What is Medicine?

Alexander V. Chalyi

What is Medicine?

Basic Principles of Physics in Medicine and Beyond

Springer

Alexander V. Chalyi
Department of Medical and Biological
Physics and Informatics
Bogomolets National Medical University
Kyiv, Ukraine

ISBN 978-3-031-64978-3 ISBN 978-3-031-64979-0 (eBook)
https://doi.org/10.1007/978-3-031-64979-0

This Springer imprint is published by the registered company Springer Nature Switzerland AG
The registered company address is: Gewerbestrasse 11, 6330 Cham, Switzerland

Preface

Nowadays, it is difficult to imagine modern medicine without the outstanding achievements of physical science. A deep understanding of complex processes in living systems that occur at the molecular level requires the use of both precise diagnostic and therapeutic methods and modern theoretical and experimental approaches. The combination of qualitative and quantitative levels of understanding of medical and biological processes in cells and in the body as a whole is, of course, the direction in which medicine of the twenty-first century will develop, based on the achievements of all natural sciences.

It is worth to remind that in ancient times and up to the seventeenth century, when science was not yet so deeply differentiated, the term "Physics" united the totality of knowledge about all nature—both living and inanimate. Later in science, differentiation processes began to accelerate more than three centuries ago during the time of Isaac Newton. Of course, these processes did not come from a "good life". Nature knows nothing about its fragmentation, which is being done so successfully to facilitate the process of studying individual sciences. In this regard, we can recall the well-known "first principles": "Nature knows better!" and "You have to pay for everything!". Indeed, you have to pay a really high price, which is caused by the loss of a holistic perception of the surrounding world.

In modern science, simultaneously with the processes of differentiation of our knowledge about the world around and inside us, integration processes are taking place and are becoming increasingly valuable, allowing us to get an idea of the objects of study as a whole. Medicine, like Synergetics, Philosophy and some other scientific approaches, is in great need and relies on such integration trends. Therefore, Medical Physics can, on the one hand, be regarded as a new branch of the living physical tree that continues to grow, while, on the other hand, it represents a necessary unifying bridge between medicine and physics, which is becoming wider and wider.

The importance of introducing this book to the dear reader is determined, not least of all, by the following definition of Medical Physics: *Medical Physics is an applied branch of physical science, in which the fundamental laws and achievements of physics are applied to describe processes in living nature, primarily with the aim of using them to successfully solve the most complex and important tasks of medicine—preservation of human health.*

This monograph is not a textbook on Medical Physics in the generally accepted sense of the word "textbook". Here, each chapter is devoted to one of the directions of theoretical or clinical medicine and consists of two parts, being written at completely different levels of their presentation. At the first level, the author makes an attempt to popularize some rather complex ideas for the dear reader, who does not have the proper level of physical and mathematical training about modern achievements of Medical Physics. At the second, more rigorous level, these same problems of Medical Physics are already considered using the appropriate mathematical apparatus and sufficiently deep physical ideas and theories. To the best of the author's knowledge, no such book on Medical Physics has yet been written. And let the dear reader forgive some of the author's immodesty (which, I assure you, is not the main feature of my character), this book is unique among competing books on this and related topics.

It should be recognized that the selection of topics for the chapters of this monograph was dictated, apparently, by such objective and subjective factors, namely: (1) their unconditional relevance for practical and theoretical medicine; (2) the author's desire to tell, if possible, interestingly and in simple words about the basic ideas, concepts, and methods of Medical Physics; (3) ensuring a sufficiently high level of physical and mathematical interpretation and substantiation of the material included in the monograph; 4) last but not least, the personal interest of the author in the selected topics of the chapters, including his own, as well as joint with colleagues, publications on these issues.

Unfortunately, I am forced to write about a tragic reason that has not yet allowed me to complete the entire intended plan of my book. Of course, this reason is the treacherous aggression of a neighboring country and the insane desire of its leader to completely kill the entire Ukrainian nation and Ukraine as a state. This war makes our lives mortally dangerous every day and hour. Holy God, help us survive and punish all those who bring death to all living things in Ukraine and other countries, violating all the sacred commandments!

There is no doubt that in the twenty-first century, the achievements of Medical Physics will have a growing influence on the development of theoretical and clinical medicine, thereby determining the level of training of highly qualified physicians. In practical terms, Medical Physics is an important integral part of all medical practice, as well as medical science and education. It is Medical Physics that provides, in the first place, a reliable basis, which is difficult to overestimate, for the creation of precision and high-tech medical equipment for diagnosis and treatment, starting from thermometers and tonometers, X-ray and cardiography devices, and ending (so far!) with the Da Vinci robotic surgical system for minimally invasive interventions, as well as a combined positron emission and magnetic resonance (PET-MRI) imaging system. Then continuing with everything that physicists, mathematicians, and specialists in the fields of biomedical engineering and IT technologies (certainly together with the most creative physicians!) will definitely invent and apply for medicine.

This book is aimed at elucidating the most important aspects of the medical physics and written for scientific researchers and general educated public as well as for students, teachers, engineers, and all those readers interested in modern problems of the medical physics.

The author would like to express his hearty thanks to my co-authors of our common publications mentioned in the lists of references, in particular to my colleagues from the Department of Medical and Biological Physics and Informatics—first of all, to my son Prof. Kyrylo Chalyy, who always supported me with his family, making my life more interesting and helping me often overcome difficulties that I certainly would not have been able to cope with on my own. I would like to convey my gratitude to Prof. Natalia Stuchynska, to whom I passed on the baton of heading the Department of Medical and Biological Physics and Informatics a year ago, which I headed for more than 40 years from March 1983 to December 2023. My special thanks to Assoc. Prof. Galyna Khrapijchuk, who helped me in the preparation of the figures. In the course of my scientific work, I had many opportunities to communicate with my colleagues Prof. Mikhail Anisimov, Prof. Savo Bratos, Prof. Leonid Bulavin, Prof. Paul Deluca, Prof. Hans-Joerg Moegel, Prof. Wolffram Schroer, Prof. Yaroslav Tsekhmister, Prof. Aleksei Vasilev, who gave me valuable remarks and friendly advice. I would also like to express my sincere and heartfelt gratitude to a Senior Editor in Physics Books, Springer Nature, Dr. Zachary Evenson; an Associate Editor, Springer-Verlag GmbH Sabine Lehr; Project Coordinator, Springer Nature, Mr. Vijay Kumar Selvaraj; and Production Supervisor - Books Production, Springer Nature, Bhagylakkshme S for their helpful advice and useful suggestions.

And last but not least, I would like to confess to everyone who hold this book in their hands and even leaf it, and maybe, to my great joy, read my book, so here is my confession: "I love my wife, I love my country, I love my doggy!" These words were said a long time ago, back in Ukrainian peacetime. I especially love my wife, without whom everything would not be the same! She is very smart and gives me surprisingly good advice, because she is wise in life and knows at the level of a Ph.D. in theoretical physics what critical opalescence of light and neutron optics are!

Among physicists, it is believed that the beginning of the development of modern Biophysics, which is close in its problems to Medical Physics, was laid by the outstanding physicist of the twentieth century, one of the founders of quantum mechanics, 1933 Nobel Prize winner in Physics Erwin Schrödinger, who wrote the book "*What is Life? The Physical Aspect of the Living Cell*" in 1944. In this book, small in scope, but extremely informative in content, the theoretical foundations of Biophysics were presented as physical laws of living nature, starting from individual molecules, cells, organs, tissues and ending with the biosphere as a whole.

Being fully aware of the certain pretentiousness of the title "*What is Medicine? Basic Principles of Physics in Medicine and Beyond*" in connection with the classic book by Erwin Schrödinger "*What is Life? The Physical Aspect of the Living Cell*", I would like to express the following two expectations: (1) sometime and someone (but certainly not the author of this monograph) will write another book "*What is Physics from the Point of View of a Physician?*" to close the peculiar "trajectory"; (2) this monograph, which focuses on the medical aspects of the application of physics to living systems, will be useful to respected readers interested in the basic ideas of Medical Physics, as well as in those outstanding achievements of medicine, which are obtained through the use of modern physical methods.

Kyiv, Ukraine Prof. Alexander V. Chalyi

Contents

Chapter 1
Brief Outline of the Development of Medical Physics

Abstract This chapter presents the most important stages of the history of the development of Medical Physics starting from the Ancient Era, continuing the the Middle Ages, and ending in the last three and a half centuries, when the differentiation of various sciences began. A special attention is also paid to the formation of the Kyiv School of Medical Physics at the St. Volodymyr Kiev University and at the Bogomolets Kiev Medical Institute (National Medical University) in the period from 1841 to the present time. The importance of Medical Physics, as an integral part of the medical science, medical education and medical industry, is emphasized. It is these directions of science, education and technology that provided a reliable basis to create precision and high-tech medical equipment for diagnostics and treatment. Medical Physics began to develop especially actively starting from the second half of the 20th century, when such medical diagnostic methods as X-ray computed tomography (CT), magnetic resonance imaging (MRI), positron emission tomography (PET), and its practical implementation became possible. It should be noted that these outstanding achievements of physics in medicine were awarded the Nobel Prizes in Physiology or Medicine in 1979 and 2003. There are two approaches to teaching the discipline "Medical Physics" in medical educational and scientific institutions of different countries of the world, The first of them is related to the teaching of Medical Physics, which includes a broad applications of physics in medicine, namely biomechanics and hemodynamics; electrodynamics and electronic medical equipment; molecular biophysics and biophysics of complex living systems; optical and quantum-mechanical methods of studying biomedical systems, as well as elements of radiation medicine. The second approach in teaching Medical Physics in a relatively narrower sense is related to the use of ionizing and non-ionizing radiation in Radiation Oncology and Nuclear Medicine. This chapter, and this book as a whole, takes a broader approach, although the Chaps. 6 and 7 touches on a number of problems in the use of tomographic methods of quantum mechanics in medicine, as well as selected problems in radiation medicine.

Keywords Medical physics · Two approaches to teaching medical physics · Medical Physics in Ancient Era · Medical Physics in middle ages · Medical Physics in the modern era · Kyiv School of medical physics

A. V. Chalyi, *What is Medicine?*, https://doi.org/10.1007/978-3-031-64979-0_1

Physics is law and everything else is a recommendation.

Elon Musk (Lex Fridman Podcast 9 Nov 2023)

1.1 Introduction

The history of medicine and physics, as well as the current trends in their development, show that Medical Physics, along with other fundamental natural sciences and educational disciplines, is very important for medical education and science and the medical field as a whole. Without a doubt, Medical Physics can be considered interdisciplinary discipline (science), because it combines two sciences (disciplines), namely: physics and medicine. At the same time, it should be emphasized that the first of these disciplines, that is, physics, provides the other with its research methods, and the second discipline, that is, medicine, provides research objects and a large base of empirical data. It is this fact that allows to transfer the results of one discipline (science), which is more precise in quantitative aspect (in our case, physics) to another discipline (science), which is much more complex and still significantly less precise (in our case, medicine), while obtaining new, including quantitative and not only qualitative results. Medicine includes the study of multi-level relationships from interatomic and intermolecular interactions that determine physical properties to intercellular interaction and synaptic transmission that determine the specifics of the functioning of organs and tissues and the human body as a whole.

In medical universities (faculties) and other scientific institutions of different countries of the world, there are, as a rule, two approaches to teaching the discipline "Medical Physics", which is also called "Medical Biophysics" or "Biomedical Physics" (see, for example, [1–19]). The first of them is related to the teaching of medical physics, which is characterized by a multifaceted approach including the following main directions (sections) of the application of physics in medicine: (1) biomechanics, biorheology and hemodynamics; (2) electrodynamics of medical and biological systems; (3) mechanisms and effects of interaction of physical fields with biological tissues; (4) electronic medical equipment; (5) elements of molecular biophysics and biophysics of membrane processes in cells; (6) thermodynamic and synergetic principles of biophysics of complex systems; (7) optical and quantum-mechanical methods of research of medical and biological systems; (8) elements of radiation medicine and the basics of dosimetry.

The following definition of medical physics in such a broad sense can be given: "***Medical physics is an applied branch of physics, in which the fundamental laws and achievements of physics are used to explain the processes occurring in medical-biological systems, in particular—in the human body, with the aim of using them to solve the problems of theoretical and clinical medicine, first of all for the sake of human health protection***". The last words in this definition make it possible to draw a line (at first glance, rather thin) between Medical Physics and Biological Physics. It should be noted that this precisely such discipline "Medical and Biological Physics" has been taught for many years in the Department of Medical and

Biological Physics and Informatics of the Bogomolets National Medical University [4–6, 17, 18]. The main goal of teaching this educational discipline is to improve the professional competences of future doctors through in-depth study and expansion of their knowledge, abilities and skills from the above-mentioned sections of medical physics, which underlie the functioning of modern (electronic) medical devices and determine their use in the diagnosis and treatment of diseases. In the future, in this chapter, our attention will be focused mainly on some features of teaching Medical Physics in such a broad and multifaceted sense.

The second approach in teaching Medical Physics in a relatively narrower sense is related to the use of ionizing and non-ionizing radiation in such medical areas as Radiation Oncology, Diagnostic Radiology, Nuclear Medicine, Radiotherapy and some other related fields [1–3, 7–11]. Note that this approach is the most developed in modern Medical Physics, it is equipped with precise and state-of-the-art medical equipment, in particular, such as X-ray computed tomographs, magnetic resonance and positron emission tomographs, etc. It is with the help of these devices that it is possible to carry out the most modern studies of the internal organs and tissues of the human body and make accurate diagnoses. This book, in Chaps. 6 and 7, will give some attention to this important approach to teaching the educational and scientific aspects of Medical Physics. However, there are many excellent textbooks and monographs devoted to the above-mentioned areas of Medical Physics. Let's name some of them [1–3, 7–9] (I apologize to the authors of many other scientific publications and textbooks that I could not refer to here). In particular, it should also be taken into account that the training of students in the mentioned medical fields takes place at the Department of Radiology and Radiation Medicine, and not at the Department of Medical and Biological Physics and Informatics of the Bogomolets National Medical University in Kyiv, Ukraine.

Turning into a joke the above difference between the two approaches to the study of Medical Physics, one can say the following: *Someone who knows more and more in smaller and smallar areas ultimately knows everything about nothing. And vise versa, someone who knows less and less in more and more areas end up knowing Nothing about everything!*

This chapter presents the most important stages of the history of the development of Medical Physics in ancient times, Middle Ages and the last three centures, when the differentiation of sciences took place. In addition, the author, who is a patriot of my native Ukraine, which is suffering so much during the last years of the war, would like to tell the respected reader also about the formation of the Kyiv School of Medical Physics (in particular, the development of Medical Physics first at the St. Volodymyr Kiev University, and then at the Kyiv Medical Institute and at the Bogomolets National Medical University) in the period from 1841 to the present time.

1.2 Main Stages in the Development of Medical Physics

It can be fairly tentatively assumed that the beginning of the development of scientific knowledge on the border of physics and medicine was laid in the Renaissance (XV-XVI centuries), although quite important results in this direction were obtained earlier, starting from the era of antiquity, namely:

In the 6th century BC, Pythagoras investigated the pitch of a tone—one of the subjective characteristics of sound—depending on the length of the string;

In the 5th century BC, Plato created the first theory of vision;

In the 4th–2nd centuries BC, first Aristotle and then Ptolemy studied the laws of refraction of light, which were later used in optical studies of biological objects;

In the III-II centuries before Christ, Archimedes and Heron of Alexandria, studying mechanical devices (lever, block, etc.), essentially create the foundations for modern biomechanics;

In the Middle Ages, in the XI century, Alhazen continued and expanded the research of Greek thinkers on physiological optics;

In the 13th century, Bacon discovered spherical aberration, which laid the foundations for the study of defects in the human eye, at the same time, the first glasses appeared first in Europe, then in Asia.

During the Renaissance, in the 15th century, Leonardo da Vinci continued research in the field of hydrodynamics, biomechanics, and optics (in particular, he created the theory of binocular vision).

In the 16th century, Mavrolik wrote a treatise on physical and physiological optics, in which he explained vision defects—myopia (nearsightedness) and hypermetropia (farsightedness), and also described the anatomy of the human eye.

At the end of the 16th century, Jansen and Anton Leeuwenhoek created the first microscope, and Galileo Galilei created the first thermometer (thermoscope).

Galileo Galilei's thermometer, being constructed in 1597, did not have a temperature scale and allowed only qualitative comparison of body temperatures. The main part of this thermometer was a glass ball in which there was air. Below this ball there was a thin glass tube, small in height, in which there was colored water, the meniscus of which was located approximately half the height of this tube. The position of the meniscus changed after the tube was placed with its end in a volume that also contained a colored liquid. This made it possible to compare temperatures: if the level of the meniscus in the tube rose (lowered), the temperature of the liquid in the vessel was higher (lower) than in Galileo's thermometer (thermoscope).

It is believed that an English physicist and mathematician Isaac Newton (1643–1727) was the first to propose in 1701 to choose the melting point of ice and the boiling point of water as two reference points for the temperature scale. As for known temperature scales, their emergence in chronological order dates back to the following years:

1709—the Fahrenheit temperature scale, which was proposed by the Polish-German physicist and chemist Daniel Fahrenheit (1686–1736). One degree in the

Fahrenheit scale is 1.8 times more than one degree in the Celsius scale, i.e., 1°F = 1.8 °C.

1730—the temperature scale of the French scientist Rene Reaumur (1683–1757) was introduced, and one degree on the Reaumur scale is 0.8 degrees on the Celsius scale, i.e., 1 °R = 0.8 °C.

1742—the temperature scale of the Swedish astronomer Anders Celsius (1701–1744) was introduced. As is well known, the temperature distance between the two reference points proposed by Newton is divided by 100 degrees in the Anders Celsius temperature scale.

1848—a temperature scale was proposed by the English physicist William Thomson (Lord Kelvin), which bears his name. The zero temperature in the Kelvin scale is taken to be the temperature of absolute zero T = 0 K, which is approximately equal to the temperature of −273 °C in the Celsius scale, because the thermal movement of atoms and molecules stops at this temperature.

It should be noted that the following relationship exists between the temperature values in different temperature scales:

$$t\left[{}^{0}\mathrm{C}\right] = (t + 273)[\mathrm{K}] = [(9/5)t + 32]\left[{}^{0}\mathrm{F}\right] = (4/5)t\left[{}^{0}\mathrm{R}\right]. \tag{1.1}$$

Here in formula (1.1), the following notations [°C], [K], [°F], and [°R] are used for temperature units in the Celsius, Kelvin, Fahrenheit, and Reaumur scales, respectively.

As consequences of formula (1.1), (a) the melting temperature of the ice occurs in different temperature scales with the following values: 0 °C, 273 K, 32 °F, 0 °R, while (b) the boiling point of water occurs at the following temperature values for different scales: 100 °C, 373 K, 212 °F, 80 °R.

The development of physics as a science began in the 17th century. Until that time, especially in ancient times, the term "physics" characterized the totality of information about all nature—living and non-living. In the future, starting from the 17th century and especially in the 19th and 20th centuries, processes of differentiation spread in science, although for quite a long time great scientists achieved success in various scientific fields (in particular, they combined the abilities of both physics and medicine).

Thus, at the very beginning of the 17th century, William Gilbert (1544–1603), the physician of the English Queen Elizabeth 1 and a professor of medicine in Cambridge, began to study the forces that act between magnets, as well as the forces that ensure the attraction of bodies to amber. He even wrote the book "Magnet", in which the foundations of magnetostatics were laid. He used magnets for healing and also studied the nature of terrestrial magnetism.

William Gilbert (1544-1603)

In the 1770s, an anatomist from Bologna, Galvani discovered the electrical properties of living beings—the so-called "animal electricity". In fact, these works initiated the development of ideas about magnetic and electric fields in biological systems and eventually led to the development of electromagnetic theory in the nineteenth century in the works of Faraday and Maxwell.

At the same time, research in the field of optics of living systems continues. In 1604, Kepler proposed his own theory of vision. In 1610–1614, after discovering the telescope, Galileo created a microscope that was more perfect than Jansen's. In the 60s and 70s of the 17th centuries, the optical microscope was further improved by Hooke and Leeuwenhoek. This gave a significant impetus to the development of anatomy, histology and other sections of medicine, in which optical methods for the study of micro-objects are increasingly used in the future. It was the English physicist Hooke who was the first to discover the cellular structure of biological objects using a microscope and wrote the book "Micrography" about it in 1665. In this book Robert Hooke introduced the term "cell" for the first time.

The famous English doctor and physicist Thomas Young (1773–1829), see his portrait above, together with the French physicist Augustin Jean Fresnel (1788–1827) are considered the creators of the wave nature of light. At first, Tomas Young studied and worked as a doctor at the St. George Hospital in London. He obtained important results in physiological optics as following: the creation of one of the first variants of color vision. In 1793, he gave an essentially modern explanation of the accommodation of the eye due to changes in the curvature of the eye's lens. Subsequently, Tomas Young became more interested in various branches of physics. Based on the wave theory of light, Tomas Young first introduced the concept of light interference and gave a correct explanation of this phenomenon. In the mechanics of deformable media, he owns another important result, which bears his name, namely the Young's modulus.

In the 18th century, Bouguer, Lambert and Behr established the laws of attenuation of light when it passes through matter, which made it possible to further develop the method of concentration colorimetry of biological objects.

In the 18th and 19th centuries, Euler, Bernoulli, Navier, Stokes, Poiseuille and other scientists discovered the basic laws of modern hydrodynamics, which allows them to be extended to the processes of blood flow through blood vessels (this section of the mechanical movement of blood is called hemodynamics).

In the 19th century, the basic laws of bioacoustics were established, which later became the basis of modern diagnostic methods in otolaryngology (Doppler, Rayleigh, Weber, Fechner, and others). In 1815, Bio establishes a law that describes the rotation of the plane of light polarization and is widely used to study the concentration of optically active biological media. At the same time, more precisely in the 40s of the XIX century, the physicist Joule, the physical chemist Helmholtz and the ship's doctor Mayer discovered the law of conservation of energy—the universal law of energy transformation in living and non-living nature. In 1852, Stokes discovered the law to which luminescent radiation obeys, which later, already in the XX century, made it possible to study the basic patterns of various types of luminescence (in particular, bioluminescence) and use them in medicine. In 1872–1873, Abbe developed the theory of image formation in an optical microscope, which has not lost its importance to our time.

The last years of the XIX century were marked by two discoveries that largely determined the subsequent development and current state of medical physics and physics in general, namely:

1. In November 1895, a famous German physicist Wilhelm K. Roentgen, the first in 1901 laureate of the Nobel Prize in Physics, has discovered the so-called X-rays, which have been used as a powerful method of diagnosis and therapy. It should be noted that a famous Ukrainian physicist, electrical engineer and one of the first translators of Bible into Ukrainian Ivan P. Puluj has invented a lamp, which produced intensive X-rays, and his activity was of great importance for the development of the physics of X-rays and its applications in medicine. The practical use of X-rays in medicine began in many countries of the world immediately in 1896.

Wilhelm Conrad Roentgen

(1845- 1923)

Ivan Pavlovich Puluj

(1845-1918)

2. In December 1900, the German physicist Max Planck introduced the concept of a quantum—a minimum portion of energy, which marked the beginning of the development of quantum mechanics—the science of the microcosm. It also made it possible to explain the laws of absolute black body radiation discovered earlier in the 70s and 90s of the XIX century by Stephan, Boltzmann and Wien, which created the scientific foundations for the development of the modern diagnostic method of thermography. More detailed information about the physical foundations of quantum mechanics, as well as the use of its methods in medicine (primarily tomographic methods of medical diagnostics) will be presented in subsequent Chaps. 6 and 7.

The XX century marked the rapid development of physics and its various applications, including in the fields of medicine and biology.

In 1901, Henri Becquerel and the spouses Pierre Curie and Marie Sklodovska-Curie first reported on the physiological effects of radioactive radiation, which had been discovered 5 years before—in 1896 by Becquerel. Henri Becquerel together with Marie Sklodovska-Curie and Pierre Curie were awarded the 1903 Nobel Prize in Physics for the discovery of the phenomenon of radioactivity. In 1911, Marie Sklodovska-Curie received another Nobel Prize, this time in chemistry, becoming the only scientist in the world who was awarded two Nobel Prizes in the different natural sciences.

Antoine Henri Becquerel (1852-1908)

Marie Sklodovska-Curie (1867-1934)
Pierre Curie (1859-1906)

In 1903, Willem Eindhoven discovered electrocardiography—a method of studying the electrical activity of the human heart, which nowadays is a powerful and precise method of diagnosing heart diseases. More precisely, Eindhoven used the ideas of Augustus Warren and his other predecessors regarding the possibility of recording the electrical activity of the myocardium and created the first electrocardiograph at the beginning of the 20th century, for which he was awarded the Nobel Prize in Medicine and Physiology in 1924.

From the very beginning of the 20th century until about the middle of the 30s of the XX century, the foundations of quantum mechanics were developed (Niels Bohr, Erwin Schrödinger, Werner Heisenberg, Louis de Broglie, Wolfgang Pauli, Paul Dirac and other physicists). This led to the development of many modern medical techniques that use the principles of quantum mechanics. Thus, in the 1920s and 1930s, based on de Broglie's discovery of the wave nature of microparticles, an electron microscope was created (Max Knoll, Ernest Ruska, Manfred Von Ardenne), with the help of which it was possible to reduce by 2–3 orders of magnitude the resolution limit, that is, the minimum distance between two points that can be seen separately under a microscope. It should be noted that thanks to the revolutionary achievements of modern optics, literally in the last two years, it was possible to reduce the resolution limit in so-called scanning optical microscopes to a value of approximately 10 nanometers, which is close to the capabilities of an electron microscope.

In 1912–1913, the English physicists—father William Bragg and son Lawrence Bragg and the crystallographer of Ukrainian origin George (Yuriy) Wulf received a formula that connects the wavelength of electromagnetic (in particular, X-ray) radiation and the parameters of the crystal lattice of inorganic and organic crystals. It was with the help of X-rays and the Bragg-Wulf formula in the early 1950s that first Rosalind Fraklin and then James Watson, Francis Crick, and Maurice Wilkins discovered the spiral structure of DNA molecules, which carry genetic information. Since then, this information has been used in the diagnosis and treatment of various diseases. Various methods, including methods of medical physics, are becoming especially promising today, at the very beginning of the 21st century, for revealing the sequence of nucleotides in the information molecules of DNA and RNA. This is the goal of the international program "Human Genome", which allows, in fact, to solve a very ambitious problem—to determine at the molecular level the cause of pathological processes in the human body and to develop effective methods of their prevention.

In 1928 and later, mathematicians and physicists created the theory of self-oscillations—undamped oscillations supported by external energy sources. Oscillations propagating in the heart muscle (myocardium) are by their nature autooscillations.

In the 40s and 50s of the 20th century, Denis Gabor developed the basics of the holographic method of recording information, which were further improved (Mykola Nakhodkin and other scientists), which made it possible to obtain optical three-dimensional images of objects with the help of light interference and to widely use the principles of holography in medicine.

In the 1950s, a laser was developed, which later became a very powerful tool in surgery and therapy. The physical basis of laser operation (optical quantum generator) was created and implemented by many scientists (Nobel Prize winners in Physics Albert Einstein, Nikolay Basov, Alexander Prokhorov, Charles Towns).

In the middle of the 20th century, resonance methods of quantum mechanics began to be successfully developed, which are now very widely used as diagnostic methods, namely: the method of electronic paramagnetic resonance, the method of nuclear magnetic resonance (Felix Bloch and Edward Parcell, 1946). All these methods, together with X-ray methods and the method of positron emission tomography, are the basis of modern computer tomography. Such a powerful research technique allows obtaining layer-by-layer images of a certain part of the object under study, that is, in essence, its three-dimensional image. The practical implementation of computed tomography became possible in the 1960s and 1970s thanks to the efforts of doctors, physicists and engineers (James Ambrose, Allan Cormick, Robert Ohlendorf, Godfrey Hounsfield, Paul Lautenbur, Peter Mansfield and other scientists). It should be noted that these outstanding achievements of physics in medicine were awarded the Nobel Prizes in 1979 and 2003.

At the second half of the 20th century, two scientists independently started research on the creation of an X-ray computer tomography, namely: the physicist and electrical engineer Godfrey Hounsfield (1919–2004), as well as the South African and American physicist Allan Cormack (1924–1998), who received the Nobel Prize in Medicine or Physiology in 1979 for the introduction of X-ray computer tomography into medicine and "for the development of computer assisted tomography".

The further history of modern methods of medical diagnostics was directly related to the Nobel Prize in Physics, which was awarded in 1902 to two Dutch physicists Peter Zeeman (1865–1946) and Hendrik Lorentz (1853–1928) for the discovery (Zeeman) and explanation (Lorentz) effect of the splitting of spectral lines in a magnetic field. It was this Zeeman effect that formed the basis for the development of magnetic resonance imaging (MRI) by the American medical physicist Paul Lauterbur (1929–2007) and the English mathematician Peter Mansfield (1933–2017), for which they were awarded the Nobel Prize in Medicine or Physiology in 2003.

1.3 Creation of the Kyiv School of Medical Physics

Medical Physics at the St. Volodymyr Kyiv University. This part of Chap. 1 will be devoted to the development of the Kyiv School of Medical Physics in the period from 1841 to the present time. Almost 80 years since the opening in September 1841 of the medical faculty at the Kyiv St. Volodymyr University and until April 1920, when the Kyiv Institute of Health Care was created on the basis of the medical faculty of the university, which in December 1920 was reorganized into the Kyiv State Medical Academy, and then exactly one year later to the Kyiv Medical Institute, there was no separate department for teaching physics to medical students.

Professor V.P. Chekhovych, who held the position of professor at the Department of General Physics until 1846, was the first lecturer who gave a physics course for medical students at Kyiv University. Prof. Chekhovych, a geologist by education, gave lectures on physics to the extent that it was taught in religious institutions, and did not go beyond the gymnasium course.

The teaching of physics improved somewhat after the arrival of Professor E.A. Knorr (1846–1858 pp.) and his successor—Professor M.I. Talizin, who headed the physics department from 1858 to 1865. On the initiative of E.A. Knorr in 1856, an observatory was built at the Kyiv St. Volodymyr University, which still exists in the center of Kyiv, on the corner of Boulvarno-Kudryavska and Observatory Streets.

Professor
Mikhail Avenarius
(1935-1995)

Conducting scientific research at the Department of Physics of Kyiv University, including at the Faculty of Medicine, which remain famous even in our time, was associated with the name of the famous scientist and teacher, founder of the first physics school in Ukraine, Professor Mikhail P. Avenarius (September 7, 1835–September 4, 1895), who headed the physics department for about 25 years, from 1865 to 1890. Professor M.P. Avenarius was a widely educated person.

He made a great contribution to the development of physical science with his tireless teaching activities and excellent scientific research.

During 1873–1877 M.P. Avenarius and his students obtained the exact values of critical parameters for many substances, which were included in the basic fund of physical quantities and remained unchanged for a long time. Although Professor M.P. Avenarius was not directly engaged in scientific research in the field of medical physics, on his initiative in 1873 the first special laboratory of medical physics was created at Kyiv University, which was headed by Doctor of Medicine A.S. Shklyarevskyi. He published in 1881–1882 pp. "Lectures on Medical Physics". Based on the analysis of the state of medicine, he advocated the need to teach physics for doctors. Shklyarevsky defended his thesis for the degree of doctor of medicine (MD) on the topic "On the passage of white blood cells through colloid membranes". He specialized mainly in pathological physiology and pathological anatomy. The main works of A.S. Shklyarevsky are devoted to the problems of inflammation, vascular wall permeability, capillary physiology. He was the first to use an experimental method to analyze the mechanism of leukocyte movement.

Professor
M. M. Schiller

The successor of Professor M.P. Avenarius as the head of the Department of Physics of Kyiv University Cv. Volodymyr for 13 years (1890–1903) was Professor M.M. Schiller (13.03.1848–23.11.1910), a specialist in the field of thermodynamics and electrodynamics, who headed the first department of theoretical physics in Ukraine and wrote about 90 scientific works. Professor M.M. Schiller graduated from the Moscow University in 1868. He worked in 1871–1874 at the University of Berlin. M. M. Schiller's scientific works relate to many branches of physics: mechanics, thermodynamics, optics, molecular physics. The scientific researches and publications of Professor M.M. Schiller were dedicated to the experimental substantiation of the main provisions of one of the most outstanding scientific achievements of the 19th century — J.C. Maxwell's electrodynamics, were very popular.

Professor Yosip Kosonogov

In 1896 a physical and medical society was founded at Kyiv University, the purpose of which was the convergence of medicine and other branches of natural science, the wide use of the results of fundamental research in medical practice. This society, which existed for about 20 years, was headed by the head of the Department of Skin Diseases, Professor M.I.Stukovenko. It is interesting to note that almost 100 years after the founding of the Physico-Medical Society, its work was resumed at the Kyiv Medical Institute named after O.O. Bogomolets thanks to the initiative of the head of the Department of Skin Diseases, Professor V.G. Kolyadenko and the head of the Department of Physics, Professor O.V. Chalyi.

Academician Y. Y. Kosonogov (31.03.1866–22.03.1922) headed the physics department of Kyiv University from 1903 to 1920. He graduated from Kyiv University of St. Volodymyr in 1889, being a pupil of M.P. Avenarius. His scientific works are devoted to the study of electrical and optical phenomena. In 1902, he discovered optical resonance in the field of visible rays and explained the reason for the bright color of heterogeneous bodies with this phenomenon. For the first time, he used an ultramicroscope to study the phenomenon of electrolysis. He was a perfectly talented experimenter and carried out very valuable scientific research devoted to the physics of dielectrics and physical optics, and paid great attention to improving the methodology of teaching physics both in higher and secondary schools. Professor Y.Y. Kosonogov published a number of textbooks and manuals on physics, including a guide to physics for university students.

Medical Physics at the Kyiv Medical Institute–Bogomolets National Medical University. The first head of the Department of Physics of the Kyiv Medical Institute, which was created, as already mentioned above, in 1920 on the basis of the medical faculty of the Kyiv University Cv. Volodymyr, became Professor B.M. Yankovych, a student of Y.Y. Kosonogov, who managed it from 1920 to 1922. He organized the first X-ray diagnostic office in Kyiv at the faculty surgical clinic of professor N.M. Volkovicha (1914), was the head of the Union of junior teachers of higher educational institutions in Kyiv. Scientific works of Professor B.M. Yankovych were devoted to the problems of theoretical mechanics.

Successors of Professor B.M. Yankovych as the head of the physics department of the Kyiv Medical Institute were professor G.S. Rudenko (headed the department in 1922–1925; 1929–1941; 1943–1948 pp.), professor O.H. Goldman (headed the department in 1925–1929 pp.) and associate professor D.M. Trubchenko (headed the department in 1941–1943; 1948–1952 pp.). They paid the main attention to the establishment of the pedagogical process, the creation of educational laboratories. The only special task that had to be solved was the preparation of future doctors for the possession of physical equipment.

Oleksandr Goldman

(1884-1971)

In solving this problem, the election of Academician O. H. Goldman (November 3, 1884–December 30, 1971) was of great importance. He graduated from Leipzig (1908) and Kyiv (1909) universities, and then was elected to the position of head of the Physics Department of the Kyiv Medical Institute in 1925. Professor Holdman was also the first director of the most famous scientific physical institution in Ukraine—the Institute of Physics of the National Academy of Sciences of Ukraine.

However, there were certain reasons that hindered the restructuring of the physics course at the Kyiv Medical Institute on a new basis. First of all, it is necessary to attribute here the very weak, especially immediately after the organization of the Kyiv Medical Institute in 1920, the supply of the department with high-quality physical equipment.

The situation changed little even in the first years after the end of the 2nd World War. The general direction of teaching physics for medical students essentially did not change until the 1950s. Only at the beginning of the 1950s, the necessity of restructuring the teaching of physics to medical students became apparent. From 1952 to 1968, the Department of Physics of the Kyiv Medical Institute was headed by Associate Professor Yu.V. Ignatovych, who graduated from the Kyiv Institute of

Public Education (1926). He was the author of a number of textbooks and manuals devoted to the teaching of the physics course at the medical institute.

Ass. Prof. Yurij Ignatovich

During this period, very important processes took place in world science, namely: biological physics took a decisive step from direct physical research methods to a broad interpretation of different physiological and pathological processes from a physicochemical point of view. Bioenergetics, neurocybernetics, molecular genetics, radiobiology, radiology, radiation hygiene and many others emerged and began to develop rapidly. Theoretical and experimental physics gave birth to such scientific concepts as the hydrodynamic theory of blood circulation, the relaxation theory of hearing, the radical-chain theory of radiation damage, quantum concepts in physiology, etc.

These achievements were directly manifested in the teaching work of the Department of Physics of the Kyiv Medical Institute. Special sections were introduced into the lecture course: X-rays, medical electronics, molecular genetics, cybernetics, such issues as the basics of the theory of relativity and quantum mechanics, paramagnetic resonance, optical quantum generators were included, and the problems of nuclear physics and semiconductors were explained in sufficient detail. All these questions were reflected in the "Lectures on Physics" published by the Department, including in the special issue "Physics and Medicine" published in 1965. At that time, from 1968 to 1982, the department was headed by Candidate of Pedagogical Sciences, Associate Professor E.O. Bezdenezhny. He was the author of more than 30 scientific works, including a number of textbooks devoted to the methodology of teaching physics in medical universities. Our veterans—participants of the 2nd World War, assistant professors H.V. Maksyutin and E.P. Chorniy, and the head of the laboratory M.P. Dudin worked very fruitfully at the Physics Department.

Ass. Prof. Boris Agapov

At the same time, those teachers who formed the core of the teaching staff of the Department of Physics (in the early 1980s, the department received a new name—the Department of Medical and Biological Physics) of the Kyiv Medical Institute (since 1995, the National Medical University) named after O. O. Bogomolets—associate professors B.T. Agapov, O.V. Govorukha, A.V. Melenevska, M.I. Murashko, N.F. Radchenko, senior teacher D.A. Makarchenko, teachers L.M. Denisenko, P.I. Ostroverkhov.

From 1982 to 1983, the Department was headed by Doctor of Biological Sciences, Associate Professor B. T. Agapov.

In order to specialize practical classes, a special physical workshop was developed for the first time at the department, where each task was solved with the help of methods and devices directly used in medical practice. In laboratory classes, students worked with ultrasound devices, radioactive isotopes and physiotherapy equipment, sound level meter and electrophotocalorimeter, etc. The results of the work on the creation of a special practicum were outlined in the textbooks "Specialized practicum in physics" and "Laboratory work in physics".

Prof. Oleksandr Chalyi

New impulse to scientific research at the Department of Medical and Biological Physics was given in March 1983 by the transfer to the Kyiv Medical Institute from the Kyiv State University and the election to the position of the head of the Department of Doctor of Physical and Mathematical Sciences, Professor O. V. Chalyi. Over the past 40 years, the Department of Medical and Biological Physics has carried out topical scientific research, which made it possible to obtain new scientific results in cooperation with scientists from the Departments of the O.O. Bogomolets National Medical University (National Medical University), a number of institutes of the National Academy of Sciences of Ukraine (Institute of Physics, Institute of Nuclear Physics, Institute of Semiconductors, Institute of Physical Chemistry, Institute of Surface Chemistry, Institute of Physiology, Institute of Materials Science, Institute of Cybernetics, Institute of Experimental Pathology, Oncology and Radiology), Taras Shevchenko Kyiv National University, Dragomanov National Pedagogical University, as well as Halle University and Freiberg Academy (Germany), Pierre and Marie Curie University (France), University of Wisconsin-Madison (USA), etc.

In November 1986, at the Department of Medical and Biological Physics of the Kyiv Medical Institute, the course "Fundamentals of Computing and Medical Informatics" was created, the head of which was appointed Associate Professor, Candidate of Physical and Mathematical Sciences Tatyana I. Zhegrii, who worked at the Department of Physics for many years. Since 1998, this course has been reorganized into the Department of Medical Informatics and Computer Learning Technologies, which is managed by Doctor of Pedagogical Sciences, Candidate of Physical and Mathematical Sciences, Professor Iryna E. Bulakh.

At the same time, in 1998, the Department of Physics received its name—"Department of Medical and Biological Physics". The staff of the department conducts extensive educational and methodological work and is a mainstay among medical and biological physics departments of higher medical educational institutions of Ukraine. The staff of the department has published a number of textbooks and teaching aids, including the first Ukrainian-language textbook "Medical and Biological Physics" (authors—Professor O.V. Chalyy, Associate Professor B.T. Agapov,

Associate Professor A.V. Melenevska, Associate Professor M.I. Murashko, Associate Professor N.F. Radchenko, Associate Professor N.V. Stuchynska, Associate Professor Ya.V. Tsekhmister), the first volume of which was published in 1999, and the second in 2001. In 2004, the 2nd expanded and revised edition of this textbook was published, and in 2010–2011 the first English- language textbooks on Medical and Biological Physics were printed. In subsequent years, updated editions of Ukrainian and English textbooks on physical and mathematical disciplines were published, which are studied by students at the departments of medical and biological physics and informatics of the higher medical and pharmaceutical institutions of Ukraine.

In the 2000–2001 academic year, the department of medical and biological physics of NMU started specialized teaching of the main academic discipline "Biophysics, informatics and medical equipment" according to the new educational programs for different specializations: "Medical Care" and "Pediatrics", "Medical and Preventive Care", "Dentistry". These profiled programs were developed by the commission under the leadership of Prof. O.V. Chalyi and approved by the Department of Education and Medical Science of the Ministry of Health of Ukraine. In 2005 and 2010, a new generation of these programs in the discipline "Medical and Biological Physics", as well as programs of other specialties, in particular "Pharmacy", "Technology of perfumery and cosmetic products, etc." were issued, for which related departments of medical and biological physics of the Lviv National Medical University and the National Pharmaceutical Academy (Kharkiv) are relevant.

In recent years, the Department of Medical and Biological Physics has significantly improved the material and technical base of the laboratory practice thanks to the purchase of modern diagnostic and physiotherapeutic devices (electrocardiographs, rheographs, UHF therapy devices, etc.), 2 computer classes have been created, in which students have the opportunity to study the disciplines of physical and mathematical profile and prepare for the "Step-1" and "Step-2" integrative license exams, using modern methods of computing and computer learning technologies. In 2018.

In 2018, a joint Departent of Medical and Biolodical Physics and Informatics was created. Currently, the scientific and pedagogical potential of the Department of Medical and Biological Physics and Informatics of the Bogomolets National Medical University is very powerful. The faculty of the department includes 24 teachers, including 3 Doctors of Sciences, 14 Candidates of Sciences and 6 highly qualified teachers. Now, the Department of Medical and Biological Physics and Informatics has the following employees: Head of the Department from March 1983 to December 2023, Corresponding member of the National Academy of Pedagogical Sciences of Ukraine, Honored Worker of Science and Technology of Ukraine, Doctor of Physical and Mathematical Sciences, Professor O.V. Chalyi; Doctor of Pedagogical Sciences, Candidate of Physical and Mathematical Sciences, Professor N.V. Stuchynska, who has been heading the Department of Medical and Biological Physics and Informatics since the end of December 2023; Doctor of Physical and Mathematical Sciences, PhD in Engineering (Japan), Professor K.O. Chalyy, 12 Associate Professors (Candidate of Physical and Mathematical Sciences O.M. Chaika, Candidate of Pedagogical Sciences A.I. Yegorenkov, Candidate of Physical and Mathematical Sciences G.V.Khrapijchuk, Candidate of Pedagogical Sciences I.P. Kryvehko, Candidate of Pedagogical Sciences I.I. Kucherenko, Candidate of Physical and Mathematical Sciences Yu.M. Litvin, Candidate of Physical and Mathematical Sciences D.V. Lukomsky, Candidate of Physical and Mathematical Sciences O.K. Lyubchik, Candidate of Physical and Mathematical Sciences I.F. Margolych, Candidate of Pedagogical Sciences P.V. Mykytenko, Candidate of Pedagogical Sciences V.V. Pashchenko, Candidate of Biological Sciences V.M. Rudneva); 2 Senior Assistant Professors (Candidate of Pedagogical Sciences N.L. Gritsenko, and A.O. Kryshtopa); 6 Assistant Professors (Candidate of Pedagogical Sciences I.M. Novikova, M.D, Andrijchuk, L.B. Desyatnyuk, O.M. Melhik, S.A. Shabatska, O.V. Tyapchenko); engineering and laboratory staff (L.B. Naida. S.V. Desyatnyuk, S.P. Rybak, Yu.V. Kuznetsov).

Teaching disciplines in Medical Physics, Biological Physics, Higher Mathematics and Medical Informatics is carried out for students who study at the magister's and bachelor's levels for the following specialties at the Department of Medical and Biological Physics and Informatics of the Bogomolets National Medical University: "Medicine", "Pediatrics", "Dentistry", "Medical Psychology", "Public Health", "Pharmacy", "Industrial Pharmacy", "Technology of Medical Diagnostics and Treatment", "Therapy and Rehabilitation".

The scientific work of the staff of the Department of Medical and Biological Physics is related to the study of modern problems of physical science, biological and medical physics, pedagogy of higher and secondary schools. Among them are the statistical theory of ordering processes in open systems of various natures, the physics of critical and metastable states, methods of the physics of phase transitions in the problem of intercellular interaction (synaptic information transmission), biophysical aspects of the interaction of electromagnetic radiation with biological objects, biophysics of muscle contraction and membrane excitability structures, quantum mechanical resonance methods of studying biological objects, problems of the diffusion of toxic substances into the environment, methods of probability theory and mathematical statistics in solving medical and biological problems, modern pedagogical aspects of teaching physico-mathematical and information disciplines, biological and medical physics, current problems of continuing professional education, etc, (Fig. 1.1).

The scientific work of the staff of the Department of Medical and Biological Physics is related to the study of modern problems of physical science, biological and medical physics, pedagogy of higher and secondary schools. Among them are the statistical theory of ordering processes in open systems of various natures, the physics of critical and metastable states, methods of the physics of phase transitions in the problem of intercellular interaction (synaptic information transmission), biophysical aspects of the interaction of electromagnetic radiation with biological objects, biophysics of muscle contraction and membrane excitability structures, quantum mechanical resonance methods of studying biological objects, problems of the diffusion of toxic substances into the environment, methods of probability theory and

Fig. 1.1 Participants of the All-Ukrainian Scientific Conference, which was dedicated to the 80th anniversary of the founding of the Department of Medical and Biological Physics of the Bogomolets National Medical University (Kyiv, Ukraine)

mathematical statistics in solving medical and biological problems, modern pedagogical aspects of teaching physico-mathematical and information disciplines, biological and medical physics, current problems of continuing professional education, etc.

The active scientific and pedagogical work of the employees of the Department of Medical and Biological Physics and Informatics has been reflected in the publication in recent years of about 150 textbooks, manuals and monographs (in particular, "Fluctuation models of self-organization processes", "Non-equilibrium processes in physics and biology", "Modern Problems of Molecular Physics " (Springer), etc.), more than 500 scientific articles (among them reviews and articles in leading domestic and foreign physics journals, such as "Ukrainian Physical Journal", "Journal of Physical Studies", "Condensed Matter Physics", "Biophysics", "Advances in Physical Sciences", "Journal of Experimental Metal and Theoretical Physics", "Theoretical and Mathematical Physics", "Journal of Physical Chemistry", "Optics and Spectroscopy", "Journal of Molecular Liquids" (Elsevier), etc.), as well as in many lectures at international and national conferences and symposia.

Nowadays, it is difficult to imagine modern medicine without the outstanding achievements of physical science and biomedical engineering. A deep understanding of complex processes in living systems that occur at the deep molecular level requires the use of both precise diagnostic and therapeutic methods and modern theoretical and experimental approaches. The combination of qualitative and quantitative levels of understanding of medical and biological processes in cells and in the body as a whole is, of course, the direction in which medicine of the 21st century will develop, based on the achievements of all natural sciences.

We will remind you that in ancient times and up to the 17th century inclusively, when science was not yet so deeply differentiated, the term "physics" united the totality of knowledge about all nature — both living and non-living. Later, especially during the last two centuries, processes of differentiation began to accelerate in science. Of course, this was done and is done not because of a "good life". Nature knows nothing about the process of its fragmentation, which man does so successfully in order to simplify the process of learning about nature. In this regard, we can recall the well-known "first principles": "Nature knows best!" and "You have to pay for everything!". You have to pay a really high price — the loss of a holistic perception of the surrounding world.

Of course, in modern science, simultaneously with the processes of differentiation, the processes of integration take place and acquire great value, which allow us to get an idea about the objects of study as a whole. Medicine, like philosophy, synergetics and some other modern scientific approaches, is in great need of and relies on such integrative tendencies. Therefore, medical physics and biomedical engineering can, on the one hand, be considered as one of the new branches of the physical and technical tree, which continues to grow, while, on the other hand, it represents a necessary connecting bridge between medicine and physics and technology, which becomes wider and wider.

There is no doubt that in the 21st century, the achievements of modern medical physics will have a growing influence on the development of theoretical and clinical medicine, thereby determining the level of training of highly qualified doctors.

1.4 Conclusions

This chapter aims to write a brief overview of the various stages in the development of Medical Physics from the ancient era to the present time. Particular attention is also paid to the history of the creation of the Kyiv School of Medical Physics since the founding in 1841 the Medical Faculty of the St. Volodymyr Kyiv University. Later in 1920, the Medical Faculty of the St. Volodymyr Kyiv University was transformed first into the Kyiv Institute of Health Care y, and then exactly in a year to the Kyiv Medical Institute, which since 1995 to the present day has been called the Bogomolets National Medical University.

Here, in Chap. 1, two approaches to the teaching and scientific research of current problems in Medical Physcs are discussed. The first approach in Medical Physics, which is used in this book and specifically in the subtitle of the book "From Thermometers to Tomography and Beyond", is reflected in a fairly broad coverage of the important and most actual problems of Medical Physics. The second approach is narrower, but admittedly more popular in Medical Physics. In this approach, all Medical Physics is reduced to problems of ionizing and non-ionizing radiation in such medical areas as Radiation Oncology, Diagnostic Radiology, Nuclear Medicine, Radiotherapy and other related fields. At the same time, understanding the significance of this approach in Medical Physics, two chapters of this book are devoted to covering the following problems: Physical principles of radiation medicine (Chap. 6); Application of quantum mechanics methods in medical diagnostics (Chap. 7).

References

1. R. Cameron, I.G. Skofronik, *Medical Physics* (Wiley, New York, 1878)
2. A. Martin, S.A. Harbison, *An Introduction to Radiation Protection* (Chapman and Hall, London, 1979)
3. Graduate Studies in Medical Physics. University of Wisconsin-Madison, School of Medicine, Madison (1996)
4. *Medical and Biological Physics* (Textbook for Students of Higher Medical Institutions), edited by A. Chalyi. Kyiv, volume 1 (1999); volume 2 (2001)
5. *Medical and Biological Physics* (Textbook for Students of Higher Medical Institutions), edited by A. Chalyi. Kyiv, 2nd edition, Knyga-Plus (2005)
6. *Medical and Biological Physics* (Textbook for Students of Higher Medical Institutions), edited by A. Chalyi, 4th edition, Vinnytsia, Nova Knyha (2020)
7. J. Van Dyk, *True Tales of Medical Physics. Insights into a Life-Saving Specialty*, 1st edition, Springer Cham (2022)
8. S. Keevil, R. Padevani, S. Tabakov, C. Lewis, *T*, 1st edn. (Greener, Introduction to Medical Physics, CRC Press, 2022)
9. - J.T. Bushberg, J.A. Seibert, *The Essential Physics of Medical Imaging. Study Guid*e, LWW (2022)
10. A.V. Chalyi, P.M. Deluka, Training in medical physics at the University Wisconsin-Madison and concepts of teaching medical physics at the higher medical school of Ukraine. *International Conference on Physics of Biologicl Systems*, Kyiv, pp. 15–16 (1998)
11. Alexander Chalyi, Paul Deluka, Fulbright program builds bridges between leading American and Ukrainian universities in the study of medical physics and biomedical engineering. Fulbright Newsletter **4**, 8–11 (1999)
12. B. Nolting, *Methods of Modern Biophysics* (Springer, Berlin, Heidelberg, 2006)
13. R. Cotterill, *Biophysics. An Introduction*. Wiley (2003)
14. K. Russel, *Hobbie* (Intermediate Physics for Medicine and Biology. Springer-Verlag, New York, Berlin, 1997)
15. A.V. Chalyi, *Nonequilibrium Processes in Physics and Biology* (Naukova Dumka, Kyiv, 1997)
16. R.S. Ledley, L.B. Lusted, Reasoning foundations of medical diagnosis. Science **130**(3366), 9–21 (1959)
17. E.I. Lychkovskiy, V.A. Tymanyuk, A.V. Chalyi, Yu.E. Lyakh, O.M. Zhyvotova. *Biophysics. Physical Methods of Analysis and Metrology*. Nova Knyga, Vinnitsia (2008)
18. P.G. Kostyuk, V.L. Zyma, I. S, Magyra, M.S. Miroshnychenko, M.F. Shuba. *Biophysics*. Kyiv University, Kyiv (2014)
19. L.A. Bulavin, L.G. Grechko, L.B. Lerman, A.V. Chalyi, *Medical Physics* (Dynamic and Statistical Models. Kyiv University, Kyiv, 2011)

Chapter 2
Richard Feynman's Classification and First Principles of Studying Natural Phenomena

Abstract This Chapter is devoted to the formulation of "Richard Feynman's Classification of Three Stages of Cognition of Natural Phenomena", as well as to the application of this classification to the following three examples, namely: (1) the phenomenon of light refraction at the air–water boundary and (2) the laws of conservation of energy, momentum and angular momentum. Particular attention is paid to the application of "Richard Feynman's Classification of Three Stages of Cognition of Natural Phenomena" to (3) the discovery of the place and hexagonal grid cells in the brain, which was awarded the 2014 Nobel Prize in Physiology and Medicine. The problem of grid cells in the brain is considered with taking into account: the experimental studies that led to the emergence of hexagons in the human and animal brain, discussion of the problem of generation and propagation an action potential along nerve fibers, physical parameters of the human brain and its medical applications in the method of hyperthermia for the treatment of malignant tumors, theoretical considerations using a certain analogy between grid cells in the brain and Abrikosov's vortex lattice in superconductors of the 2nd type, and hexagonal graphene and dimensional crossover. Moreover, the main purpose of this Chapter is to use the fluctuation theory of phase transitions to formulate the first principle of the emergence of hexagonal grid cells in the brain and, generally speaking, the first principle for the hexagon formation in different objects of inanimate and living nature. Our original theoretical findings are the following: (a) Polyakov's conformal invariance hypothesis is violated for a system of grid cells in the brain; (b) the grid cells system in the brain belongs to the universality class including the 3D Ising model in a magnetic field as well as a real classical liquid–vapor system; (c) to formulate the first principle for a reliable theoretical justification of the emergence of hexagonal grid cells in the brain, it is necessary to use the fluctuating part of Gibbs thermodynamic potential (the Ginzburg–Landau Hamiltonian) for a system with chemical reactions.

Keywords Richard Feynman's classification · Light refraction · Fermat's principle of least time · Principal conservation laws · Discovery of grid cells in the brain · First principle of emergence of hexagonal grid cells

A. V. Chalyi, *What is Medicine?*, https://doi.org/10.1007/978-3-031-64979-0_2

...the real glory of science is that we can find a way of thinking such that the law is evident [1]

Richard Feynman, 1965 Nobel Prize winner in Physics

2.1 Introduction

When we start a conversation about the so-called "first principles of nature", we must remember the statements of the 1965 Nobel Prize in Physics, Richard Feynman, which are contained in the famous "Feynman Lectures on Physics" [1]. In article [2], these statements were called "Richard Feynman's Classification of Three Stages of Cognition of Natural Phenomena". The corresponding three stages of studying natural phenomena are as follows: the 1st experimental stage is the observation and obtaining experimental data; the 2nd theoretical stage is the creation of a theory that describes the experimental data. Finally, the 3rd stage is the most important and requires the formulation of a basic law or first principle underlying the theory that explains the available experimental data. Emphasizing the importance of the 3rd stage for a deep understanding and explanation of the main causes of natural processes, Feynman used the following words: "...the real glory of science is that we can find such a way of thinking as to make the law obvious", which are given in the epigraph to the this chapter.

As a 1st example to illustrate the study of a certain natural phenomenon at the level of all three stages mentioned above, consider the phenomenon of light refraction, as proposed by Richard Feynman [1]. The beginning of the 1st stage of the study of this phenomenon was associated with the research of the ancient Greek scientist Claudius Ptolemy, who around 140 AD experimentally established the connection between the angles of incidence and refraction of light at the air–water interface. At the 2nd stage, the Dutch physicist Snell obtained in 1621 the well-known theoretical formula for the law of refraction of light, which relates the ratio of the sines of these angles to the index of refraction of water relative to air. And, finally, at the 3rd stage, the French mathematician, physicist and lawyer Pierre de Fermat formulated in 1662 the first principle, which now bears his name "Fermat's Principle of Least Time", according to which a light beam chooses such a trajectory between two points, to move along which the minimum time is required (see some details in Sect. 2.2.1, as well as physical and mathematical justification in Sect. 2.3.1).

As a 2nd example, it will be shown that the first principles of nature, which explain the laws of conservation of energy, momentum and angular momentum, are determined by their connections with the symmetrical properties of space and time. It turns out that the law of conservation of energy is a direct consequence of the principle of time homogeneity, the law of conservation of momentum is a direct consequence of the principle of homogeneity of space, and the law of conservation of angular momentum is a direct consequence of the principle of isotropy of space. These connections were discovered by the mathematician Emma Noether in her "Noether theorem" [3] and successively proved by the 1962 Nobel Prize winner in Physics, the eminent theoretical physicist Lev Landau and his co-author, Prof. Yevgeny Lifshits, in the 1st volume "Mechanics" [4] of the classic course of theoretical physics by L.D. Landau and E.M. Lifshits (see details in Sects. 2.2.1 and 2.3.1).

This chapter will also consider the 3rd example of the application of "Richard Feynman's classification of 3 stages of studying natural phenomena" to explaining the hexagonal grid cells in the brain discovered by the Norwegian neurophysiologists May-Britt Moser and Edward Moser in 2005 and receiving the 2014 Nobel Prize in Physiology and Medicine "...for their discovery of cells that constitute a positioning system in the brain", together with the American and British neurophysiologist John O'Keefe discovered place cells in 1971. Both place and grid cells create a coordinate system in the brain for determining the space position, angular orientation and velocity of movement of humans and animals. To explain the emergence of hexagonal grid cells in the brain, Edward Moser who also has a scientific degree in mathematics, statistics and programming, proposed in the Nobel lecture [5] to apply the similarity between grid cells and vortex structures resulting from fluctuating phase transitions in superconductors of the 2nd type in magnetic fields [6], discovered by Alexei Abrikosov, 2003 Nobel winner in Physics.

Here, another theoretical explanation for the emergence of grid cells in the brain will be proposed in [2], using a synergetic approach [7–10] for a deeper understanding of the formation process of hexagons near bifurcation (critical) points in various living and inanimate systems. The cellular network system in the brain is synergetically similar (isomorphic) to the biochemically reacting system with a scalar order parameter, which is the degree of completeness (coordinate) of the reaction, and the conjugate (in the thermodynamic sense) value is the affinity of the reaction [11–13]. This approach allows us to transfer the theoretical results of Hermann Haken [7, 8] to study the physical system of Benard cells [14] on the neurophysiological system of the cellular network. One of our important findings is the establishment of the fact that the system of grid cells in the brain near the bifurcation point of formation of coherent ordered structures belongs to the same universality class as the real liquid–vapor system near the critical point, and also the 3-dimensional Ising model in an external magnetic field [15, 16]. To prove this statement, such main results of the fluctuation theory of phase transitions and synergetics as the Ginzburg–Landau Hamiltonian [17], the hypotheses of scaling [18–23] and conformal [24] invariance, the method of collective variables in the theory of phase transitions [25–29], as well as Haken's approach [7, 8] to Benard cells were used. As a result, the consistent use of powerful and well-tested methods of the fluctuation theory of phase transitions, in particular the concept of classes of universality [30–32], allows to formulate the basic conditions of similar behavior of systems of different nature near their bifurcation (critical) points. This makes it possible to obtain a universal first principle from the point of view of "Richard Feynman's classification of three stages of studying natural phenomena", which explains the appearance of hexagonal structures in the brains of humans and animals as a manifestation of the nonlinear interaction of fluctuations of the relevant order parameters in living and non-living nature [2].

2.2 Section A

2.2.1 *Richard Feynman's Classification of Three Stages of Cognition of Natural Phenomena and Examples of the Applications of Richard Feynman's Classification*

An outstanding American physicist Richard Feynman, receiving the 1965 Nobel Prize in Physics, has essentially discussed three stages in the study of natural phenomena in a famous textbook "The Feynman Lectures on Physics" [1]. These lectures were delivered by Richard Feynman in 1961–1963 to students at the California Institute of Technology (Caltech), taped and prepared for publication by Prof. Robert Leighton and Prof. Matthew Sands.

Richard Phillips Feynman (1918-1988)

As **the first example** of the three stages of studying natural phenomena, Prof. Richard Feynman chose the optical phenomenon of light refraction at the air–water interface, considered in [1]. These three stages are proposed to be called «***Richard Feynman's Classification of Three Stages of Cognition of Natural Phenomena***" in [2], being as follows:

1st stage: ***Accumulation and systematization of experimental data***. The optical phenomenon of light refraction was studied experimentally by Claudius Ptolemy (Claudius Ptolemaeus, c. AD 100–c. AD 170). Richard Feynman cited a table of experimental data from Ptolemy's book "Optics" written about 140 AD, which presented the relationship between the angles of incidence and refraction of a light beam at the boundary of air and water (see first and second columns in Table 2.1).

Table 2.1 Comparison of Ptolemy's experimental data and theoretical (Snell's law) results

Angle of incidence α of the light beam in air (Ptolemy's experimental data)	Angle of refraction β of the light beam in water (Ptolemy's experimental data)	Angle of refraction β of the light beam in water (theoretical results according to Snell's law)
10	8	7.5
20	15.5	15
30	22.5	22
40	28	29
50	35	35
60	40.5	40
70	45	48
80	50	49.5

2nd stage: ***Finding a law (creating a theory) explaining experimental data***. The law connecting incident and refraction angles was found by the Dutch scientist Willibrordus Snellius (Snell van Royen) in 1621. As is well known, the Snell's law states that the ratio between the sine of the angle of incidence ($\sin\alpha$) and the sine of the angle of refraction ($\sin\beta$) of a light beam is equal to the relative refractive index (n) of the medium, in which the refracted ray propagates, relative to the medium, in which the incident beam propagates (see physical and mathematical details in Sect. 3.2, Sect. 3.2.1).

Claudius Ptolemaeus (c. AD 100 – c. AD 170)

It should be noted that there is a good quantitative agreement between the experimental data of Ptolemy and theoretical data according to Snell's law for the angles of refraction of a light beam in water, taking into account their "distance" in time, which is almost 1500 years. Dear reader, you can compare the experimental and theoretical data presented in Table 2.1 of this chapter, which actually combines Table 26.1 and Table 26.2 from the textbook [1] already mentioned above.

It would seem that with such agreement between experiment and theory, the problem of light refraction at the interface between two media can be considered practically solved. However, Richard Feynman believed that another final stage of studying and cognition was needed to fully understand this problem.

Willibrordus Snellius (1580-1526)

3rd stage: ***Formulation the first principle underlying the theory that explains experimental data***. For the optical phenomenon of light reflection, this first-principal law was discovered in 1662 by the French mathematician, physicist and lawyer Pierre de Fermat (1601–1665), and is called "Fermat's principle of least time". The physical meaning of this principle is that a ray of light passing from point A of one medium (for example, air) through the interface to point B of another medium (for example, water) chooses such a trajectory, along which the travel time will be minimal.

Pierre de Fermat (1601-1665)

Following the classical example from geometric optics discussed by R. Feynman and connected with the phenomenon of light refraction at the boundary of two media, let us consider other physical examples that illustrate the manifestations of "Richard Feynman's Classification of Three Stages of Cognition of Natural Phenomena" and especially the implementation of the third stage, which requires formulations of "first-principal laws" for a theoretical explanation of the observed data.

As **the second example** of the three stages of studying natural phenomena, we will turn to the laws of conservation of energy, momentum and angular momentum. At the first stage associated with the accumulation and systematization of the observed data (in the sense of Feynman's classification), these laws used a lot of experimental data, and not only of a physical nature. In this regard, the contribution of the German physician and subsequently the physicist Julius Robert Mayer, who is considered one (and possibly the first) of the founders of the law of conservation of energy, is of interest.

In the early 40 s of the XIX century, Julius Robert Mayer worked as a ship's doctor and used the method of treating diseases with the help of leeches (*hirudo*—in Latin) which was widespread at that time. The leeches' saliva contains the enzyme hirudin, having anticoagulant properties and being discovered in 1884. In the process of treating sailors' diseases with leeches (in more modern terms—by the method of hirudotherapy), Robert Mayer visually detected a change in the color of the patients' blood from bright red to darker red, which is characteristic of arterial and venous blood. Considering this fact, Mayer came to the conclusion that there must be a relationship and equivalence between the amount of heat that the human body receives (and blood, including), and work related to oxygen consumption in the human body.

At that time, the concept of "energy" was associated with the concept of "vital forces". Julius Robert Mayer published his article "A note on the forces of inanimate nature" (Annals of Chemistry and Pharmacy, 1842) on the essence of the law of conservation of energy (his first article on this topic was sent to print in 1841, but was not published), before the publications of James Joule and Hermann Helmholtz, who are also considered by right the founders of the law of conservation of energy. Figure 2.1 contains portraits of scientists, including Julius Robert Mayer (far right), who made significant contributions to both medicine and physics [33].

To these scientists should be added the German professor of medicine and physics Hermann Ludwig Helmholtz, who co-authored the law of conservation of energy due to his book "On Conservation of Force" published in 1847, as well as the great English physicist James Prescott Joule (his article on the measurement of the mechanical equivalent of heat was published in 1843). Another universal scientist was the French physicist and physician, academician of the French Medical Academy Jean Louis Poiseuille, who was the first to measure blood pressure with a mercury manometer in 1828 and who experimentally proved a hydrodynamical (hemodynamical) law in 1841 that was formulated by a German physicist and hydraulic builder Gotthilf

Fig. 2.1 Title cover of "Medical and Biological Physics. Practice" with portraits of D. Bernoulli, T. Young and J.R. Mayer, being physicians and physicists [33]

Hagen two years earlier in 1839. In accordance with this law, known as the Hagen-Poiseuille law, the volume of fluid (blood) V, flowing per unit of time t (the so-called volumetric velocity $Q = V/t$, , or minute blood flow) is directly proportional to the fourth power of the radius of the cylindrical tube (blood vessel). It is interesting that the volumetric velocity (minute flow rate) of blood in the aorta of adults is 5–6 L per minute in a calm state and up to 20 L per minute in a state of maximum excitement [34].

The laws of conservation of energy, momentum and angular momentum, which have received numerous experimental and theoretical justifications, are one of the most fundamental laws of nature. From the view point of Feynman's classification, one could say that these laws of conservation are themselves the first principles of nature, being necessary to explain the phenomena and processes occurring in animate and inanimate nature. However, this statement is not true, since the question remains as follows: what initial positions underlie the laws of conservation of energy, momentum and angular momentum, or in simpler words, where these laws come from? The answer to this question is given by symmetry considerations, which were first formulated by a German mathematician Emmy Noether in [3] (the Noether's theorem) and were consistently proved in the first volume "Mechanics" of the famous textbooks "Theoretical Physics" by Lev Landau and Eugenii Lifshits [4].

Below in Sect. 2.3.1 it will be shown that the laws of conservation of energy, momentum and angular momentum are based on the following first principles: (1) the law of conservation of energy is a consequence of the first principle which is related to the homogeneity of time, (2) the law of conservation of momentum is a consequence of the first principle which is related to the homogeneity of space, (3) the law of conservation of momentum is a consequence of the first principle which is related to the isotropy of space.

2.2.2 Discovery of Grid Cells, Physical Parameters of the Brain, Analogy Between Grid Cells and Vortex Lattice in Superconductors

Here in this paragraph, as **the third example** of the three stages of the studying natural phenomena, we will consider the following problems: (1) discovery of place and grid cells in the brain (2014 Nobel Prize in Physiology or Medicine) [5], (2) generation and propagation of the action potential (AP) along nerve fibers (1963 Nobel Prize in Physiology or Medicine) [35], (3) physical parameters of the human brain and its medical applications [2, 34, 36, 37], (4) Edward Moser's idea of analogy between hexagonal grid cells in the brain and Abrikosov's vortex lattice in superconductors of the 2nd type [5, 6, 38–40].

Discovery of place and grid cells in the brain. The 2014 Nobel Prize in Physiology or Medicine has been awarded to American and British physiologist John

O'Keefe from the University College London and to Norwegian physiologists May-Britt Moser and Edward Moser from the Norwegian University of Science and Technology "for their discoveries of cells that constitute a positioning system in the brain" [5]. Prof. John O'Keefe, being a founder of these studies, has discovered a special type of "place cells" in 1971 that are capable to create a positioning map in the hippocampus of animals (rats) [41]. Prof. Edward Moser and Prof. May-Britt Moser have found the "grid cells" in 2005, being an additional component of the orientational system that is locating in the entorhinal cortex of the brain [42]. Later, grid cells were found not only in the brains of rats and other animals, but also in the brain of humans [43]. Such studies have opened up new possibilities for understanding processes taking place in the brain [44–51].

An experiment on detecting the grid cells [42] was carried out by May-Britt Moser, Edward Moser, and their students T. Hafting, M. Fyhn, and S. Molden as follows. They implanted electrodes in the entorhinal cortex of the rat's brain. Such an operation made it possible to fix the action potential (AP) occurrence or, in other physiological words "firing of AP" in neurons, as a result of spatial movements of the rat in a confined space. It turned out that the spatial distribution of grid cells is characterized by a virtual 6-sided lattice with a periodic hexagonal symmetry (see Fig. 2), which creates a coordinate navigation system and allows animals and humans to find their position in space, as well as direction and speed of movement.

It is important to mention that the discovery of grid cells and the study of the features of their functioning is directly related to the problem of treating Alzheimer's disease associated with disorders of the entorhinal cortex of the human brain in the early stages of this disease [44]. All the above applies to the first "experimental" stage from the point of view of Feynman's classification.

Generation and propagation of the action potential. Here, we will focus our attention on the electrical mechanism of the problem of intercellular interaction caused by AP generation and propagation along the axon. A discussion of some of the physical and chemical aspects of the problem of cell-to-cell communication and synaptic transmission with taking into account the porocytosis hypothesis and synaptomers having a hexagonal geometry, as well as a quantitative description of this problem based on nonlinear kinetic models can be found, for example, in [52–54] (see also these and other details in Chap. 6).

Let us discuss first some theoretical considerations related to the description of grid cells in the brain at the 2nd "theoretical" stage of Feynman's classification. In this regard, we will briefly review the classic studies on the AP emergence and propagation along the nerve fibers by Alan Hodgkin and Andrew Huxley. These studies were awarded the 1963 Nobel Prize in Physiology or Medicine along with John Eccles, "for their discoveries concerning the ionic mechanisms involved in excitation and inhibition in the peripheric and central portions of the nerve cell membrane" [35].

As is known (see, for example, [34, 35, 37, 46, 50, 51]), the AP is generated as a result of a certain external stimulus of electrical, mechanical, or some other nature, which causes the depolarization phenomenon inside an axon. The process of depolarization of the axoplasm is associated with a local shift of the electrical potential from its value $\varphi_i^k \approx -70mV$ for the equilibrium Nernst electric potential for potassium ions to the threshold potential value $\varphi_{thr} \approx -(40-50)mV$. As a result, sodium channels open and a significant increase (approximately 500 times in a giant squid axon) of the permeability for Na^+ ions occurs in 1–2 ms. The depolarization process leads to an increase in the internal potential of the axoplasm to the value $\varphi_i^{Na} \approx +40mV$ of the equilibrium Nernst electric potential for sodium ions. Thus, a quantitative estimate of the AP amplitude φ_{AP} gives such an approximate value $\varphi_{AP} = \varphi_i^{Na} - \varphi_i^K \approx 110mV$ characterizing the firing of AP in neurons.

John Eccles (1903-1997)

Alan Hodgkin (1914-1998)

Andrew Huxley (1917-2012)

Physical parameters of the human brain. In living nature, unique electrical processes are realized in nerve fibers (axons), depending on whether the axons are covered by the myelin sheath or not. So, the nerve bundles of the white matter of the brain have, for the most part, a myelin sheath, which prevents the diffusion of the main inorganic ions of potassium and sodium and thus the continuous propagation of the action potential along the axon.

However, in the process of biological evolution, nature has created a mechanism for the transmission of an electrical signal in jumps through the so-called Ranvier interceptions, which are implemented on sections of axons not isolated from the external environment with a width of about 1 μm and at distances from each other up to 1.5 mm, which do not have a myelin sheath and contain a large number of voltage-gated sodium channels in these places. Such a saltatory (jumping) mechanism of electrical signal transmission using Ranvier's interceptions has a significantly higher speed of about $25 \div 100$ m/s, than the speed of signal propagation in unmyelinated axons, which turns out to be 1–2 orders of magnitude lower and reaches $0.7 \div 2.3$ m/s.

Here, we will consider some physical parameters of the substance of the human brain that are useful for further consideration. Speaking about the neurons of the brain and without going into the anatomical details of its structure, it should be noted that we are talking primarily about the neurons of the gray matter of the brain, since the white matter of the brain consists mainly of bundles of nerve fibers (axons) covered by the white myelin sheath.

First, we will estimate the average density of the human brain matter, using known statistical data of average mass and volume of the human brain [34, 36]. The average mass of the human brain is 1350 ± 50 g, while the average volume of the human brain is within 1400 ± 200 cm^3. Therefore, the average density of the human brain matter is 0.96 ± 0.17 g/cm^3, i.e., slightly less than the density of water $\rho_{water} \approx 1\, g/cm^3$, the most common and mysterious liquid in nature. As is shown in [34, 36], the numerical values of the relative dielectric permeability ε for such substances as water, blood, gray matter of the brain (gmb) and white matter of the brain (wmb) are almost close to each other and equal: $\varepsilon_{water} = 81$, $\varepsilon_{gmb} = \varepsilon_{blood} = 85$, $\varepsilon_{wmb} = 95$.

The electric conductivity of biological tissues may vary in a very wide interval [34]. Such biological fluids as cerebrospinal fluid (crbf) and blood (bld) are substances with high electric conductivity and, correspondingly, low electric specific resistance: $\rho_{crbf} \approx 0.55\,Ohm.m$, $\rho_{bld} \approx 1.66\,Ohm.m$, At the same time, such dense biological tissues as dry skin and bone, on the contrary, have a low electric conductivity and a very high electric specific resistance: $\rho_{skin} \approx 10^5\,Ohm\,m$, $\rho_{bone} \approx 10^7\,Ohm\,m$. The brain tissue has more or less intermediate value of the electric specific resistance: $\rho_{brain} \approx 14\,Ohm\,m$ (see Table 2.2]).

It should be taken into account that the above low and high values of the specific resistivity of biological tissues are of fundamental importance in the choice of physiotherapeutic methods for the treatment of oncological diseases. Conducted studies (see for more details, for example [26–30]) confirm that the use of the hyperthermia method in oncology, based on heating malignant tumors, leads to significant damage or even to the complete destruction of tumor cells and, as a result, the volume of the tumor.

Hyperthermia, as a method of treating malignant tumors, is used at temperatures higher than 39.5 °C (usually at temperatures 41 °C–45 °C) with a choice of time and temperature intervals depending on the type and location of tumors. Usually, hyperthermia in oncology is used in conjunction with methods such as chemotherapy and radiation therapy, increases the effectiveness of a number of chemotherapy drugs, as well as provides a higher sensitivity of tumor cells to the effects of radiation during radiation therapy.

Analogy between hexagonal grid cells and vortex lattice of superconductors. In conclusion of this section, we will focus on the theoretical considerations which can be considered as an initial attempt to formulate the first principle of the emergence of hexagonal grid cells in the brain in the spirit of Feynman's classification of studying natural phenomena. In his Nobel lecture "Grid Cells and the Entorhinal Map of Space" [5], Edward Moser suggested to explain the appearance of hexagonal structures in the brain using the results of the magnetic properties of superconductors [6]. He proposed drawing an analogy between grid cells in the brain and the vortex lattice of superconductors by Alexei Abrikosov (1928–2017), 2003

Table 2.2 The specific resistance of the human biological fluids and tissues with respect to direct electric current [34]

Biological fluids/tissues	Specific resistance [Ohm m]
Cerebrospinal fluid	0.55
Blood	1.66
Muscles	≈ 2
Brain tissue	≈ 14
Adipose tissue	≈ 33
Dry skin	10^5
Bone without periosteum	10^7

Nobel Prize winner in Physics. Actually, A. Abrikosov obtained fundamental physical results on the magnetic properties of superconductors of the second group that preserve their superconducting properties in strong magnetic fields [6]. These theoretical results are also confirmed by the experimental studies of U. Essmann and H. Trauble [38]. The idea of such an analogy between seemingly so distant fields as theoretical physics of superconductors and neurophysiological studies of the human and animal brain could be explained, as it seems to me, by the fact that Edward Moser received his first scientific degree in mathematics, statistics and programming in 1985, and five years later he received two more degrees in psychology and neurophysiology.

The problem of superconductivity has a long, interesting, and sometimes tragic history that goes far beyond the scope of this Chapter. Here it is to emphasize the fact [39] that the first experimental studies in the field of type II superconductors were carried out in Kharkov, at the Ukrainian Institute of Physics and Technology (UFTI) by L.V. Shubnikov, V.I. Khotkevich, Yu.D. Shepelev and Yu.N. Ryabinin [40]. It is on the basis of these experimental studies that A.A. Abrikosov developed the theory of type II superconductors [6], which not only qualitatively, but and quantitatively confirmed the experimental results obtained by Shubnikov (1901–1937) et al. [40].

Lev Shubnikov (1901-1937)

It is worthy to note that Alexei Abrikosov's scientific superviser was L.D. Landau (1908–1968), who together with V.L. Ginzburg (1916–2009) created the theory of superconductivity [17] or, as it is sometimes called, the theory of type I superconductors. Thus, the truth lies in the fact that at first the experimental data of L.V. Shubnikov with his collaborators were obtained [40], then 20 years later the theory of type II superconductors was created by A.A. Abrikosov [6], describing these experimental data, and 10 years later, additional experimental data by U. Essmann and H. Trauble [38] appeared, which essentially confirmed the experiments [40] and the theory [6]. Today, physicists around the world often use the term "Shubnikov phase" for type II superconductors, introduced by Pierre-Gilles de Gennes, 1991 Nobel Prize winner in Physics for achievements in the field of soft matter physics which showed that "methods developed for studying order phenomena in simple systems can be generalized to more complex forms of matter, in particular to liquid crystals and polymers".

The creators of the theory of superconductivity, the great theoretical physicists L.D. Landau, V.L. Ginzburg and A.A. Abrikosov became the winners of the Nobel Prize in Physics: Lev Landau—in 1962, Vitaly Ginzburg and Alexei Abrikosov—in 2003. Unfortunately, the great experimental physicist Lev Vasilyevich Shubnikov lived a very short 36-year life. He was shot in 1937 on false charges of espionage for inviting foreign scientists to the Ukrainian Institute of Physics and Technology in Kharkov [39].

In connection with the above, it seems necessary for the author to note the undoubtedly decisive influence of the Kharkov period of the work of Lev D. Landau and the Kyiv period of the work of Mykola M. Bogolyubov, as well as the Ukrainian theoretical schools created by them for the further development of physical and mathematical researches in the scientific centers of Kyiv, Kharkov, Lviv, Odessa, Dnieper and other cities known far beyond the borders of Ukraine, in particular, in the field of studying the influence of fluctuation effects on phase transitions in various media (see, for example, [26–29, 55–64]).

Lev Landau(1908-1968)

Mykola Bogolyubov (1909-1992)

2.3 Section B

2.3.1 *Theoretical Justification of Examples Illustrating "Richard Feynman's Classification of Three Stages of Cognition of Natural Phenomena"*

Snell's Law. First consider an example that was chosen by Richard Feynman in [1] and which illustrates the phenomenon of refraction of a light beam at the interface of two air–water media. Snell's Law represents the second stage of the study of the phenomenon of light refraction according to "Richard Feynman's Classification of Three Stages of Cognition of Natural Phenomena" and in fact provides a theoretical basis for the experimental results of Claudius Ptolemy, which were obtained in the first stagy of the study of the light refraction.

The formula for Snell's law, which characterizes the refraction of a light beam at the boundary of two media, can be written in the following form:

$$\frac{\sin\alpha}{\sin\beta} = n_{21} = v_1/v_2 \tag{2.1}$$

Here α is the angle of incidence of the light beam in medium 1 (in the example under consideration, α is the angle between the incident beam AC in the air and the perpendicular to the air–water interface at the point of incidence C);β is the angle of refraction of the light ray in medium 2 (in this case, this is the angle between the refracted ray BC in water and the perpendicular to the air–water interface at the refraction point C;$n_{21} = 4/3$ is the relative refractive index of the second medium, i.e. water, relative to the first medium, i.e. air; v_1 is the speed of light in air, which is close in value to the speed of light in vacuum $c = 3 \cdot 10^8$ m/s; $v_2 = v_1/n_{21} = 2.25 \cdot 10^8$ m/s is the speed of light in water.

Figure 2.2 illustrates this example of the refraction of light at the interface between two media, air–water, which was described by Richard Feynman in "The Feynman Lectures on Physics" [1].

Fermat's principle of least time. In mathematical language, Fermat's principle means that the following integral, which determines the time of movement of the light beam along the trajectory AB (see Fig. 2.2), should take the minimum possible value:

$$\int_A^B nds/c = t_{\min\cdot}, \tag{2.2}$$

where ds is the element of the trajectory length, c is the light speed in vacuum, n is the refractive index. In the previous case of light refraction, the role of n was played by the relative refractive index n_{21} of water–air.

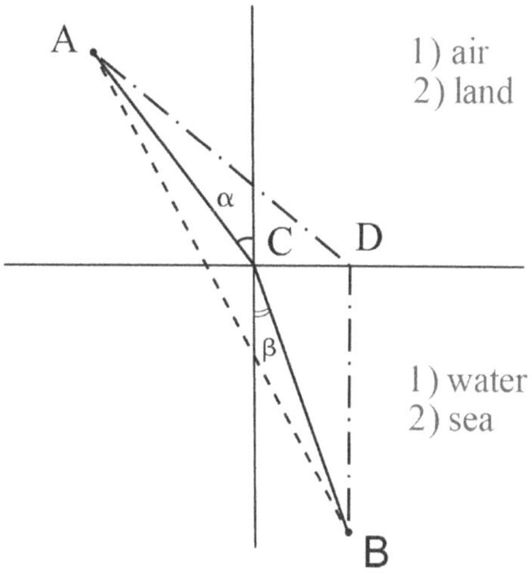

Fig. 2.2 An illustration of Snell's law of light refraction at the interface between two media, air–water (first example) and Fermat's principle of least time (second example)

Another equivalent definition of "Fermat's principle of least time" is the "Principle of least optical length", according to which a ray of light between two points A and B chooses a trajectory along which the optical length takes the minimum possible value. "Principle of least optical length" can be written in the form

$$\int_A^B ndS = c\int_{t_A}^{t_B} dt = ct_{\min} = [AB]_{\min}. \tag{2.3}$$

Here, the notation for optical length $[AB] = \int_A^B ndS$ is used [65]. To illustrate "Fermat's principle of least time" we will use Fig. 2.2 for the second time with slightly modified conditions compared to those used by Richard Feynman in [1]. Now the upper part of the picture corresponds to the land on which at point A there is a young man with outstanding athletic talents. He runs the 100 m in 9.58 s, equal to the world record held by a Jamaican sprinter Usain Bolt, and also swims the 100 m freestyle in 46.86 s, equal to the world record held by a Romanian swimmer David Popovici. Our young man is faced with a very difficult task to save a beautiful girl who cannot swim at all and has fallen from a boat into sea water. Such a disaster occurred at the point B at the bottom of Fig. 2.2, and the beauty will definitely drown if our record holder does not save her in the shortest possible time. Fortunately, the young man turned out to be not only a world sports record holder, but also endowed with mathematical abilities that allowed him to solve problems using variational principles. He very quickly calculated that his biathlon would not bring the desired result when moving along a straight-line connecting points A and B, as well as along the ADB trajectory, along which most of the AD is covered by land at a speed almost 5 times greater (or, to be more, accurate, then 4.89 times faster) than on water along the minimum length segment BD. In addition, our hero had a phenomenal memory, remembered "Fermat's principle of least time" and the proof of this principle given by Richard Feynman in [1], according to which the ratio of the sines of the angles of incidence and refraction at the boundary of two media is directly proportional to the ratio of the velocities in these environments. As a result, everything ended more than happily to the sounds of Mendelssohn's wedding march.

Thus, in full agreement with "Richard Feynman's Classification of Three Stages of Cognition of Natural Phenomena" (1) experimental data for the phenomenon of light refraction were obtained by Claudius Ptolemy at the first stage of cognition, (2) these experimental data were quantitatively described by the theoretical law of light refraction by Willibrord Snellius at the second stage of cognition, and (3) the final chord of the third stage of cognition was the formulation by Pierre de Fermat of the first principle of nature—"Fermat's principle of least time", which made these experimental and theoretical results obvious and was the undoubted pride of science.

It should be noted that the experimental stage of knowledge of natural phenomena should not necessarily precede the theoretical stage of knowledge. Quite often in physics, medicine and other sciences the situation is the opposite, i.e., theoretical achievements and discoveries precede and stimulate the conduct of corresponding experimental studies. As for the third stage of understanding natural phenomena, unfortunately, it is not always possible to realize it. Establishing the root cause or, in other words, the first principle for experimentally observed and/or theoretically explained natural phenomena is, as a rule, a great success. Later in this chapter we will consider several more such examples for which it is possible to formulate the first principles, i.e., implement the third stage of cognition in accordance with "Richard Feynman's Classification of Three Stages of Cognition of Natural Phenomena".

2.3.2 *First Principles for the Laws of Conservation of Energy, Momentum and Angular Momentum*

As noted above in the Introduction, the first principles of nature that underlie these fundamental conservation laws are related to the homogeneity of time, as well as the homogeneity and isotropy of space. Let us dwell in more detail on the connection of these conservation laws with the symmetry properties of space and time, following the presentation of this problem in the first volume of "Mechanics" [4] of the famous multi-volume course of theoretical physics created by Lev Landau and his students, among whom Evgeniy Lifshitz should be highlighted first of all.

The law of conservation of energy and its connection with the uniformity of time. A direct consequence of the homogeneity of time is the following expression for the total derivative of the Lagrange function L with respect to time t for a closed system or a system that is in a constant external field, which does not depend on time [4]:

$$dL/dt = \sum_i (\partial L/\partial q_i)\upsilon_i + \sum_i (\partial L/\partial \upsilon_i)a_i \tag{2.4}$$

Here q_i, υ_i, a_i are the coordinates, velocities and accelerations of all s particles of the system ($i = 1, 2, \ldots, s$). The absence in (2.4) of the partial time derivative $\partial L/\partial t$ indicates that the Lagrange function $L = T\text{–}U$, which was introduced by the French mathematician and physicist Joseph Louis Lagrange (1736–1813), does not depend explicitly on time precisely because homogeneity of time. Here the quadratic velocity function $T = \sum_i m\upsilon_i^2/2$ is the kinetic energy, and the function $U(q)$, which depends on the coordinates of all particles, is the potential energy.

Using Lagrange's equation

$$\frac{\partial}{\partial t}\frac{\partial L}{\partial \upsilon_i} - \frac{\partial L}{\partial q_i} = 0 \tag{2.5}$$

to replace the derivative $\partial L/\partial q_i$ in the first term on the right side of Eq. (2.4), we can obtain the following formula:

$$\frac{d}{dt}\left[\sum_i (\partial L/\partial \upsilon_i)\upsilon_i - L\right] = 0 \tag{2.6}$$

The result obtained means that the expression that is in the square brackets of formula (2.6) does not change over time.

Then, using the theorem on homogeneous functions, which was proven by the Swiss mathematician and physicist Leonhard Euler (1707–1783), we obtain the following result:

$$\sum_i \upsilon_i(\partial L/\partial \upsilon_i) - L = \sum_i m\upsilon_i^2/2 + U(q) \tag{2.7}$$

It follows that the value

$$E = \sum_i \upsilon_i(\partial L/\partial \upsilon_i) - L = \sum_i m\upsilon_i^2/2 + U(q) = const \tag{2.8}$$

is the time-conserving energy of the system E, which is equal to the sum of the kinetic energy T, depending on the square of the velocities, and the potential energy U, depending on the coordinates of all particles of the closed system. It should also be noted that the above discussion did not take into account the relativistic effects of Albert Einstein's special theory of relativity.

The law of conservation of momentum and its connection with the homogeneity of space. Taking into account the spatial homogeneity of a closed system with the parallel transfer of all coordinates of particles of the system r_i to a constant and velocity-independent small vector $\vec{a}$ leads to a change in the Lagrange function by the value [4]

$$\delta L = \sum_i (\partial L/\partial \vec{r}_i)\delta \vec{r}_i = \vec{a} \sum_i (\partial L/\partial \vec{r}_i) \tag{2.9}$$

The requirement that the Lagrange function be invariant $\delta L = 0$, taking into account Eq. (2.4) and the arbitrariness of the vector $\vec{a}$, leads to the following condition:

$$\sum_i (\partial L/\partial t)(\partial L/\partial \vec{\upsilon}_i) = (\partial/\partial t) \sum_i (\partial L/\partial \vec{\upsilon}_i) = 0 \tag{2.10}$$

The equality to zero of the time derivatives for the sum of the velocity derivatives of all particles from the Lagrange function in accordance with (2.10) means, taking into account the differentiation of the Lagrange function $L = T–U$, that the vector sum of the total momentum

$$\vec{\mathrm{P}} = \sum_i (\partial L/\partial \vec{\upsilon}_i) = \sum_i m_i \vec{\upsilon}_i = const \tag{2.11}$$

is a constant value, i.e., persists over time for a closed system.

The law of conservation of angular momentum and its connection with the isotropy of space. In [4] it is shown that, due to the condition of isotropy of space, the requirement that the Lagrange function remain unchanged under an infinitesimal rotation of the vector $\delta\vec{\varphi}$, which is directed along the axis of rotation, and its magnitude is equal to the angle of rotation, is reduced to the following relation:

$$\delta L = \sum_i (\partial L/\partial \vec{r}_i)\delta \vec{r} + \sum (\partial L/\partial \vec{\upsilon}_i)\delta \vec{\upsilon} = 0_i \tag{2.12}$$

After substituting into (2.12) the expressions for the increment of the radius vectors $\delta\vec{r}$ and the increment of the velocity vector $\delta\vec{\upsilon}\delta\vec{\upsilon}$

$$\delta\vec{r} = \left[\delta\vec{\varphi} \times \vec{r}\right], \tag{2.13}$$

$$\delta\vec{\upsilon} = \left[\delta\vec{\varphi} \times \vec{\upsilon}\right], \tag{2.14}$$

as well as replacement of derivatives

$$\partial L/\partial \vec{\upsilon}_i = \vec{\mathrm{P}}_i,\ \partial L/\partial \vec{r}_i = \partial \vec{\mathrm{P}}_i/\partial t, \tag{2.15}$$

one can get the following result:

$$\delta\vec{\varphi}(\partial/\partial t) \sum \left[\vec{r}_i \times \vec{\mathrm{P}}_i\right] = 0 \tag{2.16}$$

Taking into account the arbitrariness of the rotation vector $\delta\vec{\varphi}$ in (2.16), we have the following relation:

$$(\partial/\partial t)\sum\left[\vec{r}_i \times \vec{p}_i\right] = 0 \tag{2.17}$$

which means that the angular momentum vector

$$\vec{M} = \sum\left[\vec{r}_i \times \vec{p}_i\right] = const \tag{2.18}$$

persists over time in a closed system.

2.3.3 Physical Aspects and the First Principle for Appearance of Hexagonal Grid Cells in the Brain (2014 Nobel Prize in Physiology or Medicine)

Here, we will sequentially consider the main provisions that are necessary to formulate the first principle of the formation of hexagonal grid cells in the brain from the point of view of the proposed Richard Feynman's classification. This first principle of nonlinear interaction of order-parameter fluctuations in living and inanimate systems, which will be proposed here to explain the hexagonal structures of grid cells in the brain, uses fluctuation models of the processes of self-organization and the formation of ordered structures in chemically reacting systems [2, 11–13, 16]. The main provisions underlying the 3rd stage of the study of hexagons in the neurophysiological system of grid cells in accordance with the Feynman's classification are the following: (a) Hermann Haken's approach [7, 8], which allows to transfer the results of studying the physical system of Benard cells [14–16] on the neurophysiological system of grid cells in the brain [5, 41–43]; (b) the method of collective variables in the theory of phase transitions and critical phenomena [25–29]; (c) the Polyakov's conformal invariance hypothesis [24]; (d) the concept of universality classes near bifurcation (critical) points [30–32]; (e) the violation of the Polyakov's conformal invariance hypothesis in real classical fluids and grid cell system[2, 15, 16]; (f) the fluctuation model based on the Ginzburg–Landau Hamiltonian for chemically reacting systems [2, 16].

Theoretical background of Benard cells emergence as the 2nd stage of Feynman's classification. The theoretical explanation for the emergence of hexagonal Benard cells was proposed apparently for the first time in 1916 by Lord Rayleigh (John William Strutt), the 1904 Nobel Prize winner in Physics [66]. The appearance of hexagonal Benard cells, or the Benard-Rayleigh effect [14, 66], is caused by hydrodynamic instability and convective flows when the Rayleigh number

$$Ra = \rho^2 h^3 \Delta T C_p g \beta / \eta \lambda, \tag{2.19}$$

becomes equal or slightly greater than its critical value Ra_{cr} [67–69]. Here in (2.19), ρ is the fluid density; h is a thickness of the fluid layer;$\Delta T = T_1 - T_2$ is the temperature difference of the lower T_1 and upper T_2 surfaces of the fluid layer; C_p is the heat capacity at constant pressure; g is the acceleration due to gravity;β is the coefficient of thermal expansion;η is the dynamic viscosity;λ is the coefficient of thermal conductivity which is related to the coefficient of temperature diffusivity χ by the relation $\chi = \lambda/\rho C_P$.

The hydrodynamic theory of convective instability and formation of Benard hexagonal cells was developed in books [67–69], being ***the second stage of Feynman's classification***. To determine a critical value of the Rayleigh number Ra_{cr}, we will further use the results presented in the books by Lev Landau, Evgeniy Lifshits [68] and Werner Ebeling [69]. It turns out to be convenient to reduce the system of appropriately dimensioned hydrodynamic equations, linearized with respect to small perturbations of temperature τ, pressure Δp, and velocity $\upsilon_i, i = x, y, z$, to the following differential equation in partial derivatives for the temperature variable $\tau(x, y, z)$:

$$\Delta^3\tau = Ra\Delta_2\tau, \tag{2.20}$$

where $\Delta = \partial^2/\partial x^2 + \partial^2/\partial y^2 + \partial^2/\partial z^2$ and $\Delta_2 = \partial^2/\partial x^2 + \partial^2/\partial y^2$ are, correspondingly, the three-dimensional and two-dimensional Laplace operator. Representing the temperature variable $\tau(x, y, z)$ in a layer of a viscous liquid located between two horizontal surfaces $z = 0$ and $z = h$ with temperature T_1 and T_2 in the form

$$\tau(x, y, z) = f(z)\exp\left[i\left(k_x x + k_y y\right)\right], \tag{2.21}$$

where k_x and k_y are the components of the wave vector $\vec{k}$ in the x, y plane, one obtains the following differential equation for a function $f(z)$, which depends only on the vertical coordinate:

$$\left(\frac{d^2}{dz^2} - k^2\right)^3 f + Ra\frac{k^2}{h^4}f = 0 \tag{2.22}$$

Finding the solution of Eq. (2.22) as $f(z) = C\exp(\mu z/h)$, the sixth-order algebraic characteristic equation with respect to μ_n ($n = 1, \ldots, 6$) was obtained

$$\left(\mu^2 - a^2\right)^3 = -(Ra)a^2, \tag{2.23}$$

where a is the thermal expansion coefficient. Such an algebraic equation has the following roots [69]:

$$\begin{aligned} \{\mu_n\} &= \left\{\pm ir_0, \pm r, \pm r^*\right\}, r_0 = a\sqrt{\tau - 1}, \\ r &= a\sqrt{1 + (1/2)\tau\left(1 + \sqrt{3,}\right)} r^* = a\sqrt{1 + (1/2)\tau\left(1 + \sqrt{3,}\right)}. \end{aligned} \tag{2.24}$$

Substituting the found solutions (2.24) into the boundary conditions, one may obtain such results: (a) for the boundary conditions in the form of free surfaces, the critical value of the Rayleigh number $Ra_{\text{crit}} = 657.5$; (b) for the boundary conditions in the form of two rigid horizontal surfaces, the critical value of the Rayleigh number turns out to be equal $Ra_{crit} = 1708$. It should be pointed out, as was indicated in the book of L.D. Landau and E.M. Lifshits [68], that the last result regarding the Rayleigh number $Ra_{crit} = 1708$ was obtained by H. Jeffrey in 1908, i.e., eight years ahead of Rayleigh's article [66]; (c) it was also found that hexagons in the form of Benard cells arise only at $Ra \geq Ra_{\text{crit}}$,, if the Rayleigh number Ra was only slightly greater than its critical value Ra_{crit}. With a further increase in Ra, cylindrical structures became more preferable than hexagonal structures.

Hermann Haken's theoretical explanation of hexagonal Benard cells. For the case of hexagonal structures in Benard cells, Hermann Haken proposed an explanation, being close to ***the 3rd stage of the Feynman classification*** and using the generalized Ginzburg–Landau equations [7, 8, 16, 17]. The consequences obtained by H. Haken from these equations are as follows: (a) the formula (8.153) from [7] for the potential function $\Phi(A_k)$ is actually equivalent to the Ginzburg–Landau Hamiltonian for the Fourier components of the order parameter; (b) the following kinetic equation for the order parameter $\psi(x, t)$ (see formula (9.5.15) from [8]):

$$\partial\psi/\partial t = \lambda\psi/A\psi^2 - B\psi^3 \tag{2.25}$$

Here in (2.25), λ, A, B are the coefficients (see its connection with the parameters of the Ginzburg–Landau Hamiltonian in Sects. 8.13 and 9.4, 9.5, correspondingly, in Haken's monographs [7, 8]). As noted by H. Haken in [8], the solution of Eq. (2.25) appearance of hexagons. At the same time, in his unpublished studies, H. Haken received at zero coefficient A, found by Greenside, Cochrane Jr. and Schraer, does not lead to the hexagonal structures at $A \neq 0$. Below in Sect. 2.3, we will consider in more detail the issue of a rigorous theoretical justification of the appearance of cubic nonlinearity in the Ginzburg–Landau Hamiltonian and quadratic nonlinearity in the kinetic equation for the order parameter. It is the inclusion of these nonlinear terms that leads to the hexagonal structures based on fluctuation models of self-organization processes describing the formation of ordered structures near critical (bifurcation) points.

Here, we would like to emphasize the fact that the presence of a quadratic term $A\psi^2$ in (2.25) is equivalent to introducing an odd cubic term of the following form into the Ginzburg–Landau Hamiltonian $H\,[\psi_k]$ for the Fourier components ψ_k of the order parameter:

$$A \sum_{\vec{k},\vec{k'},\vec{k''}} \psi_{\vec{k}}\psi_{\vec{k'}}\psi_{\vec{k''}}\delta_{\vec{k}+\vec{k'}+\vec{k''},0} \tag{2.26}$$

The summation over the wave vectors should be carried out over an equilateral triangle under the conditions $\vec{k}_1 + \vec{k}_2 + \vec{k}_3 = 0$ and $\left|\vec{k}_1\right| = \left|\vec{k}_2\right| = \left|\vec{k}_3\right| = |k_0| = const.$

Obviously, these conditions lead to a coordinate lattice consisting of equilateral triangles, which in turn creates equilateral hexagons.

The method of collective variables in the theory of phase transitions. The method of collective variables in the theory of phase transitions was developed by I.R. Yukhnovskii and his co-authors [25–29]. Using this method and the Jacobian of transformation (see [26] and references there) from microscopic variables to collective variables, the authors of these works consistently obtained the fluctuation Ginzburg–Landau Hamiltonian. This Hamiltonian was commonly postulated in the renormalization group approach in the theory of phase transitions, for which the American physicist Kenneth Wilson received the Nobel Prize in Physics in 1982.

The starting point in the method of collective variables for describing second-order phase transitions is the following expression for the Hamiltonian:

$$H = -\frac{1}{2}\sum_{\vec{i},\vec{j}} \Phi\left(r_{\vec{i}\vec{j}}\right)\sigma_{\vec{i}}\sigma_{\vec{j}} - h\sum_{\vec{l}} \sigma_{\vec{l}}, \tag{2.27}$$

describing magnetic systems in a three-dimensional Ising model in an external magnetic field. In formula (2.27), the following designations are used: $\Phi\left(r_{ij}\right)$ is the interparticle interaction potential, $r_{ij} = \left|\vec{r}_{\vec{i}} - \vec{r}_{\vec{j}}\right|$ is the distance between the sites $\vec{i}$ and $\vec{j}$ of the lattice, which for definiteness will be assumed to be cubic, the variables $\sigma_{\vec{i}} = \pm 1$, and $h = \mu_B H'$, where μ_B is the Bohr magneton, H' is the magnetic field intensity. To simplify further calculations, the interparticle interaction potential is chosen in the form

$$\Phi\left(r_{ij}\right) = A\exp\left(-r_{ij}/b\right), \tag{2.28}$$

where A = const, b is a quantity describing the interaction radius.

Further, a fundamentally important transition fulfils from microscopic individual variables $\sigma_{\vec{l}}$ to N nodal collective variables $\rho_{\vec{k}}$ in the statistical sum

$$Z = \int \exp\left(\frac{1}{2}\sum_{\vec{k}} \beta\Phi(k)\rho_{\vec{k}}\rho_{-\vec{k}}\right) J_h(\rho)(d\rho)^N \tag{2.29}$$

where $J_h(\rho)$ is the corresponding Jacobian of the transition in variables $\sigma_{\vec{l}}$ to variables $\rho_{\vec{k}}$ [26]. Omitting further, generally speaking, rather complicated calculations of integral (2.29) and using the explicit form of the Jacobian $J_h(\rho)$, it can be shown that the final expression for the partition function Z of the three-dimensional Ising model, which describes the phase transition of a magnetic system in an external magnetic field, contains the following contributions. The first contribution corresponds to the Gaussian approximation (or the Ornstein–Zernike approximation) containing the squares of the Fourier components of collective variables. The second contribution corresponds to cubic nonlinearity and contains the products of three Fourier components of collective variables. The next contribution corresponds to the biquadratic nonlinearity (or the approximation of the biquadratic distribution in the terminology used in the method of collective variables), which is characterized by the presence of the product of four Fourier components of collective variables. Obviously, all these combined contributions correspond to the Ginzburg–Landau fluctuation Hamiltonian. Note that going beyond the biquadratic distribution approximation allows one to take into account higher even and odd contributions from nonlinear terms containing higher products of the Fourier components of collective variables. In addition, one should pay attention to the fact that the coefficients in terms containing odd degrees of the Fourier components of collective variables are proportional to the external field h. For $h \to 0$ all odd contributions disappear.

Polyakov's conformal invariance hypothesis of the theory of phase transitions. To rigorously substantiate the Haken's statement that Benard cells have a hexagonal geometry only in the presence of a quadratic nonlinearity in the kinetic Eq. (6) for the order parameter, one should turn to the conformal invariance hypothesis, first formulated by A.M. Polyakov in [24]. The physical meaning of this fundamental hypothesis of the theory of phase transitions and critical phenomena is reduced to the orthogonality (statistical independence) of anomalously fluctuating quantities (for example, order parameters) φ_1 and φ_2, which have different scale dimensions $\Delta\varphi_1 \neq \Delta\varphi_2$. In other words, for such statistically independent quantities, the pair correlator $\langle\varphi_1\varphi_2\rangle$ approaches a zero value: $\langle\varphi_1\varphi_2\rangle = 0$. As an example of the application of the conformal invariance hypothesis, let us consider the behavior of the pair correlation function of order parameter fluctuations and energy fluctuations near critical (bifurcation) points for two different systems: (1) a real classical liquid–vapor system and (2) a magnet in the lattice-gas model.

For the first case of a classical fluid, V.L. Pokrovskii showed in [70] that the pair correlator $\langle\varphi_1\varphi_2\rangle$ turns out to be nonzero and is characterized by the following temperature dependence in the vicinity of the critical isochore:

$$\langle\phi_1\phi_2\rangle = \Delta N \Delta E_{V=V_c} \sim \Delta\rho\Delta E_{V=V_c} \sim \tau^{-(\alpha+\gamma)/2} \approx \tau^{-0.674}. \tag{2.30}$$

Here, φ_1 and φ_2 are the fluctuations of the order parameters, being the deviations of the number of particles $\varphi_1 = \Delta N = (N - N_c)/N_c$ or the deviations of density $\varphi_1 = \Delta\rho = (\rho - \rho_c)/\rho_c$, and the deviations of energy $\varphi_2 = \Delta E = (E - E_c)/E_c$ from its critical values N_c, ρ_c, E_c, correspondingly; $\tau = (T - T_c)/T_c$ is the temperature deviation from the critical value T_c, $\alpha = 0.110$ and $\gamma = 1.237$ are the critical exponents, the numerical values of which are taken from the articles of M.A. Anisimov and his co-authors [22, 71–73]. The result obtained means that for a classical liquid the conformal invariance hypothesis is not fulfilled, and fluctuations of the number of particles ΔN (or density $\Delta\rho$) and energy fluctuations ΔE are not orthogonal, being statistically dependent quantities. Moreover, the pair correlator $\langle\Delta N \Delta E\rangle \sim \langle\Delta\rho\Delta E\rangle$ turns out to be anomalously fluctuating quantity as the critical temperature is approached. Note that taking into account the effects of spatial dispersion (spatial nonlocality of fluctuations) and/or temporal dispersion (temporal nonlocality of fluctuations) should provide a natural physical result—the finiteness of this quantity directly at the very point of the phase transition [74].

For the second system of the lattice-gas model, the situation turns out to be significantly different [16, 24, 72]. Fluctuations of the number of particles (or density) and energy fluctuations are statistically independent at the critical isochore, which corresponds to the zero value of the pair correlator

$$\langle \phi_1 \phi_2 \rangle = \Delta N \Delta E_{V=V_c} \sim \Delta \rho \Delta E_{V=V_c} = 0. \tag{2.31}$$

In other words, for the fluctuating quantities given in (2.31) in the lattice gas model, Polyakov's conformal invariance hypothesis is satisfied [36].

The considered lattice gas model is isomorphic (or belongs to the same universality class) of the three-dimensional (3D) Ising model in the absence of a magnetic field ($h = 0$). Therefore, the analogue of formula (2.31) for the 3D Ising model at $h = 0$ is the following relation [24]:

$$\langle \phi_1 \phi_2 \rangle = \Delta M \Delta E_{h=0} = 0, \tag{2.32}$$

where such a fact is taken into account that the role of the fluctuating quantity φ_1 for the Ising model is played by fluctuations of the magnetization ΔM. Formula (2.32) means that the fluctuations of magnetization ΔM and energy fluctuations ΔM are statistically independent. Therefore, in accordance with the conformal invariance hypothesis, such fluctuations for the 3D Ising model at $h = 0$ are orthogonal. Note again that the vanishing of the paired correlator is an approximate result which assumes the neglect of the effects of spatial and/or temporal dispersion. As indicated in [74], the terms describing the effects of spatial and/or temporal dispersion should naturally be added: (1) to those quantities (for example, the diffusion coefficient, the speed of sound, a pair correlator for a magnet in the lattice-gas model, etc.), which vanish at critical (bifurcation) points or points(lines) of second-order phase transitions; (2) to the inverse values of those quantities (for example, susceptibility, isobaric and isochoric heat capacities, etc.) which anomalously increase with approaching the critical (bifurcation) points or points (lines) of second-order phase transitions.

The concept of universality classes near bifurcation (critical) points. Let us now consider the concept of universality classes which is essential for the study of systems, generally speaking, of different nature, demonstrating the same behavior of physical properties in the vicinity of bifurcation (critical) points, as well as points (lines) of second-order phase transitions.

For ***bulk systems*** with linear dimensions L significantly exceeding the correlation radius (correlation length) ξ of the characteristic order parameter, the concept of a universality class includes the following main features which must be the same: (1) the same spatial dimension d, (2) the same number n of order-parameter components (for example, $n = 1$ for the scalar order parameter in the Ising model, $n = 2$ for the two-component order parameter in the plane rotator model, $n = 3$ for the vector order parameter in the Heisenberg model), (3) the same type (short- or long-range) of intermolecular interaction in comparison with the intermolecular distance, (4) the same symmetry of the fluctuation Ginzburg–Landau Hamiltonian, or fluctuation part of the Gibbs thermodynamical potential [19–23, 30–32].

For ***spatially bounded systems***, for which $L \approx \xi$, to this list of identical features, the following should be added: (5) the same type of geometry of the system or, in other words, the lower crossover dimension d_{cros} ($d_{cros} = 2$ for a slitlike, plane-parallel pore, $d_{cros} = 1$ for a cylindric pore, $d_{cros} = 0$ for a bounded sphere or cube), (5) the same type of boundary conditions (hydrophilic, hydrophobic, partially wetting), (6) the same physical properties under consideration [2, 16, 60–64, 72–77].

To explain additional conditions for universality classes in spatially limited systems, such features should be taken into account. In bulk systems, when a strong inequality $L \gg \xi$ is satisfied, the only critical values of thermodynamic parameters exist for a certain substance (for example, a single critical temperature $T_c = 647.3K$ for water). When critical values of these thermodynamic parameters are reached in bulk liquids, physical properties such as isothermal compressibility, isobaric and isochoric heat capacities, sound speed, self-diffusion coefficient, etc., take on infinite or zero values in an approximation that does not take into account the effects of spatial and temporal dispersion. In spatially limited systems, when such a condition $L \leq \xi$ is valid, the physical properties of the same substances (for example, the same water) are known to drastically change their critical behavior. The numerical values of the physical properties mentioned above become finite and non-zero at certain extremum points. It turns out that the shifts of the critical values of such thermodynamic parameters of bounded systems as the critical temperature $T_c(L)$, the critical density $\rho_c(L)$, the critical pressure $P_c(L)$ with respect to the values of the same critical parameters $T_c(\infty)$, $\rho_c(\infty)$, $P_c(\infty)$ in bulk systems cease to be universal and are different for each physical property.

To illustrate what has been said, consider the following example [78]. The shift of the critical temperature $T_c(L)$ in a bounded system with a linear dimension L relative to the critical temperature $T_c(\infty)$ in a bulk system is described by the following universal formula:

$$\Delta T_C = T_C(\infty) - T_C(L) = KL^{-1/v(L)} \tag{2.33}$$

Here, $v(L)$ is the critical index of the temperature dependence of the correlation radius $\xi(L)$ in bounded liquids, which should depend on the linear size L of the system in the direction of its spatial limitation, when taking into account the effects of the dimensional crossover [79]; and K is a nonuniversal coefficient, the appearance of which in formula (2.33) is associated with the presence of nonuniversal parameters of a particular substances in the condition of an extremum for a certain physical property.

The violation of the Polyakov's conformal invariance hypothesis in real classical fluids and grid cell system. Let us return to the connection between the Polyakov's conformal invariance hypothesis and Haken's statement that hexagonal structures in Benard cells are observed only in the presence of quadratic nonlinearity $A\psi^2$ in kinetic Eq. (6). As was noted in Sect. 2.2, this nonlinearity in (2.25) is completely equivalent to the presence of the cubic nonlinearity of the form (2.26) in the fluctuation Ginzburg–Landau Hamiltonian. In turn, the appearance of the odd (cubic) term (2.26) in the Ginzburg–Landau Hamiltonian is directly related to the Polyakov's conformal invariance hypothesis or, more precisely, is a direct consequence of the violation of the conformal invariance hypothesis in real classical fluids.

Thus, it can be argued that the two systems considered above—the real classical liquid and the magnetic in the lattice gas model—belong, generally speaking, to different universality classes due to the different symmetry of the Ginzburg–Landau Hamiltonians (the fluctuation part of the Gibbs thermodynamic potentials). Hexagons arise only in those systems located near critical (bifurcation) points for which the Ginzburg–Landau Hamiltonian is not invariable with respect to the inversion of the order parameter $\varphi \leftrightarrow -\varphi$ and an odd (cubic) nonlinearity arises in the Ginzburg–Landau Hamiltonian.

In Sect. 2.2.2, we discussed the physical properties of the human brain matter. It turned out that according to some parameters (average density, relative dielectric permittivity) the gray matter of the brain is close to water. The discovery of hexagonal grid cells in the entorhinal cortex of brain by Edward Moser and May-Britt Moser in 2005 makes the system of grid cells in the brain even closer to a real liquid–vapor system. Indeed, the main and fundamental explanation for this fact lies in the fact that both of these systems belong to the same universality class of the 3D Ising model in an external magnetic field, for which the Polyakov's conformal invariance hypothesis is violated.

It should be noted that the violation of the Polyakov's conformal invariance hypothesis is also a reason for the appearance of a ***singularity of the temperature derivative*** for the so-called ***"diameter of the coexistence curve (CC)"*** of the liquid–vapor system. As shown in a number of theoretical works (see, for example, [15, 16, 20, 22, 79–81]), the diameter of the CC, defined as the half-sum of the average densities on the liquid and gas (vapor) branches of the CC $\rho_d(\tau) = (\rho_{liquid} + \rho_{vapour})/2$, is characterized by the same singularity of its temperature derivative as the isochoric heat capacity C_V near the critical point of the classical liquid–vapor system:

$$d\rho_d/dT \sim C_V \sim |\tau|^{-\alpha} \tag{2.34}$$

The first experimental confirmation of the singularity of the diameter was obtained at the Department of Molecular Physics of the Taras Shevchenko Kyiv National University in [82], and then confirmed in a number of other experimental studies (see, for examples, [83–85]). Table 1 illustrates simultaneously (a) the similarity (isomorphism) between system of grid cells in the brain and a real liquid–vapor systen, as well as (b) the principal difference in the physical properties of a real liquid–vapor system and a lattice-gas model.

The comparison of the physical properties of three different systems, presented in Table 2.3, allows us to draw the following conclusion: the violation of the Polyakov's conformal invariance hypothesis is the reason for the appearance of hexagons in the system of grid cells in the brain, as well as the singularity of the temperature derivative of density diameter along the CC in a real classical liquid–vapor system.

Ginsburg-Landau Hamiltonian and fluctuation model for systems with chemical reactions. The first principle for grid cells in the brain. To theoretically describe the hexagonal grid cells in the brain, one should write the fluctuating part of Gibbs thermodynamic potential $\Delta G(\xi)$ in a system with chemical (biochemical) reactions, using the Ginsburg-Landau Hamiltonian in the following generalized form [2, 11–13, 16]:

$$\Delta G(\xi) = \int d\vec{r}\big[(1/2)\lambda\xi^2 + (1/2)C_2\big(\nabla\xi^2\big) + (1/3)C_3\xi^3 + (1/4)C_4\xi^4 - A^*\xi\big]. \tag{2.35}$$

The choice of the Gibbs thermodynamic potential $G(T, p, N_i)$ is determined by the real isothermal and isobaric conditions for living objects. Here in (2.35), ξ is the reaction coordinate (or the degree of completeness) which is the order parameter of a thermodynamical system with chemical reaction, being determined by the formula:

$$d\xi = dN_i/\nu_i, \tag{2.36}$$

where dN_i is the change in the number of moles of an i-component, while ν_i are the stoichiometric coefficients of the chemical reaction.

The conjugate quantity (in the thermodynamic sense) for the reaction coordinate ξ is the affinity

Table 2.3 Comparison of grid cells system, real liquid–vapor system and lattice-gas model

System	Universality class	Odd terms in Ginzburg–Landau Hamiltonian	Polyakov's hypothesis of conformal invariance	Singularity of diameter temperature derivative	Hexagons in Benard cells and grid cells in the brain
Grid cells in the brain	3D Ising model in magnetic field	Present	Not realized	–	Present
Real liquid–vapor system	3D Ising model in magnetic field	Present	Not realized	Present	Present
Lattice-gas model	3D Ising model without magnetic field	Absent	Realized	Absent	Absent

$$A^* = \sum_i \nu_i \mu_i, \tag{2.37}$$

depending on the chemical potentials μ_i of all components of a chemical reaction.

Formula (2.35) takes into account not only the Gaussian quadratic term $\lambda\xi^2$ and the fourth-order nonlinear fluctuation contribution $C_4\xi^4$, being usually accepted for the standard form the Ginsburg-Landau Hamiltonian, but also the cubic nonlinear interaction $C_3\xi^3$ of the order-parameter fluctuations in the chemically reacting system. Such an odd term appears in (2.35) due to the violation of Polyakov's conformal invariance hypothesis [24]. Note also that the last term in (2.35), containing the product of the reaction coordinate and the affinity, has the meaning of the interaction of the order parameter ξ with the external field A^*. It should be noted that some examples of the interaction mechanisms of order parameter fluctuations in systems with chemical (biochemical) reactions were considered in the kinetic nonlinear models studied in the articles [47, 52–54].

As a result of calculating the functional derivative of the fluctuation part of the Gibbs thermodynamic potential $\Delta G(\xi)$ with respect to the order parameter ξ, we obtain the following kinetic differential equation for the coordinate (degree of completeness) $\xi(\vec{r})$ of the biochemically reacting system of grid cells:

$$\frac{\partial \xi}{\partial t} = -\Gamma\left(\lambda\xi - C_2\Delta\xi + C_3\xi^3 + C_4\xi^4 - A^*\right) \tag{2.38}$$

Nonlinear kinetic Eq. (2.38) for the order parameter ξ and the Ginzburg–Landau Hamiltonian (the fluctuation part of the thermodynamic Gibbs potential) (2.35) for a chemically reacting system of neurons in the brain have the same functional form as the nonlinear kinetic equation for the order parameter (9.5.15) from [8] and expression (8.153) for the potential function from [7] in the Haken's theory describing the occurrence of hexagonal Benard cells. Therefore, all statements by Hermann Haken about the formation of hexagons in Benard cells can be almost completely transferred to the system of grid cells in the brain.

Thus, conducted research allows us to formulate the following first principle: **the formation of hexagonal structures in the system of grid cells in the brain occurs due to the nonlinear interaction of fluctuations in the degree of completeness (coordinate) of chemically interacting systems of neurons**. In addition, an obligatory condition for the nonlinear interaction of fluctuations of the order parameter in the system of grid cells in the brain is the violation of the Polyakov's conformal invariance hypothesis, a direct consequence of which is the appearance of a term cubic in fluctuations of the degree of completeness in the Ginzburg–Landau Hamiltonian.

2.4 Conclusions

This chapter is aimed at formulating "Richard Feynman's classification of three stages of the study of natural phenomena" and, first of all, the most important 3rd stage of cognition connected with understanding the "first principles", which explain the reasons for the appearance of laws describing the experimental data. Here, Edward Moser's idea of a certain analogy between grid cells in the brain and the Abrikosov's vortex lattice in type II superconductors was considered. In addition, we paid special attention to the experimental studies that led to the discovery of the hexagonal structures of grid cells in the human and animal brain, as well as the main theoretical provisions that are necessary to take into account to formulate the first principle related to the proposed Feynman's classification. Namely, physical parameters of the human brain and its medical applications, features of the process of generation and propagation of the action potential, the electrical and chemical nature of intercellular interaction. It is these essential factors that will later enable us to show that the grid cells in a chemically interacting system of neurons rather belong to the universality class of a real liquid–vapor system with a scalar order parameter. In addition, to consistently substantiate the above results, the fluctuation theory of phase transitions was used for studying the process of the appearance of hexagonal grid cells as a necessary component of the positional system in the brain of humans and animals (2014 Nobel Prize in Physiology or Medicine).

Three important conclusions necessarily follow from the above studies:

- The use of the Ginzburg–Landau Hamiltonian and the associated nonlinear kinetic equation for the order parameter of a chemically reacting system means accepting the following hypothesis. The neurophysiological system of grid cells in the brain, in which hexagons appear near the points of structural bifurcation transitions, actually belongs to the same universality class as the 3D Ising model in the presence of an external magnetic field near the second-order phase transition, as well as the real liquid–vapor system near the critical point.
- As a direct consequence of this hypothesis, the first principle of the formation of hexagonal structures in the grid cell system must be associated with the nonlinear interaction of fluctuations of the degree of completeness (coordinate) of chemical reactions in the brain.
- A rigorous theoretical substantiation of this first principle of the emergence of hexagons requires the mandatory involment of the fact that the Polyakov's conformal invariance hypothesis for fluctuations of the order parameters is violated for the grid cells s in the brain, and in a more general case, fluctuations of the corresponding order parameters in systems of living and inanimate nature near its bifurcation (critical) points and boundary of stability.

References

1. R.P. Feynman, R.B. Leighton, M. Sands, *The Feynman Lectures on Physics*, vol. 1, Chapter 26, California Institute of Technology (2013); http://www.feynmanlectures.caltech.edu
2. A.V. Chalyi, Synergetic dialogue "physics-medicine": Hexagones in living and inanimate nature. J. Mol. Liquids **329**, 114249 (2021)
3. E. Noether, Invariant variation problems. Transport Theory Statist. **1**(3), 183–207 (1971); arXiv: physics/0503066v3 [physics.hist-ph]
4. L.D. Landau, E.M. Lifshits, *Theoretical Physics*, v.1, Mechanics (Pergamon Press, Oxford, 2000)
5. J. O'Keefe, M.-B. Moser, E.I. Moser, The Nobel prize in physiology or medicine. *Nobel Prize. org. Nobel Media* AB (2014); http:/nobelprize.org/prizes/medicine/2014/summery
6. A.A. Abrikosov, On the magnetic properties of superconductors of the second group. JETP **5**, 1174 (1957)
7. H. Haken. *Synergetics. An Introduction. Nonequilibrium Phase Transitions and Self-Organization in Physics, Chemistry and Biology*. (Springer, Berlin, Heidelberg, New York, 1978)
8. H. Haken, *Advanced Synergetics. Instability Hierarchies of Self-Organizating Systems and Devices*. (Springer, Berlin, Heidelberg, New York, Tokyo, 1983).
9. A.V. Chalyi, *Synergetic Principles of Education and Science* (Bogomolets National Medical University, Kyiv, 2000)
10. A.V. Chalyi, a) Synergetics, b) Synergetical approach in education. In: V.G. Kremen' (Ed.) *Encyclopedia of Education* (Yurincom Inter, Kyiv, 2021)
11. Th.E. de Donder, *Thermodynamic Theory of Affinity: A Book of Principles* (Oxford University Press, Oxford, 1936)
12. G. Nicolis, I. Prigogine, *Self-Organization in Nonequilibrium Systems: From Dissipative Structurers to Order Through Fluctuations* (Wiley, New York, 1977)
13. A.V. Chalyi, I.V. Nezhinskii, V.M. Sysoev, The fluctuation mechanism of the formation of spatio-temporal structures in chemically active systems, in: *Proceedings of II International Workshop on Nonlinear and Turbulent Processes in Physics*, Bogolyubov Institute for Theoretical Physics, National Academy Sciences of Ukraine (Gordon and Breach, New York, 719–730, 1984)
14. H. Benard, Les tourbillans cellulaires dans une nappe liquide. Revue generale des sciences, pares et appliquees. **11**(1261–1271), 1309–1328 (1900)
15. A.V. Chalyi, K.A. Chalyy, E.V. Zaitseva, A.A. Kryshtopa. Feynman's classification of natural phenomena and physical aspects of 2014 Nobel Prize in Physiology and Medicine, Ukr J. Phys. **67**(10), 736 (2022). https://doi.org/10.15407/ujpe67.10.736
16. A.V. Chalyi, K.A. Chalyy, E.V. Zaitseva, E.N. Chaika, I.P. Kryvenko, Physical aspects of 2014 Nobel prize in physiology and medicine: 2. The first principle and universality class for grid cells in the brain. Ukr J. Phys. **68**(7), 462 (2023). https://doi.org/10.15407/ujpe68.7.462
17. V.L. Ginzburg, L.D. Landau, On the theory of superconductivity. JETP **20**, 1064 (1950)
18. M.E. Fisher, *The Nature of Critical Points, Lectures in Theoretical Physics* (University of Colorado Press, Colorado, 1965)
19. L.P. Kadanoff, Critical behavior, universality and scaling, *Proceedings of International School Physics* (Acad. Press, New York, London, 1971)
20. A.Z. Patashinskii, V.L. Pokrovskii, *Fluctuation Theory of Phase Transition* (Pergamon Press, Oxford, 1979)
21. H.E. Stanley, *Introduction to Phase Transition and Critical Phenomena* (Oxford University Press, Oxford, 1987)
22. M.A. Anisimov, *Critical Phenomena in Liquids and Liquid Crystals* (Gordon & Breach, Philadelphia, 1991)
23. L.D. Landau, E.M. Lifshits, *Theoretical Physics, Statistical Physics, Part 1*, vol. 5 (Pergamon Press, Oxford, 2013)

24. A.M. Polyakov, Conformal symmetry of critical fluctuations. JETP Lett. **12,** 381 (1970). Dumka, Kyiv (1984)
25. I.R. Yukhnovsky, *Phase Transition of the Second Order. Method of the Collective Variables* (Naukova Dumka, Kyiv, 1984)
26. I.R. Yukhnovsky, M.F. Golovko, *Statistical Theory of Classical Equilibrium Systems* (Naukova Dumka, Kyiv, 1980)
27. I.R. Yukhnovsky, M.P. Kozlovsky, I.V. Pilyuk. *Microscopic Theory of Phase Transitions in Three-Dimensional Systems.* (Eurosvit, Lviv, 2001)
28. M.P. Kozlovsky. *Influence of an External Field into the Critical Behavior of Three-Dimensional Systems.* (Galician Druker, Lviv,2012)
29. *Order, Disorder and Criticality. Vol. 6. Advanced Problems of Phase Transitions Theory.* Edited by Yu.V. Holovatch (World Scientific, Singapore, 2020)
30. M.E. Fisher, Renormalization of critical exponent by hidden variable. Phys. Rev. **176**, 257 (1968)
31. M.A. Anisimov, E.E. Gorodetskii, V.D. Kulikov, A.A. Povodyrev, J.A. Sengers, A general isomorphism approach to thermodynamic and transport properties of binary liquid mixtures near critical points. Physics A **220**, 277 (1995)
32. A.V. Chalyi, L.A. Bulavin, V.F. Chekhun, K.A. Chalyy, L.M. Chernenko, A.N. Vasilev, E.V. Zaitseva, G.V. Khrapijchuk, A.V. Severin, M.V. Kovalenko, Universality classes and critical phenomena in confined liquid systems. Condensed Matter Phys. **16**, 23008 (2013)
33. *Medical and Biological Physics. Practical Guide.* Edited by A.V. Chalyi (Kyiv: Knyga Plus, 2003)
34. *Medical and Biological Physics.* Edited by A.V. Chalyi, 4th edition (Vinnytsia: Nova Knyga, 2020)
35. J. Eccles, A.L. Hodgkin, A.F. Huxley, *Nobel Lectures, Physiology or Medicine 1963–1970* (Elsevier Publishing Company, 1972)
36. V.A. Berezovsky, N.N. Kolotilov, *Biophysical characteristics of human tissues. Reference Book* (Naukova Dumka, Kyiv,1982)
37. D. Hubel, C. Stevens, E. Kandel, et al. *The Brain, Scientific American* № 10 (1979)
38. U. Essmann, H. Trauble, The direct observation of individual flux lines in type II superconductors. Phys. Lett. **24A**, 526–527 (1967)
39. V.V. Slezov, I.I. Papirov, A.G. Shepelev. Brief history of the discovery of type II superconductors. *Questions of atomic science and technology.* №6, 86–92 (2009)
40. L.V. Shubnikov, V.I. Khotkevich, Yu.D. Shepelev, Yu.N. Ryabinin, Magnetic properties of superconducting metals and alloys. Sov. J Exp Theor Phys **7**(2), 221–237 (1937)
41. J. O'Keefe, Place units in the hippocampus of the freely moving rat. Experimental Neurology **51**, 78–109 (1976)
42. T. Hafting, M. Fyhn, S. Molden, M.-B. Moser, E.I. Moser, Microstructure of a spatial map in the entorhinal cortex. Nature **436**, 801–806 (2005)
43. J. Jacobs, C.T. Weidemann, J.F. Miller et al., Direct recording of grid-like neuronal activity in human spatial navigation. Nat. Neorosci. **16**, 1188–1190 (2013)
44. L. Kunz, T.N. Schroder, H. Lee et al., Reduced grid-cell-like representations in adults at genetic risk for Alzheimer's disease. Science **350**, 430–433 (2015)
45. P.G. Kostyuk, O.A. Kryshtal, *Mechanisms of Electrical Excitability of Nervous System* (Nauka, 1981)
46. A.V. Chalyi, L.M. Chernenko, Phase transitions in finite-size systems and synaptic transmission, in *Dynamic Phenomena at Interfaces, Surfaces and Membranes*, D. Beysens, N. Boccara, G. Forgacs (Eds.), (Nova Sci. Publ., New York, 1993, 457–464)
47. J. Nolte, *The Human Brain* (Mosby, Missouri, 2002)
48. V. Tsymbalyuk, V. Medvedev, *Man, and his Brain*, vol.1 (Nova Knyga, Vinnytsia, 2022)
49. P.G. Kostyuk, V.L. Zima, I.S. Mogura, M.S. Miroshnichenko, M.F. Shuba, *Biophysics* (Kyiv University Press, Kyiv, 2008)
50. L.A. Bulavin, L.G. Grechko, L.B. Lerman, A.V. Chalyi, *Medical Physics* (Dynamic and statistical models (Kyiv University Press, Kyiv, 2011)

51. A.V. Chalyi, A.N. Vasilev, E.V. Zaitseva, Synaptic transmission as a cooperative phenomenon in confined systems. Condensed Matter Phys. **20**(1), 13804 (2017)
52. A.V. Chalyi, E.V. Zaitseva, Strange attractor in kinetic model of synaptic transmission. J. Phys. Studies **11**(3), 322 (2007)
53. A.V. Chalyi, E.V. Zaitseva, A kinetic model of synaptic transmission on intercell interaction. Ukr. J. Phys. **54**(4), 366–370 (2009)
54. M.E. Kriebel, B. Keller, R.B. Silver, G.D. Pappas, Porocytosis: a transient pore array secrets the neurotransmitter packet. The Anatomical Record. Advances in integrative anatomy and evolutionary biology (2005); https://doi.org/10.1002/ar.b.20051
55. *Bogolyubov Institute for Theoretical Physics, NAS of Ukraine. 1966—2016* Edited by A. G. Zagorodniy (Akademperiodika, Kyiv, 2015)
56. B. Lev, A. Zagorodny, *Applications of Field Theory Methods in Statistical Physics of Nonequilibrium Systems* (World Scientific, Singapore, 2021)
57. I.M. Mryglod, V.V. Ignatyuk, Yu.V. Golovach. *Mykola Bogolyubov and Ukraine* (Lviv: Evrosvit, 2009)
58. I.Z. Fisher, *Statistical Theory of Liquids* (Fizmatgiz, 1961)
59. L.A. Bulavin, K.O. Chalyy, *Neutron Optics of Mesoscale Liquids* (Naukova Dumka, Kyiv, 2006)
60. L.A. Bulavin, A.V. Chalyi, O.I. Bilous, Anomalous propagation and scatterig of sound in 2-propanol water solution near its singular point. J. Mol. Liquids **235**, 24–30 (2017)
61. A.V. Chalyi, A.N. Vasilev, Correlation properties, critical parameters and critical light scattring in finite-size systems. J. Mol. Liquids **84**(2), 203–213 (2000)
62. K.A. Chalyy, K. Hamano, A.V. Chalyi, Correlating properties of a simple liquid at criticality in a reduced geometry. J. Mol. Liquids **92**(1–2), 153–164 (2001)
63. K.A. Chalyy, L.A. Bulavin, A.V. Chalyi, Dynamic scaling and central component width of critical opalescence spectrum in liquids with restricted geometry. J. Phys. Studies **9**(1), 66–70 (2005)
64. V.G. Boiko, H.J. Moegel, V.M. Sysoev, A.V. Chalyi, Features of metastable states in liquid-vapor phase transitions. *Soviet Physics—Uspekhi* **34**, 141–159 (1991)
65. M. Born, E. Wolf, *Principles of Optics* (Pergamon Press, Oxford-London-Edinburgh-New York-Paris-Frankfurt, 1968)
66. Lord Rayleigh, On convection currents in a horizontal layer of fluid, when the higher temperature is on the under side. London Edinburg Dublin Phylosophical Mag. J. Sci. **32**, 529 (1916)
67. S. Chandrasekhar, *Hydrodynamic and Hydromagnetic Stability* (Clarendon, Oxford, 1961)
68. L.D. Landau, EM Lifshits, *Fluid Mechanics*, 2nd edn. (Elsevier, Butterworth-Heinemann, Amsterdam, Boston, 2012)
69. W. Ebeling, *Struturbildung bei irreversible Prozessen. Einführung in die Theorie dissipativer Strukturen.* (Teubner-Verlag, Leipzig, 1976)
70. V.L. Pokrovskii, Feasibility of experimental verification of conformal invariance hypothesis. JETP Lett. **17**, 156 (1973)
71. D.A. Fuentevilla, M.A. Anisimov, Scaled equation of state for supercooled water near the liquid-liquid critical point. Phys. Rev. Lett. **9**, 149904 (2007)
72. C.E. Bertrand, M.A. Anisimov, Peculiar thermodynamics of the second critical point in supercooled water. J. Phys. Chem. B **115**, 14099 (2011)
73. V. Holten, C.E. Bertrand, M.A. Anisimov, J.V. Sengers, Thermodynamics of supercooled water. J. Chem. Phys. **136**, 094507 (2012)
74. V.M. Sysoev, A.V. Chalyi, Correlation functions and dynamic structure factor of non-isotopic systems. Theor. Math. Physics **26**, 126 (1976)
75. M.E. Fisher, The theory of critical point singularities. In: M.S.Green (Ed.), *Critical Phenomena, Proceedings of the International School of Physics "Enrico Fermi.* (Academic Press, New York, 1971, 1–99)
76. K. Binder, Phase transitions in reduced geometry. Annu. Rev. Phys. Chem. **43**, 33 (1992)
77. A.V. Chalyi, Surface tnsion in bulk and bounded liquids. J. Mol. Liquids **288**, 110873 (2019)

78. A.V. Chalyi, Dimensional crossover in liquids in reduced geometry, in L. Bulavin, N. Lebovka (Eds.) *Physics of Liquid Matter: Modern Problem,* (Springer International Publishing, Switzerland, 31–49, 2015)
79. A.V. Chalyi, Inverse correlation radius expansions and the "rectilinear diameter" singularity. Ukr. J. Phys. **21**(3), 474–477 (1976)
80. J. Wang, M.A. Anisimov, Nature of vapor-liquid asymmetry in fluid criticality. Phys. Rev. E **75**(5), 051107 (2007)
81. O. Bakai, M. Bratchenko, S. Dyuldya, On the singularity of the liquid-gas curve dimeter. Ukr. J. Phys. **65**(9), 802 (2020)
82. L.M. Aptyukhoskaya, E.T. Shimanskaya, Yu.I. Shimanskii, Heptane coexistence curve near the critical point. JETP **63**(3), 2153 (1972)
83. M. Nakata, H. Kawahara, Coexistence curve for polystyrene-cyclohexane near the critical point. J. Chem. Phys. **62**(11), 430349 (1975)
84. B.S. Maccabee, J. White, A non-symmetrical parametric quation of state for the extended critical region of carbon dioxide. Phys Lett **60A**(3), 179 (1977)
85. L.A. Bulavin, Yu.I. Shimanskii, The singularity of the coexistence-curve diameter of etheane. JETP Lett. **29**(8), 482 (1979)

Chapter 3
Basic Principles of Synergetics and its Medical Applications

Abstract The origin of living matter in the process of evolution of the world around and inside us, the emergence of ordered structures in living and nonliving nature—phenomena that continue to remain among the most amazing and difficult to understand within the framework of materialistic science. Synergetics is a relatively new direction in science and education, which studies the processes of self-organization and the formation of ordered structures in animate and inanimate worlds. The synergetic approach uses the fundamental similarity between main features of the phenomena under investigation occurring in open, nonlinear, complex systems of various nature. Most of the achievements of the synergetic approach are largely appear due to the possibility of transferring the results of one science (for example, physics) that is, as a rule, more "simple" and accurate because of omitting many details which at first glance can be considered inessential, to the objects of another science (for example, medicine) that are much more complicated and which, therefore, can be described rather qualitatively than quantitatively. This approach is based on such well-tested and precise methods as the theory of phase transitions and critical phenomena, the theory of nonlinear oscillations and autowaves, kinetic models of "brussellator", "oregonator" and synaptic trasmission, catastrophe and fractal theories and others. Dear reader, this chapter can be considered as a short excursion into the beautiful and diverse synergetic world.

Keywords Synergetics · Synergetic principles · Phase transitions · Critical phenomena · Self-organizing criticality · Medical applications

A. V. Chalyi, *What is Medicine?*, https://doi.org/10.1007/978-3-031-64979-0_3

Centuries and millennia passed until human thought was able to note the features of a single mechanism in the picture of nature, which seemed chaotic.

Volodymyr Vernadskyi, the first President of the Ukrainian Academy of Sciences (1919–1921)

3.1 Introduction

Synergetics is an interdisciplinary scientific and educational direction that studies integrative trends and processes of self-organization and ordering in open, non-linear, dynamic, complex systems of various nature. This branch of science and education, for which the Pofessor of the University of Stuttgart (Germany) Hermann Haken proposed the name "synergetics", began to take shape in its modern form about 50 years ago [1–3].

Explaining the choice of this term, Professor Hermann Haken mentioned that he chose the word "synergetics" at the time, because Greek terms were fixed for many disciplines in science. He was looking for a word that would correspond to the joint activity, the common energy to create something, because systems are self-organizing, and so it can seem that they tend to create new structures. Prof. Haken turned then for advice to his school friend Haus Christoph Wolff, who knew Greek well. While discussing various concepts, Haken aimed to set in motion a new branch of science dealing with the above problems. Even then he saw that there was a strange similarity between completely different phenomena (for example, between laser radiation and sociological processes or evolution) and that this must be only the tip of the iceberg. True, at that time Professor Hermann Haken had no idea that this field could make an impact on such distant fields of research as, in particular, psychology and philosophy.

Prof. Herman Haken (right) and Prof. Alexander Chalyi, at the International Conference devoted to celebrating 25 years of the Crafoord Prizes*) in Mathematics and Astronomy, Biology, Geosciences and Polyarthritis in Lund (Sweden), April 2007. (With heartfelt gratitude to Professor Hermann Haken, from the author's personal photo archive).

---

*) This prize was established by the Swedish philanthropist Golger Crafoord and his wife Anna-Greta Crafoord for outstanding achievements in 4 disciplines, for which the Nobel Prize is not awarded, as well as for the treatment of polyarthritis, which G. Crafoord suffered from in the last years of his life. The Crafoord Prize is worth US$500,000 (about half of the Nobel Prize), awarded by the Royal Swedish Academy of Sciences and presented by the King of Sweden.

The meaning of the term "synergetics" lies in the semantics of the Greek word "συνεργος"», which means acting together, or joint collective influence, or the phenomenon of strengthening one property by adding another. This term itself historically arose in ancient times, in the developed civilizations of Greece and Rome, primarily in connection with the understanding of the joint action of drugs. Until today, the concepts of "synergism" and "synergy" had rather a medical meaning. It should be mentioned that a famous English neurophysiologist Sir Charles Scott Sherrington (1857–1952) used this word more than hundred years ago [4]. He received the 1932 Nobel Prize in Physiology or Medicine for his studies of synaptic communication between neurons.

Charles Scott Sherrington (1857-1952)

The modern development of science and education has enriched synergetics with a new, deeper meaning. Another definition can be given: ***synergetics is a science that describes the processes of the emergence of order from disorder (chaos) and vice versa, as well as integration processes associated with changes in the structural features of the studied systems due to their internal connections and interactions with the environment***. Most of the modern achievements of synergetics are due to the possibility of integrating and transferring the results of one educational field or science (for example, physics) to similar objects of another educational field or science (for example, medicine). At the same time, physical results are, as a rule, more precise and quantitative, while due to the significant complexity of medical objects, their description, without using such a synergetic analogy, becomes possible, most likely, only at a qualitative level [1–3, 5–9].

It is obvious that the unification of various phenomena becomes possible not for all, but only for certain phenomena that occur in systems that satisfy the following important conditions:

(a) they are ***open***, that is, they exchange information, matter and energy with the environment;
(b) they are ***non-linear***, that is, there is a mutual influence and an inverse relationship between the values that describe the state of these systems;
(c) they are ***dynamic***, that is, various non-equilibrium processes occur in them, which change in time and space;
(d) they are ***complex***, that is, they can include a large number of interacting subsystems.

The uniqueness of synergetics lies in the possibility of uniting on its basis various phenomena of nature and society, which were previously the competence of different branches of scientific knowledge. The development of science at the end of the 20th century made it possible to develop, together with philosophy, another new integrative approach that largely compensates for the differentiation processes that have been so characteristic of the development of science since the time of Isaac Newton, that is, during the last more than three centuries.

The educational system is a synergetic (that is, open, non-linear, dynamic and complex) system. The importance of synergetics for science and education is connected with the possibility of creating and more effective studying an interdisciplinary dialogue, as well as the use of this new intersubjective direction for a deeper understanding of the unity of the laws of nature and society, and therefore the development of the individual as the ultimate goal of all educational activity.

Let us dwell in more detail on highlighting the features of ***the synergetic approach in science and education***. The synergetic approach is a set of principles, ideas and methods used by synergetics as a science that studies integrative trends and processes of self-organization and ordering in open, non-linear, dynamic, complex systems of various nature. Unlike other scientific and educational specialties that arise on the border of two sciences or educational disciplines, when one provides its subject, and the other—its research methods (for example, biophysics, biochemistry, mathematical linguistics, etc.), the synergetic approach is based on similarities the main features of processes that occur in systems of different nature, and models that describe these processes. This allows to transfer the results of one, more precise science (educational discipline) to similar and much more complex objects of another science (educational discipline).

The methodological basis of the synergetic approach are modern methods used to study the problems of self-organization and ordering in open systems of different nature (physical and chemical, medical and biological, educational, etc.). This approach is based on such well-tested and precise methods as the theory of phase transitions, the theory of nonlinear oscillations and autowaves, kinetic models "brussellator", "oregonator", catastrophe theory, fractal theory and others. Here is a far from complete list of sciences (educational fields) and processes for which a synergetic approach is effectively used: ***medicine***—reverberators (spiral waves) in the myocardium, hexagonal structures on the retina; synaptic transmission; carcinogenesis; ***physiology***—hexagonal structures in the brain that are responsible for the spatial orientation of man (animal); ***ecology***—fluctuating dynamics of population size, the emergence of epidemics (pandemics); ***psychology***—hysteresis and bistability in the visual perception of objects; ***computer technology***—biocomputers and parallel computing systems; ***sociology and political sciences***—the formation of public opinion; stability of political systems, etc.

To understand the uniqueness of the synergetic approach, the so-called "***first principles***" defining the basic laws of nature and society are of particular importance. In this regard, consider the following classification of the three stages of studying natural phenomena, which was actually described in [10] by Richard Feynman, winner of the 1965 Nobel Prize in Physics:

Stage 1—accumulation of experimental data. Feynman cited as an example the experiments of Claudius Ptolemy on the phenomenon of refraction, which were carried out by him about 140 AD. It should be noted that there are enough sciences (educational fields) that are in the 1st stage, associated with the accumulation and systematization of experimental data.

Stage 2—creating a theory of the phenomenon under study. At this stage, Willibrord Snellius in 1621 discovered the well-known from the school course of physics "law of refraction", which connects the angles of incidence and refraction of light with the relative refractive index of two media.

Stage 3—understanding the "first principles" that explain the reasons for the laws that describe experimental data. Richard Feynman used the following words for this crucial stage: "The true glory of science is that we must find a way of thinking that makes the law obvious." For the refraction of light, this principle was discovered by Pierre de Fermat in 1662 and called the "Fermat's principle of least time", which states that a ray of light chooses a trajectory whose travel time should be minimal.

This classification was proposed to be called for the first time in [11] as "***The Feynman Classification of the Three Stages of the Study of Natural Phenomena***" (see in more detail the previous Chap. 2).

For a synergetic approach, the first principle "***order arises as a result of the interaction of fluctuations at critical (bifurcation) points***" belonging to the Nobel laureate Ilya Prigogine is essentially important [5]. This principle makes it possible to understand fluctuating (probabilistic) processes of self-organization and the emergence of new structures at these points. Another first principle of "generalized Darwinism" asserts: **spatial, temporal and spatio-temporal structures arise as a manifestation of collective fluctuation processes, their interaction and the selection of those with a longer relaxation time** [6]. The longest fluctuating processes are described by variables called ***order parameters*** or ***collective modes***. As a result of the interaction of different modes in systems with many degrees of freedom, several order parameters must be chosen to investigate the further evolution of such systems (see Table 3.1 [6]). Using the ideas and methods of the synergetic approach creates

Table 3.1 The order parameters and conjugate fields for different systems near critical (bifurcation) points and points (lines) of phase transitions

System	Order parameter	Conjugate field (in the thermodynamical sense)
Single-component fluid	Deviation of density from its critical value	Deviation of temperature and pressure from its critical value
Binary liquid mixture near critical mixing point	Deviation of concentration from its critical value	Deviation of chemical potential from its critical value
System with chemical (biochemical) reactions	Degree of completeness (coordinate) of a chemical reaction	Affinity of a chemical reaction

a more complete implementation of the basic necessary conditions for the educational process that is based on such main principles as the connection of theoretical knowledge with experiments and practical applications, an understanding of the "first principles" underlying similar phenomena in systems of different nature.

3.2 Section A

3.2.1 Physics of Phase Transitions and Critical Phenomena, Main Ideas

The problem of phase transitions and critical phenomena has been and remains one of the most complex and important problems of physical science. Great contributions to the development and solution of this problem were made by studies of the Nobel Prize winners J. van der Waals (1910), H. Kammerlingh-Onnes (1913), T.D. Lee (1957), C.-N. Yang (1957), L. Landau (1962), J. Bardeen (1972), L Cooper (1972), J. Schrieffer (1972), I. Prigogine (1977), P. Kapitsa (1978), K. Wilson (1982), G. Bednorz (1987), A. Muller (1987), P.-G.de Gennes (1991), D. Osheoff (1996), R. Richardson (1996), A. Abrikosov (2003), V. Ginzburg (2003), A. Leggett (2003), D. Thouless (2016), D. Haldane (2016), M. Kosterlits (2016) and many other famous scientists.

Second half of the previous century passed under the sign of two outstanding achievements of physics: (1) the revolutionary discoveries and achievements in nanosciences (including nanomedicine), which will determine the future face of the XXI century; (2) solution of problem of the 2nd order phase transitions. It is difficult to overestimate the consequences of modern picture of cooperative processes in phase transitions and critical phenomena.

The essential progress in physics of phase transitions and critical phenomena of various nature is achieved because of a deeper understanding of the role of correlation effects of order-parameter fluctuations at large distances and times. The creation of the theory of superconductivity by Vitaliy Ginzburg and Lev Landau, the Nobel Prize winners in Physics [12], the development of fundamental ideas of the fluctuation theory of scaling invariance [13–17] and renormalization group approach by Kenneth Wilson (1936–2013), 1982 Nobel Prize winners in Physics [18], understanding the conditions under which the Landau mean-field theory of phase transitions is valid [19, 20], and a number of precise experiments (see, e.g. [21] and references there) have revealed a decisive role of order-parameter fluctuations near the phase transition and critical (bifurcation) points. These ideas have been also successfully used to study the specific features of phenomena and physical properties in the metastable state of substances, like the nucleation and other first-order phase transitions [22, 23].

In recent decades, a real scientific revolution has taken place, which was caused by remarkable achievements in understanding the nature of phase transitions and critical phenomena. It was this that played a primary role in creating synergy due to these circumstances. First, the study of phase transitions and various critical phenomena is a necessary part of the study of the general problem of finding peculiarities in the behavior of nonlinear environments in which, under certain conditions, the formation of ordered structures occurs. Secondly, the concept of phase transition, which until now was mainly related to physical systems (more precisely, to the physics of the condensed state), organically entered other sections of science and technology, became necessary for solving the problems of chemistry and biology, medicine and ecology, sociology and politics, etc. In recent years, a very important fact of the universality of the behavior of systems of a completely different nature near their points of phase transitions and critical phenomena has become clear. It is the universality, the common features of systems in which phase transitions occur, that allows us to apply uniform scientific approaches to them, which makes the use of the modern theory of phase transitions and critical phenomena one of the most fundamental methods of studying the characteristics of the behavior of various natural and social phenomena.

A significant majority of synergistic problems are related to the study of the characteristics of the behavior of complex multifunctional systems near their critical (bifurcation) points. The result of these studies was the understanding that such nonlinear systems near the points of phase transitions are characterized by a spontaneous violation of the previously existing symmetry of the system and the appearance of a completely different, new symmetry, which in modern mathematics and theoretical physics of phase transitions is called the symmetry of the "renormalization group" [18].

It should be noted here that the formation of modern physics of phase transitions and critical phenomena has gone through a number of main stages, which can be associated with the development of the following approaches: (a) model theories such as van der Waals theory for a classical one-component liquid; (b) Curie-Weiss theory for magnetic systems; (c) Bragg-Williams theory of ordering binary alloys and a number of others, which were generalized by L. D. Landau, winner of the 1962 Nobel Prize in Physics, when he created the thermodynamic theory of phase transitions or, as it is called in modern physical literature, the "mean field" theory in 1935–1937 [17]; (d) statistical models of systems that demonstrate phase transitions, such as the two-dimensional and three-dimensional Ising models, the Heisenberg model, the spherical model, the Baxter model, and others; (e) the theory of scale invariance ("scaling" theory) or the fluctuation theory of phase transitions of the second kind [13–17]; (f) theoretical approaches based on the ideas and methods of the renormalization group [18] and collective variables [25–27].

The main provisions of the modern theory of phase transitions of the second kind were already laid down in the works of L. D. Landau, in which the phase transition was considered as a process with a spontaneous violation of the symmetry of the system. It was L. D. Landau who introduced a fundamental quantity into the theory—the so-called order parameter, which could be used to describe the change in system symmetry. The choice of the order parameter depends on the system under study. For example, for a fluid near the critical point, the order parameter is the difference between the average density and the fluid density at the critical point. For a ferromagnet near the Curie point, the order parameter is the magnetic moment calculated per node or per unit volume. The content of the order parameters and their associated external fields for some other systems was previously shown in Table 3.1.

However, the following two essential assumptions were used in the theory of L.D. Landau, namely: (a) all thermodynamic functions (thermodynamic potentials and their derivatives, which gave physical properties observed in the experiment, such as heat capacity, susceptibility of the system to external influences, etc.) were considered analytical functions, i.e., those that at the most critical point or the point of the phase transition of the second order had distributions in power series (Taylor series) by temperature and order parameter; (b) fluctuations (deviations from the average value) of the order parameter were considered so small that they could be neglected.

Unfortunately, these assumptions did not stand the test of time, although such a fairly simple (mathematically and physically) theory gave results that were in good agreement with experimental data (although not for all objects, but in particular for ferroelectrics, low-temperature superconductors and some others). In 1944, a Norwegian and American physical chemist Lars Onsager (1903–1976) was awarded by the Nobel Prize in Chemistry. He managed to find an exact solution of the Ising two-dimensional model, as a result of which the logarithmic feature of the isochoric heat capacity C_V was discovered depending on the temperature, more precisely on the temperature distance to the critical point, which is conveniently represented by the dimensionless variable $\tau = (T - T_c)/T_c$. It was for this work, as well as for the creation of the theoretical foundations of the thermodynamics of irreversible processes, that Lars Onsager received the Nobel Prize. The anomalous growth of the heat capacity of the two-dimensional Ising model at the critical point according to the dependence $C_V \sim \ln|\tau|$ is in contradiction with the result of the Landau self-consistent field theory, in which the heat capacity should take a finite value at the critical point with a jump when the temperature approaches the critical value from above ($T > T_c$) and from below ($T < T_c$). Experiments on measuring the temperature dependence of heat capacity, which were carried out in the early 1960s by Aleksandr Voronel' and his colleagues in liquid argon, confirmed the fact that the heat capacity increases significantly as it approaches temperature to a critical value and is approximately described by a logarithmic dependence on the temperature variable τ.

Lars Onsager (1903-1976)

Another fundamental result indicating the peculiarity of the state of matter at critical points and points of phase transitions was obtained by the laureates of the 1957 Nobel Prize in Physics Tsung-dao Lee and Chen-Ning Yang. They proved that the statistical properties of systems in the vicinity of phase transition points have significant differences compared to their regular behavior far from these points. All these and other results, which confirmed the discrepancies between the real critical behavior and the predictions of the Landau theory, being associated with strongly developed order- parameter fluctuations that should be properly accounted.

The second (by the order parameter) fundamental value in the physics of phase transitions is the correlation radius or correlation length—this is the distance within which the fluctuations of the order parameter are highly correlated. To be more precise, the correlation radius characterizes the distance at which the interaction between fluctuations decreases by approximately e = 2.7 times. It is often noted in the scientific literature that the concept of the correlation radius (correlation length) was introduced for the first time by L. Ornshtein and F. Zernike in 1914–1916 in their studies devoted to the calculation of density fluctuations near the critical point. However, as the author of this monograph was told by a famous scientist in the field of phase transitions and critical phenomena in liquids and liquid crystals, Prof. Mikhail Anisimov, this result was previously obtained by Johannes Diderik van der Waals, winner of the 1910 Nobel Prize in Physics. The main idea of these works was that the spatial derivatives of the order parameter were taken into account, since they allow better consideration of the influence of the environment on the physical properties of the system at a certain spatial point. Implementing this idea, in addition to the Gaussian term proportional to the square of the order parameter (this was done by A. Einstein in 1910, creating the thermodynamic theory of fluctuations), a second term proportional to the square was also introduced into the distribution of the fluctuating part of the free energy of the system near the critical point gradient (spatial derivative) of the order parameter. The ratio of the coefficient in front of the square of the gradient of the order parameter and the coefficient in front of the square of the order parameter itself determines the square of the correlation radius, which grows abnormally as it approaches the bifurcation (critical) point. It is the correlation of fluctuations of the order parameter over large distances that is a characteristic feature of the behavior of various systems near the points of phase transitions and critical points. It turns out that the direct consequence of the increase in the correlation radius of fluctuations of the order parameter is a very slow decay of the so-called correlation function, which determines the interaction between a pair of fluctuations at two different points of the system.

The next important stage in the development of the theory of phase transitions and critical phenomena was related to the formulation of the hypothesis of similarity or, as it is called, the theory of scaling invariance. This was done mostly independently and almost simultaneously in the works of A. Z. Patashynsky, V. L. Pokrovsky and L. Kadanov in 1966. The main idea of the hypothesis of similarity (scaling) was that near the critical point, a substance (liquid, magnet, ferroelectric, polymer mixture, liquid crystal, etc.) can be represented as a set of regions (clusters) with linear dimensions of the order of the correlation radius. The laws of anomalously (strongly) fluctuating quantities in the vicinity of a phase transition when moving to a new length scale—the correlation radius—make up the content of the modern theory of phase transitions and critical phenomena—the theory of scale invariance or the fluctuation theory of phase transitions. This theory made it possible to deeply understand the role of collective effects caused by the strong interaction of fluctuations of the order parameter on large spatial and temporal intervals. This fundamental fact is the reason for the singular behavior of the physical properties of various systems near their critical points and points of phase transitions.

The application of the theory of similarity (scaling) gave universal results for all phase transitions of the second kind in terms of so-called ***power laws***. The exponents of powers in these laws are called ***critical indices*** (***critical exponents***), which was already mentioned earlier [13–18]. It turned out that it is useful to introduce the so-called ***classes of universality*** [28–30], which should include systems with the same spatial dimension (that is, two-dimensional and three-dimensional systems belong to different classes of universality), with the same dimension of the order parameter (that is, systems with a scalar order parameter, such as a liquid, a magnet in the Ising model, etc., and systems with a vector order parameter, such as a magnet in the Heisenberg model, etc., fall into different classes of universality). The radius of action of the intermolecular potential should also be attributed to the characteristics of a certain class of universality (that is, depending on whether the intermolecular interaction is short-range or long-range, the system falls into one or another class of universality). Objects belonging to the same class of universality have the same singular behavior of physical properties near the phase transition points, in particular, the same critical indices. Definitions and numerical values of the main critical indices for thermodynamic and correlation properties of liquids are given in Table 3.2 [6, 14–18, 21].

The scaling theory has been confirmed in a large number of experiments [21]. However, the theoretical calculation of critical indices remained a rather difficult problem for a long time. In the early 1970s, significant progress was made in this direction thanks to the work of Kenneth Wilson and Michael Fisher [31]. Kenneth Wilson later received the 1982 Nobel Prize in Physics for his work on phase transitions, which used the so-called renormalization group method, well known in quantum field theory. It should be noted that the methods of quantum field theory in the theory of phase transitions were used before K. Wilson in the works of N. N. Bogolyubov, A. Z. Patashynskii, V. L. Pokrovskii, A. A. Migdal, A. M. Polyakov, I.R.

Table 3.2 Critical indices of the scaling theory of phase transitions

Definitions of critical indice	Connections with scaling dimensions	The value of critical indices				Equalities between critical indices
		Landau theory	Ising model d=3, n=1	RG (ε-expan-sioins)	Experiment	
$C_V \sim \tau^{--\alpha}$	$2 - \frac{d}{\Delta_\tau}$	0 (jump)	0,125± 0,015	0,077	0,07÷ 0,14	$\alpha+2\beta+$ $+\gamma = 2$
$\Delta\rho \sim \lvert\tau\rvert^{\beta}$	$\frac{\Delta_\eta}{\Delta_\tau} = \frac{d-\Delta_h}{\Delta_\tau}$	1/2	0,312± 0,003	0,340	0,34± 0,02	$\beta(\delta -1)=\gamma$
$\beta_T \sim \tau^{--\gamma}$	$\frac{2\Delta_h-d}{\Delta_\tau} = \frac{\Delta_h-\Delta_\eta}{\Delta_\tau}$	1	1,250± 0,030	1,244	1,23± 0,02	$\alpha + \beta\ (\delta +$ $+1)=2$
$\Delta\mu \sim \Delta\rho^{\delta}$	$\frac{\Delta_h}{d-\Delta_h} = \frac{\Delta_h}{\Delta_\eta}$	3	5,0÷ 5,2	4,46	4,5± 0,5	$d\nu =2-\alpha$
$R_c \sim \tau^{-\nu}$	$\frac{1}{\Delta_\tau}$	1/2	0,638	0,626	0,60± 0,03	$(2-\eta_0)\nu=$ $=\gamma$
$R_c \sim \Delta\rho^{-\zeta}$	$\frac{1}{\Delta_\eta} = \frac{1}{d-\Delta_h}$	1	2	1,841	1,77±0,05	$d\zeta=\delta -1$
$G(r) \sim \frac{1}{r^{d-2+\eta_0}}$	$2+d-2\Delta_\eta=$ $2-d+2\Delta_\eta$	0	0,055± 0,010	0,037	0÷0,1	$(2-\eta_0)\zeta=$ $\delta -1$

Note d is a spatitial dimension, n is the number of components of the order parameter, $\varepsilon = 4$—d

Yukhnovsky and other theoretical physicists who fruitfully worked on this problem in various scientific centers of the former Soviet Union.

The application of the renormalization group method marked a new qualitative stage in the development of modern physics of phase transitions and critical phenomena. The renormalization group transformation is a complex nonlinear transformation affecting the coefficients of the expansion of the fluctuation part of the free energy. As a result of its application, the temperature dependence of these coefficients changes (renormalizes). Thus, the method of the renormalization group made it possible to calculate various critical indices using the so-called ε-expansions, where $\varepsilon = 4-d$ is the deviation of the spatial dimension d of the system from 4. It is interesting to note that the thermodynamic Landau theory of phase transitions, which neglects the fluctuation effects, turns out to be valid in space with $d = 4$. Such a method of ε-expansions was proposed by K. Wilson and M. Fisher in 1972 [31]. Saving the first few terms of the series of ε-expansions for the critical indices made it possible to obtain results that are in quantitative agreement (up to 2–3%) with the experimental data.

Interesting and valuable results for the theory of phase transitions were obtained using the method of collective variables by a famous Ukrainian theoretical physicist and politician, Academician and Professor Igor Yukhnovsky (1925–2024), who created the world-known Lviv school of theoretical physics of the condensed state of matter.

Igor Yukhnovsky (1925-2024)

In particular, Prof. Igor Yukhnovsky and his scientific colleagues were able to rigorously obtain (see, e.g. [24–27] and links there) the Ginzburg-Landau fluctuation Hamiltonian, which is postulated in the scaling invariance theory and in the renormalization group approach. To prove such an important result, a strict transformation was used with taking into account the corresponding Jacobian transition from the coordinates and momentum of all particles to the so-called collective variables that are related in a certain way to density fluctuations. This approach made it possible to obtain important expressions for three-dimensional Ising model. The method of collective variables is essentially one of the options for implementing the renormalization-group approach, which made it possible to develop a microscopic method of calculating critical indices.

We dwelled in sufficient detail on the current state of the physics of phase transitions due to the fact that this theory allows studying fluctuation effects at a rigorous level. In turn, fluctuation effects significantly affect the synergetic processes of formation of spatial, temporal and spatio-temporal ordered structures near critical (bifurcation) points.

3.2.2 Self-organized Criticality: Universal Dynamic Evolution of Systems to a Critical State

About 45 years ago, at the end of the 70s, P. Buk, C. Tang and K. Wiesenfeld) discovered the phenomenon of self-organized criticality [32], the essence of which can be formulated as follows: ***systems of various nature consisting of a large number of interacting elements naturally evolve to a critical state, in which a rather insignificant event can cause a chain reaction that leads to catastrophic changes in this system***. In other words according to the concept of self-organized criticality, catastrophic changes or chain reactions affecting a large number of system elements are as inherent a feature of dynamic behavior as minor events. Moreover, it can be argued that both small and large (catastrophic) events have the same nature.

Let us consider several examples of phenomena that confirm the concept of self-organized criticality.

Earthquakes. In 1956, scientists-geologists B. Gutenberg and C. Richter (the latter introduced the well-known scale of earthquake strength, which now bears his name) established that the number of earthquakes N, during which energy E is released, is inversely proportional to the energy E that rises in degree 1.5. In the language of a mathematical formula, this law is written as follows:

$$N \sim 1/E^{1.5}. \tag{3.1}$$

The index of the power of 1.5 is universal and it should be called the critical index, by analogy with similar exponents in the theory of phase transitions and synergetics, and the above formula (3.1) should be called ***the scaling Gutenberg-Richter law of***

earthquakes. Based on the Gutenberg-Richter law, it follows the well-known fact that low-energy earthquakes occur much more often than catastrophic earthquakes. For example, if $N_1 = 10000$ earthquakes with energy $E_1 = 10$ (in relative units) occur in a certain area during the year, then $N_2 = 10$ earthquakes with energy $E_2 = 1000$ (in the same relative units) should occur in the same area during the same period of time units. Or in some other place, for $N_3 = 1000$ "weak" earthquakes with an energy $E_3 = 1$ (in relative units), $N_4 = 1$ "strong" earthquake with an energy $E_4 = 100$ (in relative units) should occur during the same time. Thus, the number of small and large earthquakes are closely related and are the consequences of the same mechanism that leads to the appearance of earthquakes in the earth's crust.

It is now generally accepted that earthquakes occur as a result of blocks of the Earth's crust sticking together with the accumulation of deformation energy, and then slippage occurs when the accumulated energy is released and spreads to other areas. Researchers in various scientific laboratories around the world recreated this process on models and established that the Gutenberg-Richter law is confirmed experimentally and theoretically. Moreover, as a detailed analysis of the proposed models showed, the Gutenberg-Richter law is an experimental confirmation of the fact that the earth's crust is in a constant state of self-organized criticality. The results of model experiments and real studies of earthquakes confirm the validity of the Gutenberg-Richter law, which in turn can be considered as evidence of the validity of the theory of self-organized criticality. Of course, it is difficult to overestimate the significance of these results for the prediction of catastrophic earthquakes, which are very dangerous and take the lives of many hundreds and thousands of people.

Snow avalanches. The study of the dynamic behavior of snow avalanches in model and real experiments once again confirms the concept of self-organized criticality. Such experiments proved that these objects, like the earth's crust, naturally, that is, arbitrarily, tend to reach a critical state. If, on the one hand, the steepness of the snow slope is less than the critical one, then the avalanche in such a "precritical" state will, of course, be smaller than the one that will be observed in the critical state. Then the steepness of such a precritical slope will increase until it reaches a critical value. If, on the other hand, the steepness of the slope is greater than the critical one, then in such a "subcritical" state, avalanches will be generated larger than those that occur in the critical state. Thus, both in the precritical and subcritical states, there is an arbitrary transition to the critical state, that is, self-organized criticality occurs.

A very unusual and interesting feature of the studied objects is their local instability (that is, the instability of a certain area of snow) and the absolute stability of their critical state, which is due to self-organized criticality. Thus, it appears that self-organized criticality is a universal property of these and many other objects.

As researches have shown, it is likely that complex dynamic processes in the economy have similar properties. Thus, Benoit Mandelbrot, who is the author of the modern theory of fractals (he owns the excellent book "Fractal Geometry of Nature" [33]), as a result of the analysis of such indicators of the securities market as the well-known Dow-Jones index, proved that the dynamics of stock market operations satisfy the regularities "flicker noise", which is a sign of self-organized criticality.

A study of the frequency f of such random events as the dynamics of earthquakes, markets, ecosystems, etc. revealed that this frequency changes over time within very wide limits and satisfies the law $1/f$. This law is known for chaotic processes, which have the name "flicker noise" and characterize systems with memory, i.e. those in which the dynamic behavior is largely determined by the past state of these systems. The appearance of "flicker noise" means that self-organized criticality is associated with the presence of chain reactions with different amplitudes and time intervals between the elements of the system.

This same important conclusion should necessarily be transferred to universal by nature processes of self-organized criticality that can occur in the human body. Anomalous increase in human susceptibility to external fields (for example, the influence of geomagnetic storms in weather-dependent people on changes in solar activity or the influence of an external electromagnetic field on the peculiarities of the flow of biochemical reactions near the state of self-organized criticality) can cause significant (sometimes decisive) effects on the cardiac activity or on the synaptic mechanisms of information transmission. Thus, if a medical-biological, economic, or political system is in a state of self-organized criticality, significant fluctuations (deviations from the average value) of its internal parameters under the influence of very minor disturbances are an inherent feature of the dynamic behavior of such systems.

Phenomenon of self-organizwd criticality in human brain. The phenomenon of self-organized criticality is of increased interest in connection with the study of temporal correlations and scaling behaviors of oscillatory processes observed in the brain [34], as well as in the creation and analysis of a brain network model [35] and a brain plasticity model [36] taking into account the phenomenon of self-organized criticality. The authors of [37], working in well-known research medical and physical centers and universities in Argentina and the USA, studied the influence of the phenomenon of self-organized criticality for the following three models: (a) the 2D ising ferromagnetic model, (b) the 3D Vicsek flocking model and (c) a small-world neural network model. As a result of their research, the following conclusion was made: "feedback from the autocorrelation function of the order parameter fluctuations shifts the system towards its critical point."

3.2.3 Peculiarities of the Bifurcation Behavior in Open Systems: The Formation of Ordered Structures as a Non-equilibrium Phase Transition

It is known that in an isolated (closed) system there is a transition from order to complete disorder with an increase in entropy. In such a system, the processes of self-organization and formation of ordered structures cannot take place. If the system is open, that is, it exchanges matter, energy, and information with the environment, then ordered structures can arise in such a system. It turns out that here, in principle, two possibilities are realized depending on how far this system is from the equilibrium state:

(1) for systems that are close to equilibrium and for which it is possible to limit ourselves to taking into account small deviations of parameters from equilibrium values, that is, for small thermodynamic forces, the laws of thermodynamics of irreversible processes apply. In this case, only equilibrium structures can arise (similar to those that appear during crystal ordering or during the transition of a crystal to a liquid state due to melting, etc.). It is these structures that are allowed from the point of view of those irreversible processes that occur in such open systems;
(2) for systems that are far from equilibrium, due to the growth of fluctuations near critical (bifurcation) points, qualitatively new dynamic structures appear, in other words, the emergence of dynamic order. Thus, the emergence of ordered structures occurs in open systems, far from equilibrium, in which the effects of nonlinearity, characterizing the mutual influence of system parameters on each other, are very significant.

In the modern synergetic approach to the problem of self-organization and ordering, the process of formation of spatial, temporal and spatio-temporal structures is considered as a non-equilibrium phase transition. The need to study the analogy between non-equilibrium phase transitions, in which ordering and self-organization occur, and equilibrium phase transitions, where fluctuations of the order parameter play a decisive role, was clearly reflected in the above-mentioned principle "order through fluctuations", which was formulated in [38]. The implementation of this well-known synergistic principle requires the creation and study of so-called "fluctuation models" of self-organization processes, which are based on the fundamental ideas of the theory of phase transitions, which underlie the hypothesis of scaling invariance and the renormalization-group approach [13–18, 21].

Let us now consider the important question of the relationship between equilibrium and non-equilibrium phase transitions, between which there is a certain analogy. Indeed, the phase transition is characterized by one or more order parameters that describe the collective behavior of systems consisting of an extremely large number of particles (molecules and atoms) in the vicinity of critical (bifurcation) points. Another feature of phase transitions of both types (equilibrium and non-equilibrium) is the increase in the susceptibility of the system to external influences. It is this circumstance that has a significant, sometimes decisive, effect on the properties of the studied systems and on the course of the phase transition itself. In other words, very weak external fields can cause a significant reorganization of the system during the phase transition. It should also be noted that near the critical (bifurcation) points and at the limit of system stability, the so-called "small" influences cease to be such, i.e., they become essentially considerable, if the system is at a "distance" from the phase transition point, which is smaller in energy units , than the external influence itself. The loss of the concept of "weak external influence" near the points of phase transitions has as a consequence the possibility of the system transition from one phase state to another under the influence of an external stimulus. For non-equilibrium phase transitions, such a state change can be associated with the appearance of coherent ordered structures. Finally, near the critical (bifurcation) points, the relaxation time increases strongly, i.e., the time during which the dynamic behavior of the system changes. In the physics of phase transitions and critical phenomena, such slowing down of the speed of processes in time is called "critical deceleration".

Let us also dwell on the question of the fundamental possibility of studying ordered structures based on an approach that uses the concept of fluctuation Hamiltonians of the physics of phase transitions, that is, one of the main concepts of equilibrium phase transitions. In fact, here it becomes possible to go beyond the equilibrium theory, which of course cannot always be applied to the study of non-equilibrium phase transitions.

There are at least two reasons for the successful implementation of this approach:

1. Changing the signs of the coefficients in the fluctuation Hamiltonians ensures system transition from a single-phase stable or metastable region to a thermodynamically unstable state. As indicated earlier, a similar transition can occur when the signs of the coefficients, which are responsible for the inverse susceptibilities of the system or for its diffusion coefficients, change. Stabilization of thermodynamically unstable states, that is, the appearance of structural stability, becomes possible precisely due to the appearance of spatial, temporal, and spatio-temporal orderly periodic (quasi-periodic) structures.
2. Use of fluctuation Hamiltonians like the fluctuation Ginsburg-Landau Hamiltonian does not necessarily mean that the system is closed. Indeed, the approach to the problem of self-organization and ordering that we use includes the case of "stationary state" according to Ilya Prigogine, which is quite common in real open systems of various nature (see paragrapf 3.3.3 for more details). This stationary state is characterized by a complete compensation of the entropy production produced inside the system due to irreversible dissipative processes and the inflow of negentropy (negative entropy) from outside the system. Since the influence of fluctuation effects must be taken into account for a consistent consideration of the processes of formation of ordered structures in space and time, then for stationary open systems the free energy contains only a fluctuation contribution, i.e. it is reduced to the fluctuation Hamiltonian of the equilibrium theory of phase transitions. Of course, such formally balanced fluctuation Hamiltonians are the basis for the study of non-equilibrium phase transitions and self-organization processes in stationary open systems based on the corresponding time-dependent equations of motion for the order parameters of the studied system. This approach will be implemented in the creation and study of fluctuation models of self-organization processes, where the results of obtaining ordered structures of various types within the framework of the proposed fluctuation models will also be presented and discussed.

To conclude this paragraph, we shall consider the general characteristics of fluctuation models of self-organization processes. The main goal will be to determine the type of the singular points of fluctuation models of self-organization and ordering, as well as the associated equations of motion for the interacting order parameters (ruling modes). These models, on the one hand, must be consistent with the main provisions of modern theory phase transitions, and, on the other hand, they can be considered as models of the emergence of coherent ordered structures in objects of various nature (medical and biological, physical and chemical, etc.). The starting points of this study are as follows:

(a) the use of so-called fluctuation Hamiltonians, which play the role of a peculiar energy characteristic of the interaction of fluctuations in the studied system. At the same time, the form of these Hamiltonians is used, which is similar (isomorphic) to the original Landau-Ginzburg Hamiltonian in the modern theory of phase transitions for systems with two interacting order parameters. As will become clear, this essentially statistical (probabilistic) approach to the study of the problem of self-organization and ordering allows taking into account, firstly, the results of the traditional Gaussian approximation, which corresponds to the model of the free field of fluctuations, as well as the very important near critical (bifurcation) points effects of non-locality (spatial dispersion) according to the van der Waals (Ornstein-Zernike) approximation and various effects of the interaction of order parameter fluctuations;
(b) it should be noted that the use of the fluctuation Hamiltonian in the form of the Landau-Ginzburg Hamiltonian for two interacting order parameters uniquely determines the structure of the kinetic equations (equations of motion) for these parameters. The equation of motion for two order parameters (controlling modes) in its functional form resembles the equations of reaction-diffusion models, which contain both a diffusion term proportional to the spatial derivatives of the order parameters describing the influence of the environment on the properties of the system at a certain spatial point, and the necessary nonlinear the relationship between the parameters of the order, which takes into account their interaction;
(c) external influences change the magnitudes and, possibly, the signs of the coefficients in the equations of motion associated with the inverse susceptibilities of the system (coefficients with Gaussian terms in the fluctuation Hamiltonian), as well as diffusion coefficients (coefficients with Ornstein-Zernikow terms containing the squares of spatial derived from the order parameters). In addition, external influences give additional contributions to the fluctuation Hamiltonian and, accordingly, to the equations of motion.

3.2.4 Laws of Thermodynamics and Steady State of Open Systems

Most objects of animate and inanimate nature are not isolated from the external environment. Having such contact with the environment, they exchange matter, energy and information with it. Such systems are called ***open***. The science that studies the general laws of energy transformation and thermal interaction between different systems is called ***thermodynamics***. To study open systems of various nature (physical-chemical, medical-biological, etc.), the laws of thermodynamics of open systems should be used. In addition to special relationships that operate in open systems, there are fundamental laws of nature (they are called, as mentioned earlier, first principles). Such fundamental laws include the first and second laws of thermodynamics. Let us formulate these laws, which are always in effect in any system (open or isolated, with or without the formation of ordered structures (see e.g. [38–40]).

The first law (principle) of thermodynamics: **Heat supplied to a system goes to change its internal energy and to the work that this system performs on external bodies**. In essence, the first principle of thermodynamics is a universal law of nature—***the law of conservation of energy***. It is interesting to note that the honor of discovering this law in the middle of the 19th century belonged not only ***to the physicists J. Joule and H. Helmholtz***, but also ***to the doctor Y. Mayer***, who studied the connection between the color of the blood of sailors and thermal effects while sailing at different latitudes (see Chap. 2, where this issue is covered in more detail).

The second law (principle) of thermodynamics: **In any isolated system, it is impossible to transfer heat from a less heated body to a more heated one.** This formulation of the second law of thermodynamics belongs ***to the German theoretical physicist Rudolf Clausius*** (1822–1888), who discovered the second law of thermodynamics in 1850.

Another very important concept of ***entropy*** belongs to him, which he introduced in 1865 and which can be formulated as follows: ***Entropy S is a function of the state of the system that determines the direction of an arbitrary process in an isolated system, and entropy increases as it approaches the equilibrium state, and in the equilibrium state it takes its maximum value.***

The important role played by entropy in thermodynamics is determined by two main reasons:

(1) the change in entropy dS characterizes the heat that the system receives or gives off during its interaction with the environment,
(2) entropy characterizes the degree of system order in accordance with the principle that bears the name of Ludwig Boltzmann's.

The famous Austrian theoretical physicist Ludwig Boltzmann (1844–1906) laid the foundations of statistical physics in 1877. In particular, he established the statistical meaning of entropy. According to Boltzmann's principle, entropy takes on a smaller value, the more ordered the system is. Conversely, when moving from a more ordered state to a less ordered state and, ultimately, to complete disorder, entropy increases and reaches its maximum value at the equilibrium state. The following important conclusion should be drawn from this: ***if ordered structures appear in the system, then the entropy of such a system decreases***.

Ludwig Boltzmann (1844-1906)

Consider ***the concept of a stationary state***, which was introduced by a Belgian physical chemist Ilya Prigogine (1917–2003), 1977 Nobel Prize winner in Chemistry. The concept of stationary state plays an important role in the thermodynamic study of open systems. The total entropy change dS of an open system consists of the sum of two terms that generally describe the following processes: (1) entropy change processes inside the system (we denote this entropy change by dS_i), (2) entropy change processes due to the interaction of the open system with the environment (denote this entropy change due to dS_e).

Note that in the case when inverse processes occur in the system, i.e. such cyclic processes, as a result of which there are no changes in the external environment, the change in entropy dS_i within the system is zero. However, in real systems, irreversible dissipative processes occur, as a result of which the entropy dS_i inside the system increases. As for the sign of the second term dS_e, which is related to the interaction of the open system with the environment, it can be arbitrary and determined by whether there is an inflow or outflow of entropy through the surface that limits the allocated volume of the studied system. The physical reason that causes this change in entropy is determined by positive or negative fluxes of particles, heat, momentum, and other transport processes.

Stationary is the state of the system in which the entropy of the entire open system is preserved, that is, the total change in entropy is zero (dS = 0). It directly follows from the condition of entropy constancy that both contributions to the total entropy of the system are mutually compensated, i.e., $dS_i = -dS_e$. Since the change in entropy dS_i as a result of dissipative irreversible processes occurring inside the system is a positive value, the change in entropy dS_e due to the interaction of the open system with its environment must be negative and equal in magnitude to the change in entropy dS_i inside the system. A similar implementation of the condition of stationarity of an open system becomes possible if the entropy produced inside the system due to the processes of diffusion, heat conduction, viscosity, chemical (biochemical) reactions and other irreversible processes completely passes through the limiting surface into the environment. In other words, it can be stated that open systems in a stationary state are fed by ***negative entropy***, which is also called ***negentropy*** and is denoted by $N = -S$.

Ilya Prigogine proved that ***in a steady state the rate of entropy change over time (the so-called entropy production) reaches a minimum value***. This statement is called "***Prigogine's theorem***". The principle of minimum entropy production in the steady state of an open system is important. It gives a quantitative criterion that allows you to determine the direction of development (evolution) of an open system, namely: if irreversible processes occur in an open system near thermodynamic equilibrium, then its transition to a stationary state can be predicted by the decrease in entropy production in such a system. In other words, ***the criterion for the system approaching a steady state is the negativity of the rate of change of entropy production, that is, the decrease of entropy production over time.***

Ilya Prigogine (1917-2003)

Prigogine's theorem also explains the fundamental stability of steady states of open systems. Indeed, if an open system arbitrarily exits its stationary state due to fluctuations, then the rate of entropy production increases in it. Then, as an inevitable consequence of such changes in the state of the system, due to the effect of Prigozhin's theorem, such processes must be implemented inside the system, during which it will again return to its stationary state. The principle of minimum entropy production (Prygozhin's theorem) plays the same role for open systems as the well-known ***La Chatelier-Brown principle***, which explains the stability of equilibrium systems: ***any external influences that bring the system out of equilibrium cause such processes in it that tend to weaken the result of these external influences***.

3.3 Section B

3.3.1 Theory of Phase Transitions and Critical Phenomena: Main Provisions and Results

In Section 3.2 (see Sect. 3.2.1) a general description of the current state of the physics of phase transitions and critical phenomena was given. Here we will consider this issue in more detail.

Numerous theoretical and experimental works have questioned the direct use of Landau "classical" phenomenological theory and similar versions of the self-consistent (mean) field theory to describe the immediate vicinity of critical points and points of phase transitions of the 2nd order. The theory of a mean field is valid only in the region of small fluctuations of the order parameter, where the Ginzburg-Lewaniuk criterion

$$\mathrm{Gi} \ll |\tau| \ll \mathrm{I}, \quad \tau = \frac{\mathrm{T} - \mathrm{T_c}}{\mathrm{T_c}}.$$

where *Gi* is the Ginzburg number [19, 20]. In the region $|\tau| \ll Gi$ fluctuations cease to be small, their amplitudes in a volume with linear dimensions of the order of the correlation radius R_c become comparable to the equilibrium value of the order parameter$\langle\eta\rangle$, which completely determines the singular behavior of physical properties in the vicinity of the critical point.

Correlation of fluctuations over long distances requires a transition from inter-atomic distances to a new unit of dimension length—the correlation radius *R*c. The laws of transformation of anomalously fluctuating physical quantities during the transition to this new scale are at the basis of the modern theory of critical phenomena—the theory of scale transformations ("scaling"), developed in the works of A.Z. Patashynskyi and V.L. Pokrovsky, L. Kadanov, B. Widom, M. Fisher and other scientists [13–21].

Among the most important results of this theory is the establishment of a scale-invariant form of abnormally fluctuating quantities at the critical point

$$\mathrm{A}(\lambda\vec{r}) \rightarrow \lambda^{-\Delta_A}\mathrm{A}(\vec{r}) \tag{3.2}$$

and related thermodynamic and correlation functions. The scale dimensions Δ_A of the main physical quantities (order parameters $\eta \rightarrow \lambda^{-\Delta\eta}\eta$, temperature $\tau \rightarrow \lambda^{-\Delta\tau}\tau$ and field $h \rightarrow \lambda^{-\Delta\eta}h$ variables, etc.) are parameters of the theory. They can be associated with critical indices (or exponents) of asymptotic power laws that satisfy a number of equalities, so that the number of independent indices in the static case turns out to be equal to two (see Table 3.2 [13–17, 21, 41–44]).

A significant achievement of the scaling invariant theory was obtaining an expression for the singular part of the free energy

$$\Delta F_s(\rho,\tau) = |\tau|^{2-\alpha} f_0^{(F)}\left(\Delta\rho/\tau^{\beta}\right) = |\Delta\rho|^{\delta-1} q_0^{(F)}\left(\tau/\Delta\rho^{1/\beta}\right). \quad (3.3)$$

This expression makes it possible to obtain formulas for all thermodynamic properties in the critical region by differentiating (3.3) with respect to the temperature variable τ and the order parameter. For the liquid-vapor phase transition, such an order parameter is the density deviation $\Delta\rho = (\rho - \rho_c)/\rho_c$ from the critical value. In particular, the equation of state has the following form:

$$\mu(\rho,\tau) = \mu(\rho_c,\tau) + \begin{cases} |\tau|^{\beta\delta} f_0^{(\mu)}\left(\Delta\rho/\tau^{\beta}\right) sign\Delta\rho, \\ |\Delta\rho|^{\delta} q_0^{(\mu)}\left(\tau/\Delta\rho^{\frac{1}{\beta}}\right) sign\Delta\rho. \end{cases} \quad (3.4)$$

Here, the scaling functions $f_0^{(\mu)}(y)$ and $q_0^{(\mu)}(x)$ have the following asymptotics along the limiting directions of the approach to the critical point:

(a) in the vicinity of the critical isochore

$$f_0^{(\mu)}(y \to 0) = \sum_{n\geq 0} a_{n+1} y^{2n+1},\ q_0^{(\mu)}(x \to \infty) = \sum_{n\geq 0} a_{n+1} x^{\gamma-2n\beta}, \quad (3.5)$$

(b) in the vicinity of the critical isotherm

$$f_0^{(\mu)}(y \to \infty) = \sum_{n\geq 0} b_{n+1} y^{\delta-n/\beta},\ q_0^{(\mu)}(x \to 0) = \sum_{n\geq 0} b_{n+1} x^{n}, \quad (3.6)$$

where a_n, b_n are the substance parameters.

For the quantitative description of an arbitrary vicinity of the critical point, it turned out to be quite convenient to use the parametric form of the scaling invariance theory:

$$H = r^{\beta\delta}\tilde{h}(\vartheta),\ M = r^{\beta}\tilde{m}(\vartheta),\ \tau = r\tilde{t}(\vartheta), \quad (3.7)$$

where $\tilde{h}$, $\tilde{m}$, $\tilde{t}$ are regular functions of the angular variable ϑ that specifies the direction of approach to the critical point, and r is the "distance" to the critical point, the dependence of which determines the nature of the singular behavior of various physical properties. The practical implementation of the parametric form of the theory of large-scale transformations in order to obtain the equation of state led to the so-called "linear model", in which $\tilde{m}(\vartheta) \sim \vartheta$. Similar results were obtained by A.A. Migdal, who wrote the equation of state in the form

$$H = \chi^{-(\beta+\gamma)/\gamma} q(m). \quad (3.8)$$

Here $\chi = (\partial M/\partial H)_T > 0$ is the susceptibility in a finite external field H, and $q(m)$ is an analytical function containing only odd powers of the scale variable $M\chi^{\beta/\gamma}$. The equation of state of the "linear model" and equation (3.8) in the first approximation

($\beta + \gamma = 3/2$, $q(m) = m + \varphi_3 m$) quite accurately (up to 5÷10%) describe the experimental data.

The requirements of scaling invariance also made it possible to establish that the multipoint correlation function (CF) $G^{A}{}_{1} \ldots {}^{A}{}_{n} (\vec{r}_1, \ldots \vec{r}_n)$ is a homogeneous function of its spatial order arguments $\Delta_{A_1} + \ldots \Delta_{A_n}$ and depends on the ratios $\left|\vec{r}_i - \vec{r}_j\right|/R_c$. Thus, a pair correlator of a spatially homogeneous medium has the following form:

$$G^{AB}(\vec{r}_1, \vec{r}_2) = \frac{C_{AB}}{|\vec{r}_1 - \vec{r}_2|^{\Delta_A + \Delta_B}} q(|\vec{r}_1 - \vec{r}_2|/R_c), \tag{3.9}$$

and in the case of fluctuations of the order parameter $\Delta_A = \Delta_B = (d - 2 + \eta_0)/2$. Here η_0 is the critical index of the anomalous dimentionsionaly, which describes the difference between the asymptotic behavior of the CF $G^{\rho\rho}$ and the predictions of the free fluctuation field theory or the Ornstein-Zernike approximation with $\eta_0 = 0$. The correlation radius Rc depends, strictly speaking, on the proximity to the critical point both in terms of temperature and order parameter:

$$R_c(\tau, \Delta\rho \to 0) = a_0|\tau|^{-\nu}, R_c(\tau \to 0, \Delta\rho) = a_0|\Delta\rho|^{-\varsigma}, \tag{3.10}$$

where the critical exponent $\varsigma = \frac{1}{\Delta_\eta} = \frac{1}{d - \Delta_h} = \frac{\delta}{\Delta_h}$.

In addition to invariance with respect to a change in scale, correlation functions can also have, as was shown by A.M. Polyakov in [45], invariance with respect to a special conformal transformation. If the scaling invariance made it possible to determine the structure of the pair correlation function with accuracy up to a constant factor of C_{AB} and the explicit form of the function $q\,(|\vec{r}_1 - \vec{r}_2|/R_c)$ from formula (3.9), which turns into a constant at the critical point, then the requirements of conformal invariance make it possible to carry out the same procedure for triple correlators, and also significantly narrow the class of functions describing higher-order correlation functions.

Another important consequence of conformal invariance is the orthogonality of abnormally fluctuating quantities A_1 and A_2, which have different scale dimensions $\Delta_{A_1} \neq \Delta_{A_2}$. To test the hypothesis of conformal invariance, V. L. Pokrovsky proposed in [46] to experimentally investigate the combination of thermodynamic derivatives

$$(\partial\rho/\partial P)_T \cdot \left[(\partial P/\partial T)_\rho - (\partial\rho/\partial T)_P\right., \tag{3.11}$$

in which the most singular contributions from the average $\langle A_i{}^2 \rangle$ were cancelled out. If A_1 and A_2 are the main basic elements of the algebra of fluctuating quantities with dimensions $\Delta_{A_1} = \Delta_\eta \approx 1/2$ and $\Delta_{A_2} = \Delta_\varepsilon \approx 3/2$ for d = 3, then in case of violation of conformal invariance, the temperature dependence of the indicated combination (3.11) on the critical isochore will be determined by the mean value of $\langle A_1 A_2 \rangle \sim |\tau|^{-(\gamma+\alpha)/2}$.

The pair correlation function $G^{\rho\rho}$ of density fluctuations determines the integral intensity of single-scattered light and X-rays, as well as the intensity of coherent neutron scattering in liquid systems in the critical region. Thus, for $q \sim \exp(-|\vec{r}_1 - \vec{r}_2|/R_c)$ from (3.9), using M. Fisher approximation with a non-zero value of the critical exponent η_0 and unpolarized incident light beam, the following formula describes the integral single-scattering light intensity [47]:

$$I^{(1)} = A(\vartheta)\beta_T/\left(1 + R_c^2 k^2\right)^{1-\eta_0/2}, \tag{3.12}$$

where the first multiplier in the right side

$$A(\vartheta) = I_0 \frac{\pi^2 V}{2L_0^2 \lambda^4}\left(\rho\frac{\partial\varepsilon}{\partial\rho}\right)_T^2 k_B T\left(1+\cos^2\vartheta\right), \tag{3.13}$$

while β_T is the isothermal compressibility, and

$$k = \frac{4\pi}{\lambda}\sin\frac{\vartheta}{2} \tag{3.14}$$

is the change in the wave vector in the process of scattering a light beam at an angle ϑ.

Note that the equalities between the critical indices

$$(2 - \eta_0)\nu = \gamma, (2 - \eta_0)\varsigma = \delta - 1, \tag{3.15}$$

follow from (3.12)–(3.15) based on the natural assumption regarding the finiteness of the scattering intensity at the critical point:

$$0 < \lim_{\tau, \Delta\rho \to 0} I^{(1)}(\tau, \Delta\rho, k \neq 0) < \infty. \tag{3.16}$$

Obtaining information about the behavior of older correlation functions in the critical region should, in particular, be associated with the rejection of the assumption of single scattering of light, X-rays or neutrons when considering the phenomena of its critical opalescence. The study of the effects of multiple scattering and the use of correlation spectroscopy methods may prove to be promising in connection with the verification of important consequences concerning the asymptotic behavior of older correlation functions when groups of particles are approaching long distances between each other. This, in turn, is fundamental for the study of collective effects that determine the specifics of the behavior of systems in which the processes of self-organization and ordering take place.

The scaling invariance theory is essentially a phenomenological theory that requires microscopic justification. In this direction, which is fundamental for a deeper understanding of the nature of critical phenomena, significant progress has been achieved, associated with the successful application of quantum field theory methods. The first attempt to calculate the sum of a number of Feynman diagrams, significant around the points of phase transitions, was made by A.Z. Patashynsky and V.L. Pokrovsky [15], who first formulated the hypothesis of scaling invariance of the correlation functions. However, the specific results obtained by them turned out to be correct only for the spatial dimension $d = 4$ (in particular, under the assumption, the scale dimension of the order parameter turned out to be $\Delta_\eta = 3/4$ instead of the required value $\Delta_\eta = (1 + \eta_0)$ for 3-dimensional space). Later, in the works of A.M. Polyakov, A.A. Migdal and other authors, with the help of exact relations of the quantum field theory (Dyson's equation, conditions of unitarity and analyticity, Ward's identity, etc.) [47], it was possible to confirm many physical ideas underlying the theory of large-scale transformations In these works, the scale and conformal invariance for CF, the identity between critical indices, and local and bilocal algebras of fluctuating quantities were microscopically substantiated. In particular, the asymptotics of the $G^{\eta\eta}$ $(\vec{r}, \vec{r\prime})$ was obtained as follows:

$$G^{\eta\eta}(\vec{r}, \vec{r\prime}) \sim m^{2\Delta_\eta - 1} e^{-m|\vec{r}-\vec{r\prime}|} / \left|\vec{r} - \vec{r}'\right|, \tag{3.17}$$

where m is a quantity with the scale dimension of the inverse length.

The studies mentioned above greatly stimulated the further development of the microscopic theory, which led to the use of the ideas of the renormalization group (RG) in the physics of phase transitions in the works of K. Wilson and M. Fisher with their co-authors. Previously, RG was used in quantum electrodynamics by M. Gell-Mann and F. Lowe and in quantum field theory by M. M. Bogolyubov and D. V. Shirkov. The simplified idea of the RG-approach, which was proposed by K. Wilson, consists in obtaining an explicit dependence of the parameters of the effective Ginzburg-Landau Hamiltonian

$$H_{GL} = \int_V d^d x\left\{(\nabla\eta)^2 + R\eta^2 + U\eta^4\right\} \tag{3.18}$$

from the dimension L associated with the magnitude of the cutting impulse, which characterizes the area of averaging of the order parameter. For this purpose, in (3.18) an approximate statistical averaging is performed over all short-wave fluctuations with $\lambda < L$. As a result, it becomes possible to obtain a system of differential equation

$$\frac{dR}{dL} = -3L^{\varepsilon-1}uR, \frac{du}{dL} = -9L^{\varepsilon-1}u^2. \tag{3.19}$$

This achieves the renormalization of the parameters of the Hamiltonian (3.18) from the values $R = R_0|\tau|$ and $u = u_0$ in the Landau theory of a mean field to the following values (in the linear approximation in $\varepsilon = 4\text{-}d$):

$$R = R_0|\tau|^{1-\varepsilon\nu/3}, U = U_0|\tau|^{\varepsilon\nu}, \nu = 1/(2 - \varepsilon/3). \tag{3.20}$$

The main result (3.20) consists in the appearance of a non-analytical temperature dependence, the direct consequence of which is the correct large-scale behavior of various physical properties in the critical region. Thus, the very problem of the phase transition is not solved with the degree of accuracy that is possible when describing model systems. However, at the same time, it is possible to find out many fundamentally important properties that are manifested under the action of the successive RG transformation of the following form:

$$H_{GL'} = -k_B T \ln \left\{ \left[\prod_{i, \frac{\vec{\Lambda}}{s} < \vec{k}' < \vec{\Lambda}} \exp(-H_{GL}/k_B T) d\eta_{i\vec{k}'} \right]_{\eta_{\vec{k}} \to a_s \eta_{s\vec{k}}} \right\} \tag{3.21}$$

The transformation (3.21), written in the symbolic form $\mu' = R_s\mu$, is carried out over the set $\{\mu\}$ of the parameters of the effective Hamiltonian (usually the H_{GL}), bringing the point μ of the parametric space closer to the so-called fixed point μ^*, which plays the role of a real critical points Note that the transformation (3.21) leads to the following important equation for the CF $G(\vec{k}, \mu)$:

$$G\left(\vec{k}, \mu\right) = s^{2-\eta_0} G\left(s\vec{k}, R_s\mu\right). \tag{3.22}$$

All further studies of critical behavior are largely reduced to the task of finding the eigenfunctions e_i and the eigenvalues s^{yi} of the linearized near μ^* operator $R_s{}^L$ of the RG transformation at $s \to \infty$. The solution of this problem turned out to be successful when K. Wilson and M. Fisher chose the dimension of space d as a free parameter, more precisely the deviation $\varepsilon = 4-d$ of the dimension from the case d = 4. At the same time, in a 4-dimensional space with logarithmic accuracy, "classical" Landau critical behavior is valid, since the contribution of fluctuations is determined by the ratio L^4 / L^d, where $L \approx R_c \to \infty$ at the point of the phase transition. Unfortunately, the method of ε-expansions leads to asymptotic series that diverge as the number of retained terms increases. However, due to the rapid reduction of the first few coefficients of ε-expansions the truncation of the series at the 3rd or 4th terms leads to a good agreement between the results of theoretical calculations and the experiment.

Another convenient parameter that allows implementing the RG-approach is the inverse dimension of the parameter of order n^{-1}. The results of the exactly solvable spherical model ($n \to \infty$), which give $\beta = 1/2$, $\delta = 5$, $\eta_0 = 0$, $\nu = 1$, $\gamma = 2$, are used as a zero approximation in n^{-1}—expansions.

Application of the renormalization group method is a fruitful approach to the problem of phase transitions. Along this path, it was possible to more deeply study the qualitative features of cooperative effects, find quantitative values of critical indices, calculate the explicit form of large-scale functions of the equation of state of various systems, and obtain a number of other important results.

Let us emphasize some basic provisions of the theory of phase transitions, which are widely used in synergistics to study the behavior of systems of various nature near bifurcation points.

1. As indicated earlier in 2.4 and 2.6, the modern statistical theory of phase transitions uses the following expression for the thermodynamic potential (***Ginzburg-Landau fluctuation Hamiltonian***) to take into account fluctuation effects:

$$H_{GL} = \int d^d x\left[(1/2)a\eta^2(x) + (1/4)b\eta^4(x) + (1/2)c(\nabla\eta(x))^2 - h\eta(x)\right]. \tag{3.23}$$

Here $\eta(x)$ is the order parameter; h is the field conjugate to the order parameter (in Table 3.1, the order parameters η and the corresponding fields h for some systems were given); a, b, c are parameters; d is the spatial dimension.

The coefficient a in (3.23) is related to the inverse susceptibility of the system $\chi^{-1} \sim dh/d\eta$. For temperature phase transitions, this value is directly proportional, according to the assumption of L.D. Landau, to the deviation of the temperature from the critical value: $a = a_0\tau$, where $\tau = (T - T_c)/T_c$, Ts is the critical temperature. As mentioned above, the transformation of the renormalization group, which is the basis of the modern microscopic theory of phase transitions, is carried out over the parameters a, b, c of the Ginzburg-Landau Hamiltonian. The renormalization of the temperature dependence $a(\tau)$ essentially means a transition from Curie's law $\chi \sim \tau^{-1}$ for the susceptibility of the system (isothermal compressibility of a liquid or magnetic susceptibility of a magnet) to a scaling law for this quantity $\chi \sim \tau^{-\gamma}$ with a critical index $\gamma = 1 + \varepsilon/6 + O(\varepsilon^2)$, where $\varepsilon = 4-d$ is the deviation of the dimension of space d from 4.

2. In the dynamic theory of phase transitions, the equation for the order parameter takes the form

$$\partial\eta/\partial t = -L * \left(a\eta + b\eta^3 - c\Delta\eta - h\right), \tag{3.24}$$

where L^* is the Onsager kinetic coefficient. It is not difficult to notice the complete analogy between equations of reaction-diffusion models (see e.g. [2, 7, 38]) and the equation (3.24). The origin of the diffusion term in (3.24) is related to the presence in (3.23) of a term containing $(\nabla\eta)^2$ and taking into account the effects of non-locality of order-parameter fluctuations near the critical (bifurcation) points.

The equilibrium value of the order parameter η_0 satisfies the relation

$$a\eta_0 + b\eta_0^3 - c\Delta\eta_0 = h, \tag{3.25}$$

that is the equation of state of the studied system. Knowledge of the equation of state opens up wide possibilities of using many reliably established results of the theory of condensed media.

3. Equilibrium and kinetic properties are related to correlation functions due to fluctuation or fluctuation-dissipation relations (theorem). One example is the compressibility theorem (or the compressibility integral)

$$\rho k_B T \beta_T = 1 + \rho \int G_2(r) d\vec{r} \tag{3.26}$$

which connects the isothermal compressibility β_T of the liquid with the pair correlation function $G_2(r)$ of density fluctuations. In the vicinity of the phase transition or the critical (bifurcation) points, this function) takes a scaling invariant form

$$G_2 = f(r/R_c)/r^{1+\eta}, \tag{3.27}$$

where η is the critical index of the anomalous dimension of the correlation function of the order parameter (see Table 3.2), and the function $f(x) \sim \exp(-x)$ in the Ornstein-Zernike approximation.

4. The dependence of the correlation length (radius) R_c on the equilibrium value of the order parameter η_0 is characterized by the following formula:

$$R_c = R_0 \eta_0^{-\nu/\beta} \tag{3.28}$$

Here R_0 is the амплітуда of the correlation length, which in order of magnitude coincides with the radius of action of the intermolecular potential $U(r)$; v and β are critical indices. Critical indices depend on the spatial dimension d of the system, the dimension (number of components) n of the order parameter η, as well as on the radius R_0 of the interaction potential. As the phase transition point (line) is approached, the value of R_c increases and can reach the linear dimensions of the system. This means that near the phase transition the state of the system is completely determined by cooperative effects caused by long-range correlated fluctuations of the order parameter.

5. The role of fluctuations is clarified by analyzing the Ginzburg number Gi [15, 19, 20]. The Ginzburg number Gi is determined by the ratio of the mean square fluctuation of the order parameter $<\Delta\eta^2>$ to the square of the equilibrium value of the order parameter $\eta_0{}^2$. It can be shown that

$$Gi \sim \eta_0^{d-4}/R_0^d. \tag{3.29}$$

It follows from (3.29) that the Ginzburg number $Gi << 1$ near the phase transition point, if only $d \geq 4$ and (or) $R_0 \to \infty$. In this case, the fluctuations of the order parameter can be neglected, and the Landau mean-field theory turns out to be valid.

3.3.2 General properties of Spatially Bounded Media Near Bifurcation Points and Phase Transition Points

The influence of spatial limitation on the phase transitions and critical phenomena, which can lead to the formation of ordered structures and self-organization, has recently attracted increased interest. Such effects are very important for various systems of not only theoretical but also practical interest, such as surfaces, transition layers (interphases), pores, biomembranes, vesicles, synaptic clefts, etc. We will return to this question in Chap. 4 when studying the process of synaptic transmission.

A system is called to be spatially bounded (limited or finite-size) in the vicinity of the phase transition, if its characteristic linear dimensions are comparable to the correlation radius of fluctuations of the order parameter, i.e. $L_i \approx R_c$. As is known from the theory of scaling invariance, the correlation radius of fluctuations of the order parameter $R_c = R_{co}\tau^{-\nu}$, where the amplitude of the correlation radius in liquids reaches the value $R_{co} \approx 0.1$–1.0 nm. In modern experiments on the study of phase transitions and critical phenomena in classical liquids, record achievements regarding the closest possible approach to the critical point in terms of temperature are characterized by the following values of the absolute deviation of the temperature from the critical value $|T\text{-}T_c| \approx 10^{-3}$ K at $T_c \approx 10^2$ K, or in relative values, the minimum temperature deviations are equal to $\tau_{min} = |T - T_c|/T_c \approx 10^{-5}$. Since the critical index of the correlation radius is equal to $\nu \approx 0{,}63$ at $d = 3$, the maximum values of the correlation radius in classical liquids can reach the following values: $R_c \approx 0.1$–1.0 nm $\cdot$ $(10^{-5})^{-0{,}6} \approx 0.1$–$1.0$ μm . (A special case is quantum liquids, where due to very small values of the critical temperature, the correlation radius can increase by another two orders of magnitude; for example, in liquid helium $\tau \approx 10^{-8}$ and $R_c/R_{co} \approx 10^5$. Thus, media with linear dimensions of the order of $L \approx (0.1$–$1.0) \cdot 10^2$ μm can be considered spatially limited near points of phase transitions in quantum liquids).

For similar spatially limited systems, a famous English and American theoretical physicist Michael Fisher (1931–2021) and other scientists in the field of phase transition physics proposed a modified hypothesis of scale invariance (scaling), which includes not only thermodynamic variables and external fields, but also linear dimensions of the system [48–52]. According to this hypothesis, the scaling formulas for the fluctuating parts of the thermodynamic potential and the correlation radius in a system with linear dimensions L have the following form:

$$\Delta F_{fl} = L^{-d} f_F\left(a\tau L^{1/\nu}, bhL^{\beta\delta/\nu}\right), \tag{3.30}$$

$$R_c = Lf_R\left(a\tau L^{1/\nu}, bhL^{\beta\delta/\nu}\right). \tag{3.31}$$

The first scale argument $x = a\tau L^{1/\nu}$ in the scaling functions f_F and f_R can be obtained from the following relations:

$$R_c \sim \tau^{-\nu}, \tau \sim R_c^{-1/\nu}, x \sim \frac{\tau}{R_c^{-1/\nu}} \sim \tau L^{1/\iota}(R_c \to L). \quad (3.32)$$

Similarly, the expression $y = bhL^{\beta\delta/\nu}$ and for the second scale argument can be obtained, bearing in mind the following formulas:

$R_c \sim \tau^{--\nu} \sim \varphi j^{-\nu/\beta} \sim h^{-\nu/\beta\delta}, \varphi \sim \tau^{\beta}, \varphi \sim h^{1/\delta}, \tau \sim \varphi^{1/\beta}, h \sim R_c^{-\beta\delta/\nu}$,

$$y \sim \frac{h}{R_c^{-\beta\delta/\nu}} \sim hL^{\beta\delta/\nu}(R_c \to L). \quad (3.33a)$$

$$R_c \sim \tau^{-\nu} \sim \varphi^{-\nu/\beta} \sim h^{-\nu/\beta\delta}, \varphi \sim \tau^{\beta}, \varphi \sim h^{1/\delta}, \tau \sim \varphi^{1/\beta}, h \sim R_c^{-\beta\delta/\nu},$$

$$y \sim \frac{h}{R_c^{-\beta\delta/\nu}} \sim hL^{\beta\delta/\nu}(R_c \to L). \quad (3.33b)$$

Let us consider ***the consequences of the scaling hypothesis for spatially bounded systems*** [48–50]:

(a) *the equilibrium value of the order parameter*

$$\varphi_0 = \{(\rho_0 - \rho_C)/\rho_c; M_0; \ldots\} = \left(\partial\Delta F_{fll}/\partial h\right)_\tau = L^{-d+b\delta/\nu} f_\varphi(x, y),$$

$$f_\varphi = b\left(f_\varphi\right)_h{}', -d + \beta\delta/\nu = -\beta/\nu(3\nu = 2 - \alpha, \alpha\beta + \gamma = 2, \beta\delta = \beta + \gamma). \quad (3.34)$$

Thus, it is possible to propose ***the following algorithm for obtaining the dependence of various properties on the size of L in a bounded system***:

$$A \sim \tau^{-n} \sim R_c^{n/\nu} \sim L^{n/\nu} \quad (3.35)$$

In other words, if in a spatially unbounded system some quantity has the peculiarity of the form $A \sim \tau^{-n}$, then taking into account the relations

$$R_c \sim \tau^{-\nu}, \tau \sim R_c^{-1/\nu}, R_c \to L \quad (3.36)$$

The dependence of this value on the linear size near critical points and points of phase transitions in a spatially limited system has the following form:

$$A \sim L^{n/\nu}. \quad (3.37)$$

So, for example, for the order parameter in an unbounded system, we have $\varphi_0 \sim \tau^{\beta}$. Then, in a spatially limited system with a linear dimension L (say, for a cube with side L), we will have ***the following dependence of the order parameter on the linear dimension***:

$$\varphi_0 \sim \tau^{\beta} \sim R_c^{-\beta/\nu} \sim L^{-\beta/\nu} \tag{3.38}$$

It directly follows from this that in a spatially unlimited system, where the linear size L and the correlation radius R_c can grow indefinitely (L, $R_c \to \infty$) upon reaching the phase transition point, the value of the order parameter goes, as it should, to zero ($\varphi_0 \to 0$).

For other properties of bounded systems we have

(b) *the susceptibility of a spatially limited system*

$$\chi = \left(\partial^2 F_{fl}/\partial h^2\right)_\tau = L^{-d+2\beta\delta/\nu} f_\chi(x, y), \tag{3.39}$$

$$f_\chi = b\left(f_\varphi\right)''_h, -d + 2\beta\delta/\nu = \gamma/\nu.$$

(c) *the heat capacity of a spatially limited system*

$$C = \left(\partial^2 \Delta F_{fl}/\partial \tau^2\right)_h = L^{-d+2/\nu} f_c(x, y), \tag{3.40}$$

$$f_c = a\left(f_\varphi\right)''_\tau, -d + 2/\nu = \alpha/\nu$$

It turns out that in spatially limited systems, phase transitions are not as sharp as in bulk (infinite) systems. In this regard, ***the concept of the temperature neighborhood (τ nbh) of a phase transition in bounded systems*** should be introduced:

$$x = a\tau_{nbh} L^{\beta\delta/\nu} \approx 1, \; \tau_{nbh} \approx L^{-1/\nu}/a \sim L^{-\theta_T}, \tag{3.41}$$

where the critical index $\Theta_T = 1/\nu = 1.59 \approx 3/2$ ($\nu = 0.63 \approx 2/3$, $d = 3$).

In a similar way, we have ***the neighborhood (h_{nbh}) of a phase transition with respect to an external field (magnetic, gravitational, etc.) in bounded systems*** as follows:

$$y = bh_{nbh} L^{\beta\delta/\nu} \approx 1, \; h_{nbh} = |H - H_c|/H_c \approx L^{-\beta\delta/\nu}/b \sim L^{\theta_H}, \tag{3.42}$$

where the critical index $\Theta_H = \frac{\beta\delta}{\nu} \approx 2,5$ ($\beta \approx 1/3$, $\delta \approx 5$, $\nu \approx 2/3$, $d = 3$).

Another new and fundamental result concerns ***the shift of critical parameters*** (temperature, density, magnetic field) of spatially limited environments in comparison with the values of critical parameters in unbounded systems.

Let us find the shift of the critical temperature Tc in bounded systems from the condition of the susceptibility extremum, for which we have

$$\chi = L^{\gamma/\nu} f_\chi(x, y) \underset{x = a\tau L^{1/\nu}}{=} \left(\frac{x}{a\tau}\right)^\gamma f_\chi(x, y). \tag{3.43}$$

This extremum condition can be written in the following form with taking into account the formula for $x = a\tau L^{1/\nu}$:

$$(\partial \chi/\partial \tau)_h = -\gamma\left(\frac{x}{a\tau^*}\right)^{\gamma}\frac{1}{\tau^*}f_\chi + \left(\frac{x}{a\tau^*}\right)^{\gamma} f_\chi{}' a L^{1/\nu} = 0, \tag{3.44}$$

from which one may obtain the dimensionless value of the shift τ^* of the critical temperature

$$\tau^* = \frac{T_c(L) - T_c(\infty)}{T_c(\infty)} = \frac{\gamma f_\chi}{a f_\chi{}'} L^{-1/\nu} = A_\chi L^{-1/\nu}. \tag{3.45}$$

We will show that the temperature shift $\tau^* = [T_c(L) - T_c(\infty)]/T_c(\infty)$ in limited systems is not the universal quantity. For this purpose, let us find the shift τ^* of the critical temperature from the extremum (maximum) condition of the heat capacity C, which gives the following result:

$$C = L^{\alpha/\nu} f_c(x, y) = (x/a\tau)^{\alpha} f_c\left(a\tau L^{1/\nu}, y\right)$$

$$(\partial C/\partial \tau)_h = -\alpha\left(\frac{x}{a\tau^*}\right)^{\alpha}\frac{1}{\tau^*}f_c + \left(\frac{x}{a\tau^*}\right)^{\alpha} f_c{}' a L^{1/\nu} = 0,$$

$$\tau^* = \frac{\alpha f_c}{a f_c{}'} L^{-1/\nu} = A_c L^{-1/\nu}. \tag{3.46}$$

Similarly, the following formula can be obtained for the shift τ^* of the critical temperature from the extremum (maximum) condition of the correlation length R_c:

$$\tau^* = \frac{\nu f_R}{a f_R{}'} L^{-1/\nu} = A_R L^{-1/\nu}. \tag{3.47}$$

Thus, we came to the following ***important conclusion***: in accordance with formulas (3.45)–(3.47), the dependence of the shift τ^* of the critical temperature on the linear size of the system ($\tau^* \sim L^{-1/\nu}$) is universal, but the numerical values of τ^*, which give extremes to various physical properties (susceptibility χ, heat capacity C, correlation length R_c, etc.), are not universal quantities, since amplitude values are different in (3.45)–(3.47): $A_\chi \neq A_c \neq A_R$. These amplitudes contain non-universal coefficients like a and b in (3.30)–(3.31), which depend on the physical properties and nature of the substances being studied.

A schematic graph of the temperature dependence of susceptibility in bulk and bounded systems is shown in Fig. 3.1. This figure illustrates the basic facts established by M. Fisher, K. Binder and other scientists [48–52], which distinguish the critical behavior of spatially unlimited (bulk) and spatially limited (bounded) systems. As can be seen from Fig. 3.1, in bounded systems there are:

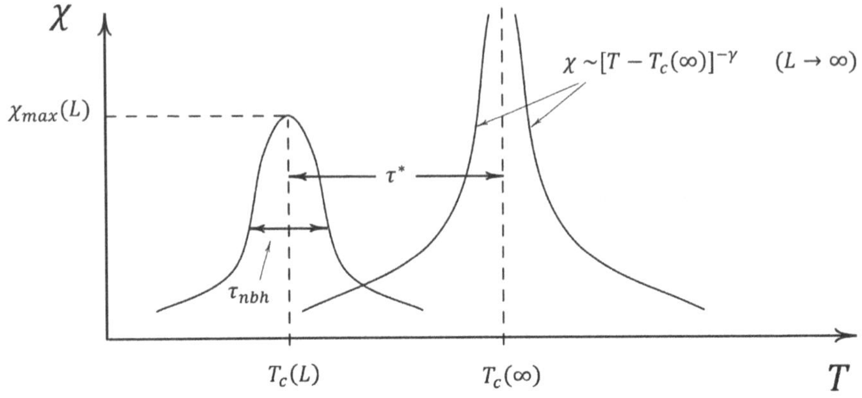

Fig. 3.1. Temperature dependence of the susceptibility of spatially limited (left) and unlimited (right) systems.

(1) shift of critical parameters (in the case under consideration—the critical temperature),
(2) a certain region of the phase transition, which narrows when moving to a bulk system,
(3) the maximum of susceptibility at the new point of the phase transition, which shifts towards higher temperatures and increases to infinity when approaching a bulk system (without taking into account the effects of spatial dispersion).

3.3.3 Stationary State of an Open System in Case of Formation of Ordered Structures

Consider an open system of stationary type [1, 6, 38]. As is known from Sect. 3.2.4, the concept of stationarity assumes that the total entropy S_{tot} of an open system is unchanged due to the fact that the increase in entropy ΔS_i, associated with irreversible dissipative processes inside the system, is fully compensated by the interaction of the system with the environment, which is accompanied by an outflow of entropy ΔS_e, is due to the surface limiting the volume of the system. Thus, there is a correlation

$$\Delta S_{tot} = \Delta S_i + \Delta S_e = 0, \tag{3.48}$$

since $\Delta S_i > 0$, and $\Delta S_e = -\Delta S_i < 0$. Here, in formula (3.48), ΔS_i denotes the positive production of entropy within the system

$$\Delta S_i = \iint \sigma dV dt > 0, \tag{3.49}$$

and ΔS_e indicates the inflow of negative entropy (negentropy) from the environment

$$\Delta S_e = -\iint J_s d\Omega dt < 0. \tag{3.50}$$

In expression (3.50), the notation standard for non-equilibrium thermodynamics is used, namely: $J_s = d^2S_e/dAdt < 0$ is the entropy flow through the surface A directed inside the system, and dA is the element of the surface, which coincides in direction with the external normal to the element of surface.

In other words, stationarity is achieved due to such an inflow of negentropy from the outside ($\Delta N_e > 0$), which is exactly equal to the decrease of negentropy inside the system ($\Delta N_i < 0$), i.e.

$$\Delta N_{tot} = \Delta N_i + \Delta N_e = 0. \tag{3.51}$$

The above justifies the well-known statement: ***an open stationary system is powered by negative entropy, i.e., negentropy $N = -S$.***

It is for such stationary open systems that Prigogine's theorem on minimum entropy production is fulfilled, which is valid for small deviations of the system from equilibrium. The term "small deviations" implies the validity of the linear thermodynamics of irreversible processes, that is, such an approximation for which the linear law is fulfilled

$$J_i = \sum_k L_{ik} X_k \tag{3.52}$$

between flows J_i and thermodynamic forces X_k, as well as the problem of entropy change

$$\Delta S = \sum_{i,k} g_{ik} X_i X_k \tag{3.53}$$

in the form of a quadratic form of thermodynamic forces, where g_{ik} are coefficients calculated in the equilibrium position.

Let us generalize the concept of an open system of the stationary type according to Ilya Prigogine to the case when the formation of ordered (spatial, temporal, or spatio-temporal) structures is possible in an arbitrary open system, where thermodynamically unstable states are realized and where the processes of spinodal decay, self-organization, and formation of ordered structures take place.

During the formation of ordered structures, the change in entropy ΔS_i within the system generally consists of the following contributions:

(1) a positive change in entropy $\Delta S_i^{(d)}$ due to dissipative processes,
(2) a negative entropy change $\Delta S_i^{(s)}$ due to the formation of ordered structures. Thus, we have

$$\Delta S_i = \Delta S_i^{(d)} + \Delta S_i^{(s)} \tag{3.54}$$

At the same time, the stationary state can be ensured in two ways:

(a) if $\Delta S_i^{(d)} < \left|\Delta S_i^{(s)}\right|$, then $\Delta S_i > 0$ and to fulfill the condition $\Delta S_{\text{tot.}} = 0$, it is necessary to ensure the inequality

$$\Delta S_e = -\left|\Delta S_i^{(d)} - \Delta S_i^{(s)}\right| < 0, \tag{3.55}$$

similar to the condition (3.48) of the stationarity of the open system according to Prigjgine's theorem;

(b) if $\Delta S_i^{(d)} < \left|\Delta S_i^{(s)}\right|$, then $\Delta S_i < 0$ and in this case an inflow of entropy from the outside ($\Delta S_e > 0$) is necessary to ensure the condition of stationarity ($\Delta S_{\text{tot}} = 0$), i.e.

$$\Delta S_e = \left|\Delta S_i^{(s)}\right| - \Delta S_i^{(d)} > 0. \tag{3.56}$$

Inequality (3.56) is called ***the condition of stationarity of open systems in the generalized sense.*** We note that for an arbitrary open system, the condition of small deviations of parameters from the equilibrium position (small gradients that determine the magnitude of thermodynamic forces) and thus the conditions of application of linear thermodynamics of irreversible processes and, accordingly, Prigogine's theorem regarding the minimum of entropy production in a steady state may turn out to be unfair.

And, finally, it is necessary to emphasize once again that when studying the processes of formation of ordered spatial or temporal structures near critical points, i.e., at Turing or Hopf bifurcations, it is necessary to consistently take into account the fluctuating contributions ΔS_{fl}. Therefore,

$$\Delta S_{tot} = \Delta S_i^{(d)} + \Delta S_i^{(s)} + \Delta S_e + \Delta S_{fl}. \tag{3.57}$$

It follows that when the stationarity conditions (3.56) or (3.57) are met, the term ΔS_{fl} becomes the determining factor for entropy. Accordingly, for the fluctuating part of the free energy of such a system we have: $\Delta F_{fl} = -T\Delta S_{fl}$.

3.4 Conclusions

The complex and diverse processes of the development of nature and society are gradually receiving more and more adequate research methods based on the achievements of synergetics and other sciences. These methods are intensively developed all the time and enriched with new ideas. On their basis, such processes as the spontaneous (fluctuating) complication of biological objects, the correlation of the processes of structure formation, the cooperative nature of the course of these processes over large spatial and temporal intervals, the significant responses to weak external influences characteristic of many systems, etc., have already found their explanation.

In the modern synergetic picture of animate and inanimate nature, which is based on the achievements of the theory of phase transitions and critical phenomena, an important role is assigned to the hypothesis of universality. According to it, systems of completely different nature belonging to the same "class of universality" exhibit similar behavior near their critical (bifunctional) points. Numerous confirmations of such universality make the aspiration of researchers to unify the phenomena of the world around us justified and necessary. It is this that gives grounds for confidence that beautiful and perfect synergetic methods will be effective for studying the complex processes of development of nature and society.

To confirm this opinion, dear readers, allow me to recall a saying that belongs to the outstanding scientist of the 20th century, Nobel Prize laureate Richard Feynman. Keeping the meaning of these bright words, which were said during Feynman's famous lectures on physics at the California Institute of Technology (USA), we could finish our story about synergetics, which of course is only an introduction to this wonderful science:

Once a poet said: "The whole world is in a glass of wine." We will probably never understand what meaning he put into it, because poets do not write in order to be understood. But there is no doubt that by carefully looking into a glass of wine, we will actually discover a whole world. It contains physical phenomena (sparkling liquid, evaporation that changes depending on the weather and your breath, the shine of glass) and atoms (about which our imagination already tells us). Glass is a purified rock; it contains the secrets of the age of the universe and the development of stars. And what a strange set of reagents this wine consists of! How did they come about? There are starters, enzymes, extracts and various other products. After all, there is a great generalization hidden in wine: all life is fermentation. Studying the chemistry of wine, one cannot help but discover, as Louis Pasteur did, the causes of many diseases. How much life is in this claret, if it imposes its spirit on our minds, if we must be so careful with it! Our limited mind for convenience divides this glass of wine, this world into parts: physics, biology, geology, astronomy, etc., but nature does not really know any division! Let's pour it all together, not forgetting what we saw. Let this glass of wine give another pleasure at the end: to drink it and forget about everything!

References

1. H. Haken, Synergetics, in H. Haken (Ed.) *Proceedings of Symposium on Synergetics*. Elmau, Teubner, Stuttgart (1973)
2. H. Haken, *Synergetics. An Introduction. Nonequilibrium Phase Transitions and Self-Organization in Physics, Chemistry and Biology*. Springer, Berlin, Heidelberg, New York (1978)
3. H. Haken, *Advanced Synergetics* (Springer, Berlin, Heidelberg, New York, Tokyo, Instability Hierarchies of Self-Organizating Systems and Devices, 1983)

4. Ch.S. Sherrington, *The Integrative Action of the Nervous System* (Oxford University Press, Oxford, 1906)
5. P. Glansdorff, I. Prigogine, *Thermodynamic Theory of Structures, Stability and Fluctuations*, Wiley, New York (1971)
6. A.V. Chalyi, *The Synergetic Principles of Education and Science*, Bogomolets National Medical University, Kyiv (2000)
7. D.S. Chernavskii, *Synergetics and Information: Dynamic Theory of Information.* Mir (2001)
8. V.I. Sugakov, *Fundamentals of Synegetics.* Oberig, Kyiv (2001)
9. *Medical and Biological Physics.* A.V. Chalyi (Ed.) Knyga Plus, Kyiv (2005)
10. R.P. Feynman, R.B. Leighton, M. Sands, *The Feynman Lectures on Physics*, vol. 1, Chapter 26, California Institute of Technology (2013)
11. A.V. Chalyi, Synergetic dialogue "physics-medicine": Hexagones in living and inanimate nature. J. Mol. Liquids **329**, 114249 (2021)
12. V.L. Ginzburg, L.D. Landau, On the theory of superconductivity. J. Exp. Theor. Phys. **20**, 1064 (1950)
13. M.E. Fisher, The Nature of Critical Points, *Lectures in Theoretical Physics* (University of Colorado Press, Colorado, 1965)
14. L.P. Kadanoff, Scaling laws for Ising models near T_c. Physica **2**(6), 263–273 (1966)
15. A.Z. Patashinskii, V.L. Pokrovskii, *Fluctuation Theory of Phase Transition* (Pergamon Press, Oxford, 1979)
16. H.E. Stanley, *Introduction to Phase Transition and Critical Phenomena* (Oxford University Press, Oxford, 1987)
17. L.D. Landau, E.M. Lifshits, *Theoretical Physics, Statistical Physics*, part 1, vol. 5. Pergamon Press, Oxford (2013)
18. K.G. Wilson, J. Kogut, Renormalozation group and ε-expansioin. Phys. Rep. **12C**, 75 (1974)
19. A.P. Levanyuk, Contribution to the theory of light scattering near the second-order phase-transition points. Zh. Eksp. Teor. Fiz. **36**, 571–576 (1959)
20. V.L. Ginzburg, Some remarks on phase transitions of the second order and the microscopic theory of ferroelectric materials. Phys. Solid State **2**, 1824–1834 (1960)
21. MA Anisimov, *Critical Phenomena in Liquids and Liquid Crystals* (Gordon and Breach, Phyladelphia, 1991)
22. V.G. Boiko, H.J. Moegel, V.M. Sysoev, A.V. Chalyi, Features of metastable states in liquid-vapor phase transitions. Phys. Sci. Adv. **34**, 141–159 (1991)
23. A.V. Chalyi, A.G. Lebed, *Non-Homogeneous Liquids near the Critical Point and the Boundary of Stability and Theory of Percolation in Ceramics* (Harwood Academic Publishers, London, 1993)
24. I.R. Yukhnovsky, M.F. Golovko, *Statistical Theory of Classical Equilibrium Systems.* Naukova Dumka, Kyiv (1980)
25. I.R. Yukhnovsky, *Phase Transitions of the Second Order. Method of Collective Variables.* Naukova Dumka, Kyiv (1984)
26. I.R. Yukhnovsky, M.P. Kozlovsky, I.V. Pilyuk, *Microscopic Theory of Phase Transiyions in Three-Dimensional Systems.* Eurosvit, Lviv (2001)
27. *Order, Disorder and Criticality. Vol. 6. Advanced Problems of Phase Transitions Theory.* Edited by Yu.V. Holovatch (World Scientific, Singapore, 2020)
28. M.E. Fisher, Renormalization of critical exponents by hidden variables. Phys. Rev. **176**, 257 (1968)
29. M.A. Anisimov, E.E. Gorodetskii, V.D. Kulikov, A.A. Povodyrev, J.A. Sengers, A general isomorphism approach to thermodynamic and transport properties of binary liquid mixtures near critical points. Physics A **220**, 277 (1995)
30. A.V. Chalyi, L.A. Bulavin, V.F. Chekhun, K.A. Chalyy, L.M. Chernenko, A.N. Vasilev, E.V. Zaitseva, G.V. Khrapijchuk, A.V. Severin, M.V. Kovalenko, Universality classes and critical phenomena in confined liquid systems. Condensed Matter Phys. **16**, 23008 (2013)
31. R.G. Wilson, M.E. Fisher, Critical exponents in 3.99 dimensions. Phys. Rev. Lett. **28**, 240–243 (1972)

32. P. Buk, C. Tang, K. Wiesenfeld. Self-organized criticality: an explanation of 1/*f* noise. Phys. Rev. Lett. **59**, 381–384 (1987)
33. B. Mandelbrot, *The Fractal Geometry of Nature* (Freeman, San Francisco, W.H, 1983)
34. K. Linkenkaer-Hansen, V.V. Nikouline, J.M. Palva, R.J. Ilmoniemi, Long-range temporal correlations and scaling behavior in human brain oscillations. J. Neurosci. **21**(4), 1370–1377 (2001)
35. D.R, Chialvo, Critical brain network. Physica A **340**, 756–765 (2004)
36. L. de Arcangelis, P. Perrone-Capano, H.J. Hermann, Self-organized criticality model for brain plasticity. Phys. Rev. Lett. 96 (2006)
37. D.R. Chialvo, S.A. Cannas, T.S. Grigera, D.A. Martin, D. Plenz, Controlling a complex system near its critical point via temporal correlations. Sci Rep. **10**(1), 12145 (2020)
38. G. Nicolis, I. Prigogine, *Self-Organization in Nonequilibrium Systems: From Dissipative Structurers to Order Through Fluctuations* (Wiley, New York, 1977)
39. A.V. Chalyi, *Non-Equilibrium Processes in Physics and Biology*. Kyiv, Naukova dumka (1997)
40. A.V. Chalyi, a) Synergetics, b) Synergetical Approach in Education. In: *V.G. Kremen' (Ed.) Encyclopedia of Education*, Yurincom Inter, Kyiv (2001)
41. A.V. Chalyi, Dynamic anomalies in confined supercooled water and bulk fluids. In: *L. Bulavin, A. Chalyi (Eds.) Modern Problems of Molecular Physics,* Chapter 12 (Springer International Publishing, Switzerland, 253–289, 2018).
42. D.A. Fuentevilla, M.A. Anisimov, Scaled equation of state for supercooled water near the liquid-liquid critical point. Phys. Rev. Lett. **9**, 149904 (2007)
43. C.E. Bertrand, M.A. Anisimov, Peculiar thermodynamics of the second critical point in supercooled water. J. Phys. Chem. B **115**, 14099 (2011)
44. V. Holten, C.E. Bertrand, M.A. Anisimov, J.V. Sengers, Thermodynamics of supercooled water. J. Chem. Phys. **136**, 094507 (2012)
45. A.M. Polyakov, Conformal symmetry of critical fluctuations. JETP Lett. **12**, 381 (1970)
46. V.L. Pokrovskii, Feasibility of experimental verification of conformal invariance hypothesis. JETP Lett. **17**, 156 (1973)
47. E.L. Lakoza, A.V. Chalyi, Multiple light scattering near the critical point. Phys. Sci. Adv. **26**, 573 (1983)
48. M.E. Fisher, The theory of critical point singularities, in: M.S. Green (Ed.), Critical Phenomena, Proceedings of the International School of Physics "Enrico Fermi", New York: Academic Press, 1–99 (1971)
49. K. Binder, Phase transitions in reduced geometry Annu. Rev. Phys. Chem. **33**, 43 (1992)
50. A.V. Chalyi, Critical phenomena in finite-size systems. J. Mol. Liquids **58**, 153–164 (1993)
51. K.A. Chalyy, K. Hamano, A.V. Chalyi, Correlating properties of a simple liquid at criticality in a reduced geometry. J. Mol. Liquids **92**, 153–164 (2001)
52. A.V. Chalyi, Dimensional crossover in liquids in reduced geometry, in *L. Bulavin, N. Lebovka (Eds.) Modern Problems of Molecular Physics,* Chapter 2 (Springer International Publishing, Switzerland, 31–49, 2015)

Chapter 4
Physical Foundations of Neurophysiology: Non-linear Models of Action Potential Propagation and Synaptic Transmission

Abstract The problem of synaptic transmission (generally saying, the problem of intercellular interaction) in neurophysiology, for which there are electrical and chemical aspects, is as important for living nature as the problem of intermolecular interaction for inanimate nature. The electrical aspect of intercellular interaction is associated with the processes of generation and propagation of the action potential along the axon. The chemical aspect is due to the chemical reactions of the interaction of mediators and receptors in the synaptic clefts (synapses) that separate the cell body, axons and dendrites. Here, in Sects. 4.2 and 4.3, the general provisions and sequential equations of the Hodgkin-Huxley model (1963 Nobel Prize in Physiology or Medicine) will be considered. In addition, this chapter will propose new nonlinear kinetic models of synaptic transmission of a nerve impulse, as well as obtain analytical solutions to the corresponding systems of differential equations and analyze the consequences that follow from the analysis of singular points in accordance with the classifications of Poincaré and Lyapunov. As an application of our theoretical approach to the problem of synaptic transmission in cholinergic synapses, the systems of two (or three) nonlinear differential equations will be proposed to find the change of order parameters (ruling modes) such as concentrations of acetylcholine (ACh), acetylcholine-receptor (AChR) complexes, and acetylcholinesterase (AChE). The main findings of our studies are as follows: the linear size of the activation zone is evaluated as the correlation length for a AChR binary mixture (Sects. 4.2.2 and 4.3.2); the stationary states and types of singular points are studied for all models of synaptic transmission (Sects. 4.2.4 and 4.3.3); the nonlinear kinetic model of synaptic transmission with three concentrations of ACh, AChR, and AChE demonstrates a strange-attractor behavior (Sects. 4.2.4 and 4.3.3).

Keywords Synaptic transmission · Action potential · Hodgkin-Huxley model · Nonlinear kinetic models · Cholinergic synapse · Hypothesis of isomorphism · Strange attractor

A. V. Chalyi, *What is Medicine?*, https://doi.org/10.1007/978-3-031-64979-0_4

Can the brain understand the brain?

Can it understand the mind?

David Hunter Hubel,

1981 Nobel Prize in Physiology or Medicine

4.1 Introduction

This chapter will be devoted to synaptic transmission of nerve impulses (generally speaking, to intercellular or cell-to-cell communication). The processes of intercellular interaction determine one of the main stages of the work of the human and animal brain—perhaps the most complex device that nature has created. A sacramental question arises: can the human brain explain how this brain itself works? It is clear that there cannot be a final positive answer to this question. But this does not mean that there are no significant achievements of such sciences as neurophysiology, medicine, biology, biophysics, biochemistry, psychology, and others on the way to understanding the brain. These achievements gradually and iteratively bring us closer to the absolute truth in understanding all the processes that determine the principles of the human brain. We will get acquainted with some significant achievements in this and other chapters of the book on the way to a deeper understanding of the decisive problem—how our nonlinear brain system organizes its thinking function [1–17].

Figure 4.1 shows the main parts of the body (soma) of a nerve cell (neuron), which consists of mitochondria, ribosomes and many other organelles such as nucleus with its genetic information; dendrites emitting synaptic information from their own soma and receiving this information from other nervous cells; axons transmitting information to dendrites of other neurons. In other words, signals of intercellular interaction originating in the axon near its junction with the neuron body are propagated along the axon and then transmitted to other cells through synapses. In turn, these cells transmit the necessary information to the neuron using a large number of nerve branches—dendrites.

Let us consider some important characteristics that underlie the neurophysiological and psychophysical processes of synaptic transmission of information [1–17]:

(1) size of a human nervous cell (neuron)—from 5 μm to 130 μm (the largest size is typical for giant pyramidal Betz cells discovered by the Ukrainian anatomist Vladymyr Betz in 1874, who worked at the Medical Faculty of the St. Vladymyr Kiev University [18]);
(2) size of synaptic cleft (thickness of synapse)—from 3 to 20 nm;
(3) diameter of ion channels in the membrane—up to 1 nm;
(4) axon length—up to 1 m; width (diameter) axons—up to 0.5 mm for giant squid axons;

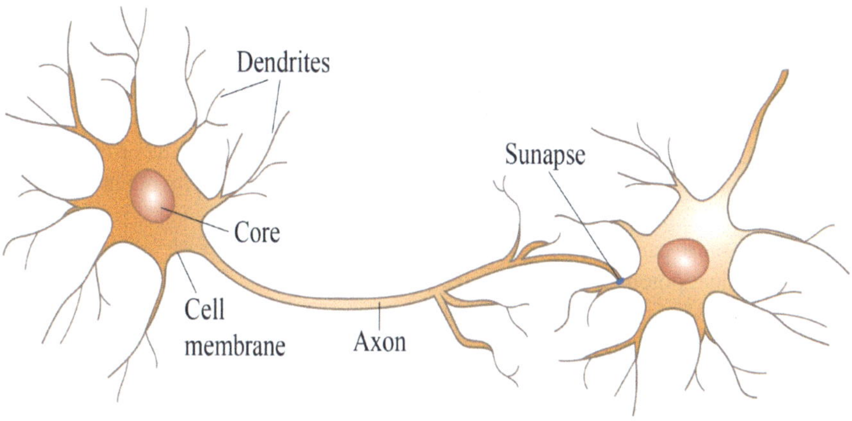

Fig. 4.1 The main components of the neuron involving in the synaptic transmission of nerve impulses

(5) number of neurons in the human brain—about 10^{11};
(6) number of contacts (synapses) between dendrites of one neuron with other neurons—up to $1.5 \cdot 10^4$;
(7) number of synapses in the human brain—about 10^{14}–10^{15};
(8) time of a signal exchange—about 10^{-3} s;
(9) number of channels in the membrane surface—about $10^2 - 10^3$ per 1 μm^2;
(10) number of impulses generated by neuron per 1 s—from 1 to 10^3;
(11) number of neurotransmitters (for example, ACh molecules contained in 1 vesicle)—about 10^4;
(12) number of ACh molecules, releasing simultaneously from the axonal presynaptic terminal with approximately 10^3 vesicles, under the influence of one nerve impulse (action potential)—about 10^7.

Below, we will focus the attention of our respected readers on a qualitative explanation of the electrical and chemical mechanisms of intercellular interaction along nerve fibers and through synaptic clefts.

4.2 Section A

4.2.1 Generation and Propagation of Action Potential Along Nerve Fibers (Nobel Prize in Physiology and Medicine)

The emergence of an action potential. As a result of electrical stimulation, the plasmatic membranes of cells are able to generate electrical impulses called ***action potentials***. Action potentials travel along nerve fibers called ***axons***. This electrical signals propagating along the axon are called ***spikes***. Depending on whether the axon is covered or not covered by the outer sheath (myelin sheath), the speed of propagation of the nerve impulse (spike) can vary within fairly wide limits, namely: for myelinated axons, which touch the external environment only in small areas of length about 1 μm (the so-called nodes of Ranvier), this speed is of the order of 25 $\div$ 100 m/s, whereas in the absence of the myelin sheath, i.e. for unmyelinated axons, it is 0.7 $\div$ 2.3 m/s.

The process of generating an action potential is associated with a sharp change in the membrane potential of nerve cells. In this case, the axon begins to produce a spike only if the potential on the plasmatic membrane of the cells exceeds a certain threshold value. Typical values of the resting membrane potential are about 70 $\div$ 80 mV, and the innermost environment of the axon (axoplasm) is negatively charged relative to the external environment ($\varphi_i \approx -75$ mV). The threshold value of the potential, at which the action potential begins to be generated, is of the order of threshold value $\varphi_i \approx -(40 \div 50)$ mV. As a result of exceeding this threshold potential value (and only in this case), an action potential develops, which at the maximum spike is equal to $\varphi_i \approx +(40 \div 50)$ mV. The above values of membrane potentials are typical for the squid axon and may change when moving to other objects. Thus, the process of the emergence of an action potential is of a critical (threshold) nature or, as they say, occurs according to the "all or nothing" principle. The generation of an action potential is associated with a change in the state of the plasmatic membrane. Let us consider what processes can lead to a change in the electrical potential of the axoplasm from the initial value $\varphi_i \approx -75$ mV to the final value $\varphi_i \approx +(40 \div 50)$ mV. Obviously, as a result of passive transport, the potassium ions K^+ can leave the cell, since there are tens of times more of these ions in the axoplasm than outside. This process can only increase the absolute value of the potential inside the axon, i. e. make it more negative (this process is called ***hyperpolarization***). A similar effect should also be caused by the chloride ions Cl^-, which, as a result of passive transport, enter the cell from the external environment. Thus, a change in the membrane potential in the desired (positive) direction, leading to a decrease in its absolute value (this process is called ***depolarization***), can only be due to the penetration of sodium Na + ions into the axon. Studies conducted by various scientists (A. Hodgkin, A. Huxley, B. Katz, K. Cole and others) have shown that when an action potential is generated, it is sodium channels that open. This leads to a sharp increase (approximately 500 times) in the permeability of the membrane to sodium ions. If at rest the ratio of membrane permeabilities (denoted by the letter P) for monovalent inorganic ions (K^+, Na^+, Cl^-) on the squid giant axon is P_K: P_{Na}: $P_{Cl} = 1$: 0,04: 0,45, then in a state of excitation, i. e. when an action potential is generated (at the maximum spike), this ratio is already P_K: P_{Na}: $P_{Cl} = 1$: 20: 0,45.

Such a significant increase in the permeability of the plasma membrane to sodium ions can be estimated through the value of the membrane potential at the maximum of the spike, retaining the contribution only from these ions, which gives $(\varphi_i)_{\max} \approx +$ 40 mV. Note that the change in sign of membrane potential due to the depolarization process from a negative value (at rest) to a positive one (at the maximum of the spike) is called ***inversion***.

It is possible to approximately estimate the amplitude of the action potential (spike) in relation to the resting potential on the membranes of nerve cells, taking advantage of the fact that at rest the main contribution to the membrane potential $(\varphi_i)_{\text{rest}} \approx -(70 \div 80)$ mV comes from potassium ions. Therefore, the amplitude $\Delta\varphi_{\text{AP}}$ of the action potential is approximately equal to

$$\Delta\varphi_{AP} = (\varphi_i)_{\max} - (\varphi_i)_{rest} \approx (40 \div 50)mV - [-(70 \div 80)mV \approx (110 \div 130)mV$$

After the membrane potential reaches its maximum value (as a result of the opening of sodium channels), these channels are blocked (inactivated) and the potassium conductivity of the membrane increases. In this case, the passive transport of potassium ions from the cell to the outside leads to the restoration of the original value of the resting membrane potential to the value $\varphi_i \approx -(70 \div 80)$ mV. This process is called membrane ***repolarization***. Thus, depolarization of the membrane and inversion of the sign of the membrane potential, i.e. the ascending part of the spike is caused by the opening of sodium channels for a time of about 1–2 ms, while the repolarization of the membrane, i.e. the descending part of the spike—by opening potassium channels for a time of 8–10 ms. The duration of the action potential deepends on the type of nerve fiber. Thus, in large skeletal nerve fibers, the duration of the action potential can be up to 5–10 ms, while in the nerve fiber of heart muscle it can be significantly longer.

Propagation of action potential along nerve fibers (Hodgkin-Huxley model). The creation of a model of the propagation of a nerve impulse is associated with research conducted in 1949–1952 by famous British neurophysiologists and biophysicists Alan Hodgkin (1914–1998) and Andrew Huxley (1917–2012). Ten years later in 1963, they received the Nobel Prize in Physiology and Medicine for this research together with famous Austalian neurophysiologist John Eccles (1903–1997) "for their discoveries concerning the ionic mechanisms involved in excitation and inhibition in the peripheric and central portions of the nerve cell membrane" [19].

Let us consider the general provisions of the Hodgkin-Huxley model:

1. The model takes into account both the capacitive current associated with a change in the potential difference on the membrane, and the currents of potassium, sodium, and chlorine through ion channels in accordance with the equivalent electrical circuit of a biological membrane shown in Fig. 4.2. In this figure, φ_e and φ_i are the electrical potentials on both sides of the membrane, C_m is the membrane capacitance, which is about 1 μF/cm^2 per unit area, the arrows indicate the electrical resistances of the sodium and potassium channels, which

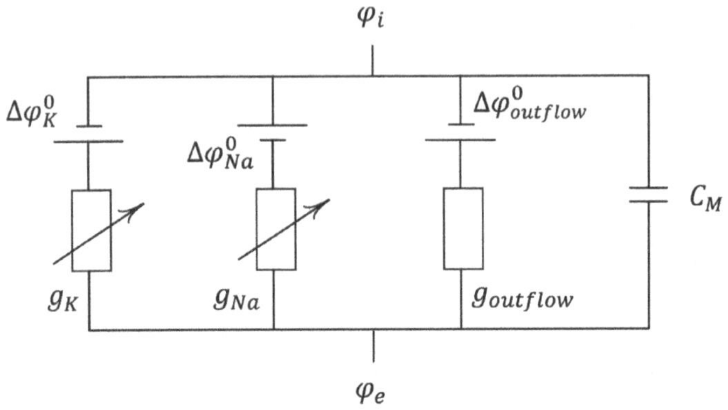

Fig. 4.2 Equivalent electrical cicuit of a biological membrane in the Hodgkin-Haxley model

change during depolarization and repolarization, g_i are the specific electrical conductivities of these channels, and the letters $\Delta\varphi_{Na}^0$, $\Delta\varphi_K^0$, $\Delta\varphi_{out}^0$ denote the electromotive forces (voltages) of sodium, potassium and chloride channels (the latter is also called the outflow channel).

2. In numerous precision experiments with perfused axons, i.e., axons whose internal environment was selectively changed relative to the external environment, Hodgkin and Huxley convincingly demonstrated that ionic conductance is directly proportional to the difference between the membrane potential and the equilibrium Nernst potential for a certain type of ion. Of course, first of all, the Hodgkin-Huxley model takes into account the main ion currents Na+ and K+, since it is these ions, as noted above, that lead to the emergence of ascending and descending branches of the action potential (spike). The remaining ions, such as Cl^- and others, the permeability coefficients of which do not change so sharply during the formation of an action potential, are combined into the so-called outflow current.
3. When constructing their model, A. Hodgkin and A. Huxley had to correctly take into account the main effect—a change in the conductivity of sodium (g_{Na}) and potassium (g_K) channels. To do this, they considered the processes of activation and inactivation of these channels as possible processes that can be described using chemical kinetics equations. This allowed the authors to model the kinetics of activation of ion channels in the membrane; more precisely, to obtain formulas for the dependence of the electrical conductivity of sodium and potassium channels on time.
4. Further experimental studies showed that the potassium channel opens only if four single-charge negative particles simultaneously approach a certain area of the membrane. Then for the conductivity of the potassium channel, we can write the corresponding relation containing a certain function to the fourth power, characterizing in this case the probability of such an event that one particle approaches the membrane, activating the potassium channel. The degree "4" is associated with the total number of activating particles and the assumption of

the independence of random events of their approach to a given section of the membrane (probability multiplication theorem). The opening of the potassium channel can also be interpreted as a result of the removal of two Ca^{2+} ions from it. The resulting expression for potassium channel conductance is an example of the so-called phenomenological equations of the Hodgkin-Huxley model, which were written based on experiments with perfused axons of the giant squid axon.

Accordingly, another phenomenological equation for sodium channel conductance was written based on the experimental fact that a sodium channel opens if three activating (negatively charged) particles enter it and one blocking (positively charged) particle is removed.

Taking into account the results of modeling the kinetics of activation of ion channels and the phenomenological equations for the conductance of sodium and potassium channels, Hodgkin and Huxley derived an expression for the electrical current flowing through the axon membrane.

5. At the final stage of constructing the Hodgkin-Huxley model, a differential equation was obtained that describes the process of propagation of a nerve impulse along the axon. Without dwelling here on the details of the derivation of the Hodgkin-Huxley equation (this issue will be discussed in Sect. 4.3), it should be noted such an important fact that once again emphasizes the universality of the laws of animate and inanimate nature: the Hodgkin-Huxley equation for the process of propagation of a nerve impulse along an axon by a structure completely coincides with the so-called telegraph equation, which describes the propagation of an electrical signal along conductors and is widely used in theoretical electrical engineering.

Further development of models for the propagation of a nerve impulse (spike) along the axon showed that the speed of spike propagation depends on the electrical and geometric characteristics of the nerve fiber, but does not depend on the shape of the initial impulse itself. A significant contribution to the study of molecular processes of excitability of biological membranes (in particular, the mechanisms of operation of ion channels) was made by scientists from the Institute of Physiology named after Academician Oleksandr Bogomolets of the National Academy of Sciences and other scientific and educational institutions of Ukraine. The results of Ukrainian scientists have received scientific recognition throughout the world.

4.2.2 Synergetic Features of Intercellular Interaction: The Chemical Aspect of Synaptic Transmission of a Nerve Impulse

In the synaptic mechanisms of nerve impulse transmission, biochemical reactions near the stability threshold play an important role. As noted earlier, features of the critical behavior of processes occurring nearby the points (lines) of phase transitions or the boundaries of stability of a homogeneous state of matter are characterized by a universal similarity. The reason is associated with the cooperative (synergetic) nature of such phenomena, determined by the strong interaction of fluctuations of the characteristic order parameter of the system under study, and their correlation over large spatial and time intervals.

It is well known that a "spike" is a wave of action potential propagating along a nerve fiber (axon), reaching the synaptic cleft, causes a certain biochemical reaction. During this reaction, a mechanism for releasing a mediator is triggered (in the future, for clarity, the cholinergic synapse and the corresponding mediator, acetylcholine, will be considered), interacting with a receptor substance—a specific hydrophobic protein contained in the postsynaptic membrane. As a result of the interaction between the mediator and the receptor, conformational changes occur in the membrane, leading to a change in the ionic conductivity of the channels. The release of acetylcholine into the synaptic cleft is quantum (cooperative) in nature: under the influence of one nerve impulse, several million acetylcholine molecules are simultaneously released from certain points near the surface of the presynaptic membrane of the axon. The end plate of the postsynaptic membrane contains several million receptor molecules with a high packing density of about 10^4 receptor molecules per 1 μm^2. The postsynaptic membrane also contains the enzyme acetylcholinesterase, which catalyzes the hydrolysis of acetylcholine. The formation of acetylcholine receptor complexes leads to the activation of channels in the postsynaptic membrane and a sharp increase in its conductivity.

Figure 4.3 shows a simplified diagram of a cholinergic synapse. The following sequence of events occurs at this synapse:

1. Synthesis of acetylcholine (ACh) by choline acetyltransferase.
2. ACh reserve in vesicles (synaptic bladders).
3. Release of ACh into the synaptic cleft.
4. The interaction reaction between ACh and specific receptors.

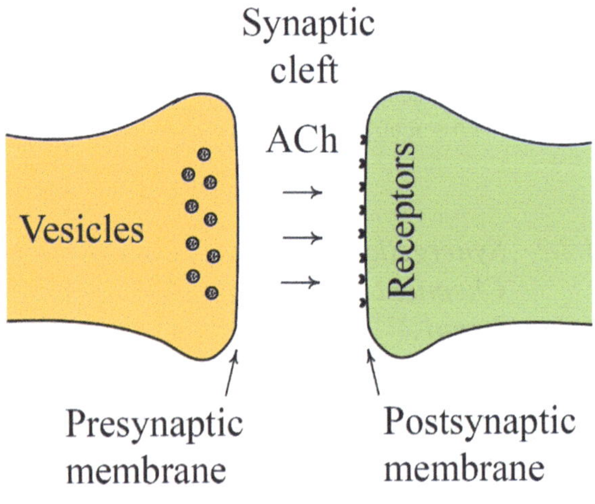

Fig. 4.3 A simlified scheme of cholinergic synapses

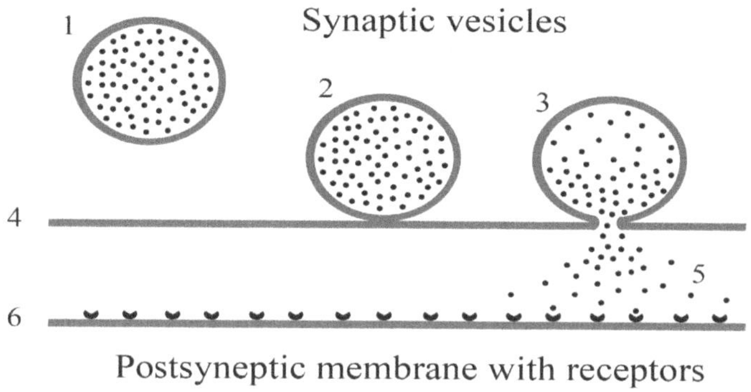

Fig. 4.4 Schematic representation of the processes of ACh synthesis in vesicles and its release into the synaptic cleft

5. Formation of mediator-receptor complexes.
6. Conformational changes in the postsynaptic membrane and changes in membrane potential.
7. Inactivation of ACh by acetylcholinesterase (AChE) or its removal from the synaptic cleft due to diffusion.

A schematic representation of the processes of storing acetylcholine in spherical vesicles and its release into the synaptic cleft is shown in Fig. 4.4 according to Charles Stevens' article [17].

As noted above, the conduction of a nerve impulse through the synaptic cleft due to biochemical reactions with the formation of acetylcholine receptor complexes is a quantum (cooperative) process. It is advisable to interpret it as a kind of phase transition. More precisely, the synchronous activation of a certain receptor zone by acetylcholine molecules, causing the opening (or closing) of channels in the postsynaptic membrane, can be considered as a phase transition of the mixing-separation type in a binary liquid mixture. The quantitative nature of the occurrence of such phase transitions at the level of biochemical reactions in the structurally stable and metastable regions will be studied in detail in Sect. 4.3.

4.2.3 *The Porocytosis Hypothesis in the Theory of Synaptic Transmission*

As already mentioned above, the problem of synaptic transmission and cell-to-cell communication is extremely important for understanding the functioning of the human brain [4–6, 10–17, 19–27]. Recent experimental studies of the mechanisms of synaptic transmission have established a significant role of the porocytosis hypothesis and synaptomers [21–23]. Using ultramicroscopic studies, it was found that synaptomers are hexagonal formations on the inner side of the presynaptic membrane, consisting of six vesicles. The sites of vesicles' attachment were located at the vertices of a regular hexagon in deepenings with a linear dimension about 10 nm. Under the influence of the action potential reaching the presynaptic membrane, calcium channels were activated causing a simultaneous contraction of all 6 vesicles for each synaptomer.

It should also be noted that hexagonal structures play an important role in the brain's operation. In addition to the example of the manifistation of hexagons in the structure of synaptomers considered here, there is another phenomenon in the brain with the participaton of hexagons in the process of spatial orientation of humans and animals. As is known (see details in Sect. 4.3), the 2014 Nobel Prize in Medicine or Physiology was awarded for the discoveries of the so-called "place cells" by J. O'Keefe in 1971 and "grid cells" by M.-B. Moser and E.I. Moser in 2005 [26]. Both types of cells have formed a positioning system in the brain, representing the spatial position, direction, and speed of movement. In particular, grid cells have multiple excitation regions creating a virtual 6-grid lattice with hexagon structures.

The porocytosis hypothesis versus the hypothesis of quantum-vesicular exocytosis. A number of experimental studies of the mechanisms of synaptic transmission have established an important role of ***the porocytosis hypothesis*** and so-called ***synaptomers having a hexagonal geometry*** [21–23]. The porocytosis hypothesis becomes an alternative to the previously existing hypothesis of quantum-vesicular exocytosis. According to the last hypothesis a distinctive feature of the mechanism of synaptic transmission is the complete release of mediator molecules into the synaptic cleft for uncorrelated vesicles in space, which merge with the presynaptic membrane. Processing the distribution of the amplitudes of postsynaptic potentials by methods of mathematical statistics showed that the theoretical value of the coefficient of variation (in other words, the mean value of the squared deviation of the amplitudes of postsynaptic potentials from their mean value, i.e., the mathematical expectation) is numerically consistent with the experimental data when using the porocytosis hypothesis. Using the hypothesis of quantum-vesicular exocytosis, the theoretical value of the coefficient of variation is almost 5 times higher than the experimental value of this coefficient.

Theoretical estimation of the number of synaptomers. Let us estimate the number of synaptomeres in the terminal bouton of a neuron near the presynaptic membrane. For this numerical estimate, we use the following geometric parameters: (a) the diameter of each vesicle is approximately $d_v \approx (40-50)\,nm \approx 45\,nm$; (b) the area of the maximum cross-section of the vesicle $A_v = \pi d_{uc}^2/4 \approx 1600\,nm^2$; (c) the radius R_S of the circle into which the hexagonal synaptomere is inscribed equals $R_S = d_v + r_v \approx 67.5\,nm$, with taking into account the approximation of close packing of all 6 vesicles of the synaptomere; (d) the area of the circle, into which all 6 vesicles of one synaptomere are inscribed, equals $A_S = \pi R_S^2 \approx 1.4 \cdot 10^4\,nm^2$; (e) the linear size (diameter) of the active zone of the axon terminal of the neuron, we assume equal to $D_B \approx 2\,\mu m$; (f) the surface area of the active zone of the terminal bouton $A_B = \pi D_B^2/4 \approx 3.1 \cdot 10^6\,nm^2$. As a final result, we obtain the following number N_s of synaptomeres in the active zone of the terminal bouton of a neuron: $N_S = A_B/A_S = 3.1 \cdot 10^6/1.4 \cdot 10^4 \approx 221$. Despite a number of significant approximations, the numerical value for the number $N_S \approx 221$ synaptomeres in the active zone of the axon terminal of a neuron shows a fairly satisfactory agreement with an accuracy of about 11% compared to the numerical value $N_S \approx 200$ of synaptomeres adopted in the hypothesis of porocytosis. The result of the sequence of these processes is the synchronous secretion into the synaptic cleft a small part of the mediator molecules for each of about 1200 vesicles. Due to the diffusion of secreted mediator molecules through the space of the synaptic cleft and their subsequent reaching the postsynaptic membrane, mediator molecules bind to specific receptors located on the outer surface of the postsynaptic membrane. This leads to conformational changes in the postsynaptic membrane, which cause the activation or inhibition of ionic channels and the transmission of an electrical signal from one neuron to another, thereby carrying out the process of cell-to-cell communication.

The reason for appearance the hexagonal structures. One of the interesting problems in the porocytosis hypothesis is the answer to the question why, in many cases, nature often chooses hexagonal formations. Typically, the focus of the answers to this question is on geometric and energetic-efficiency considerations (see other details in [25]). Here we will make numerical estimates and discuss the following results on hexagon emergance:

(a) Hexagons have a fairly large ratio of their surface area A to the perimeter P, the value of which

$$(\mathrm{A/P})_{\mathrm{hex}} = \frac{6\sqrt{3}}{4}\mathrm{r}^2/6\mathrm{r} \approx 0.43\mathrm{r}$$

is close to the highest value

$$(\mathrm{A/P})_{\mathrm{cir}} = \pi\mathrm{r}^2/2\pi\mathrm{r} = 0.50\mathrm{r}$$

for a circle at the same characteristic size r of these geometric formations, where r is the radius of the circle or the length of the hexagon side.

(b) Hexagonal structures in various systems have no free space between neighboring cells. This fact explains their greater energetic efficiency compared to circular or cylindrical geometry, respectively for 2-dimensional or 3-dimensional cases, where such vacant places are available.

(c) For another form of polygons, the energetic efficiency of hexagonal cells is maintained. So, for structures in the form of squares or equilateral triangles, for which there are also no empty spaces between cells, the A/P ratio is less than for hexagons, namely:

$$(\mathrm{A/P})_{\mathrm{sqr}} = \pi\mathrm{r}^2/4\mathrm{r} = 0.25\mathrm{r}$$

for squares and

$$(A/P)_{tri} = r/4\sqrt{3} \approx 0.14r$$

for equilateral triangles. For the right polygons with more sides n than $n_{hex} = 6$, the ratio

$$(A/P)_{plg} > (A/P)_{hex},$$

but the area of free space between neighboring cells grows with n increase. This requires a greater mass production of natural building materials.

(d) In the early stages of hexagons formation, their shape has circular or cylindrical geometry. The number of other cells surrounding the fixed cell has a decisive influence on the further change in the shape of fixed cells. Just this number determines the hexagonal structure of cells. From a physical point of view, an important role in the creation of hexagones belongs to the forces of surface tension of natural building materials at the triple junctions in each of the 6 hexagon vertices. For a deeper understanding of the role of the surface tension coefficient, we can recommend a review paper [27].

To conclude this paragraph of the Sect. 4.2, let us formulate the following important assertion. Here, we consider a new mechanism of porocytosis with taking into account hexagonal formations, which are called synaptomers and are a new component of the general problem of synaptic transmission and cell-to-cell communication. The number of synaptomeres N in the terminal bouton of a neuron near the presynaptic membrane is theoretically estimated, being equal to $N_{theor} \approx 221$. This theoretical estimate is in a good quantitative agreement with the experimental numerical value $N_{exp} \approx 200$ of synaptomeres adopted in the hypothesis of porocytosis.

4.2.4 *Poincaré Classification of Singular Points, Phase Portraits, Lyapunov Exponents*

The famous French mathematician and physicist Henri Poincaré (1854–1912) was the creator of a method for studying a system of nonlinear differential equations for a qualitative description of the process of the emergence of dynamic structures (in other words, time-dependent dynamic order). He introduced the concept of singular points for such systems of differential equations and proposed a classification of singular points, the essence of which reduces to the following provisions:

1. At the first stage of studying singular points, stationary solutions of a system of nonlinear differential equations are found, for which all time derivatives of dynamic variables, which are order parameters (in terms of the theory of phase

transitions and bifurcation phenomena) or control modes (in terms of the synergetic approach), are assumed to be zero. It is these stationary solutions that characterize the singular points for the order parameters (controlling modes).

Henri Poincaré (1854 – 1912)

2. At the second stage, the system of nonlinear differential equations is linearized, which means that they retain small deviations of the order parameters from stationary values only to the first degree.
3. At the third stage, the system of linearized differential equations for time-dependent small deviations of the order parameters is reduced to a characteristic algebraic equation, which is, for example, quadratic or cubic for two or three order parameters, respectively.
4. At the last fourth stage, depending on the specific values of the roots of the characteristic agebraic equation, it is possible to classify singular points, as well as study the phase portraits of a system of nonlinear differential equations.

Table 4.1 contains information about the real (*Re*) and imaginary (*Im*) parts of the roots λ of the characteristic algebraic equation. Depending on the signs of $Re\ \lambda$ and $Im\ \lambda$, the following singular points are distinguished: stable and unstable nodes and foci, saddles and centers (limit cycles).

Table 4.1 Types of singular points according to Poincaré classification

Singular point type	Signs of roots of the characteristic equation	
	Real part	Imaginary part
Stable node	$Re\ \lambda_1 < 0$ $Re\ \lambda_2 < 0$	$Im\ \lambda_1 = 0$ $Im\ \lambda_2 = 0$
Unstable node	$Re\ \lambda_1 > 0$ $Re\ \lambda_2 > 0$	$Im\ \lambda_1 = 0$ $Im\ \lambda_2 = 0$
Saddle	$Re\ \lambda_1 < 0$ $Re\ \lambda_2 > 0$	$Im\ \lambda_1 = 0$ $Im\ \lambda_2 = 0$
Stable focus	$Re\ \lambda_1 < 0$ $Re\ \lambda_2 < 0$	$Im\ \lambda_1 \neq 0$ $Im\ \lambda_2 \neq 0$
Unstable focus	$Re\ \lambda_1 > 0$ $Re\ \lambda_2 > 0$	$Im\ \lambda_1 \neq 0$ $Im\ \lambda_2 \neq 0$
Center (limit cycle)	$Re\ \lambda_1 = 0$ $Re\ \lambda_2 = 0$	$Im\ \lambda_1 \neq 0$ $Im\ \lambda_2 \neq 0$

Let us first consider the phase portraits of singular points that correspond to stable and unstable nodes. These phase portraits, as in the case of a saddle-shaped singular point, are characterized by the presence of only the real part and the absence of the imaginary part of the roots of the characteristic algebraic equation of the dynamical system. A stable node characterizes such a movement of the representing point in the phase portrait, in which it approaches its stationary state in both parameters. And, conversely, for an unstable node, the representing point moves away from the singular point in both respects. When the real roots of the characteristic equation have different signs, then according to one parameter the representing point approaches, and according to another parameter it moves away from the singular point, which is called the saddle. In all of the listed cases of singular points (stable node, unstable node and saddle), the dynamic system is characterized by nonperiodic behavior.

Periodic behavior of a dynamic system occurs in the case of the obligatory presence of the imaginary part of the roots of the characteristic equation for singular points, which are the stable and unstable foci, as well as the center (limit cycle). In the case of a stable focus, the phase portrait has the form of a spiral, which winds around a special point in both parameters with a negative real value of two roots. This behavior is characterized by damped oscillations of two order parameters of a dynamic nonlinear system. In the case of an unstable focus, the real part of the roots of the characteristic equation is positive, and the phase portrait is a spiral that unwinds and corresponds to oscillations of two order parameters with an amplitude increasing in time. In the latter case of the center (limit cycle), the real part of the roots of the characteristic equation is equal to zero, which corresponds to undamped oscillations (self-oscillations) of two order parameters of the dynamic system.

Figure 4.5 illustrates phase portraits for dynamical systems of nonlinear differential equations with two order parameters in accordance with Henri Poincaré's classification for various types of singular points. Here, from left to right and from

Fig. 4.5 Phase portraits of various types of singular points

top to bottom, the following phase portrauts are demonstrated: (1) stable focus, (2) stable node, (3) saddle, (4) unstable node, (5) unstable focus, (6) center (limit cycle). The phase portrait of a strange attractor is located at the very bottom of Fig. 4.5.

The behavior of solutions to a dynamic system of nonlinear differential equations may turn out to be unstable and chaotic when certain conditions are met. In such cases, the movement of the point representing the temporal behavior of solutions of a dynamical system in the phase space of order parameters ceases to be deterministic. Such a state corresponds to the phase portrait, which is shown below in Fig. 4.5 and called the term "strange attractor".

It should be emphasized that the singular points considered above in the form of a persistent node, persistent focus and center (limit cycle) are simple attractors. All trajectories in the phase portraits of a dynamical system in the vicinity of the three indicated singular points tend to these points or, in other words, are attracted to them. In the case of a strange attractor, all trajectories in its phase portrait diverge and chaotically mix with one another.

An effective method for studying the stability or chaoticity of the trajectories of a representative point that describes the state of a dynamic system using its phase portrait is the study of exponents that bear the name of the famous mathematician Alexander Lyapunov (1857–1918). Let us consider the case of a three-dimensional dynamic system of nonlinear differential equations, for which we have three main order parameters and the same number of Lyapunov's exponents. As was shown by A. Lyapunov, there are the following criteria for establishing the type of singular points when studying simple and strange attractors:

(a) if all three Lyapunov's exponents are negative, then the attractor is simple and represents a stable focus;
(b) if two Lyapunov's exponents are negative and the third Lyapunov's exponent is zero, then the attractor is simple and represents a center (limit cycle);
(c) if one Lyapunov's exponent is negative and the other two Lyapunov's exponents are equal to zero, then the attractor is simple and represents a stable torus;
(d) if one of the Lyapunov's exponents is positive, then the movement of the representing point can become chaotic. In such cases, with a slight change in the initial conditions, a strange attractor arises, and neighboring trajectories begin to quickly diverge.

One of the first examples of a dynamic system with the properties of a strange attractor was the hydrodynamic (meteorological) model proposed by the American mathematician and meteorologist Edward Lorenz (1917–2008) in his article [28]. This model is described by an initial system of three nonlinear differential equations, which are the Navier–Stokes equation, the equation of continuity, and the equation of thermal conductivity, which are well known from hydrodynamics (see, for example, [29]). As a result of certain transformations, Lorentz wrote down a system of three ordinary nonlinear differential equations for time dependent functions X, Y, Z, which describe the convective movement of a fluid, which is heated from below.

It should be noted that the Lorentz model for describing the convective movement of a viscous fluid in a horizontal layer with a vertical vector of the temperature gradient directed downwards is a definite approximation for a more consistent description of the hydrodynamic instability during the formation of hexagonal structures known as Bénard cells (see details in Chap. 2). Another note concerns explaining the popularity of the Lorenz model due to its use in creating accurate weather forecasts. Solving a system of nonlinear differential equations in the form of a strange attractor, which exhibits a chaotic behavior, makes accurate weather prediction impossible due to a number of circumstances. For example, due to the high sensitivity of solutions of the system of differential equations of the Lorentz model to the inaccuracy of the initial conditions for meteorological parameters. It was this effect that was apparently the reason for the appearance of the scientific term "deterministic chaos" and the literary term "butterfly effect" in science fiction.

4.3 Section B

4.3.1 The Hodgkin-Huxley Theory of Propagation of a Nerve Impulse Along the Axon

Let us discuss in more detail the main theoretical principles underlying the Alan Hodgkin and Andrew Huxley model of the propagation of nerve impulses along axons [19, 30]. The initial equation for the density J_m of the electric current flowing through a biological membrane has the following form (see, e. g., [5]):

$$J_m = C_m(d\Delta\varphi/dt) + \sum_i J_i \quad (4.1)$$

Here $\Delta\varphi = \varphi_e - \varphi_i$ determines the membrane potential, and the first term on the right side in (4.1) specifies the capacitive current associated with the change in the membrane potential difference, while the second term in (4.1) is the currents due to the presence of ion channels (see equivalent electrical circuit of the membrane in Fig. 4.2).

Temporal kinetics of ion channels. In experiments with perfused axons, i.e. when the concentration of ions in the axoplasm changes, A Hodgkin and A Huxley established that ion currents J_i are directly proportional to the difference between the membrane potential $\Delta\varphi$ and the equilibrium Nernst concentration potential $\Delta\varphi_i^0$ for a given tepe of ions in accordance with the following phenomenological formulas:

$$\begin{aligned} J_{Na} &= g_{Na}(\Delta\varphi - \Delta\varphi^o_{Na}), \\ J_K &= g_K(\Delta\varphi - \Delta\varphi^o_K), \\ J_{outflow} &= g_{outflow}(\Delta\varphi - \Delta\varphi^o_{outflow}). \end{aligned} \quad (4.2)$$

Here, first of all, the main ion currents of sodium Na^+ and potassium K^+ are taken into account, and the remaining ions (chlorine Cl^- and others) are combined into the last term, called the outflow current. Coefficients g_i characterize the conductivity of ion channels.

When taking into account the main effect—changes in the conductivity of sodium and potassium channels—it was considered that the processes of activation and inactivation of channels are probabilistic processes that can be described using chemical kinetics equations. The basic diagram of these processes is as follows:

$$\text{Unactivated channel} \underset{\beta_n}{\overset{\alpha_n}{\rightleftarrows}} \text{Activated channel}$$

Here α_n, β_n are constants characterizing the probability of activation (inactivation) of channels per unit time, i. e. rate constants of activation and inactivation of channels. Let C be the number of activated channels, and C_0 be the total number of channels, then C_0–C be the number of non-activated channels. Using the above diagram, we can write the following differential equation for the number C of activated channels:

$$dC/dt = \alpha_n(C_0 - C) - \beta_n C. \quad (4.3)$$

Let us divide Eq. (4.3) by C_0 and introduce the relative number $n = C/C_0$ of activated channels, which gives the following first-order inhomogeneous differential equation for the time dependence of the relative number $n(t)$ of activated channels:

$$dn/dt = \alpha_n(1 - n) - \beta_n n, \quad (4.4)$$

or

$$\mathrm{dn/dt} = -(\alpha_\mathrm{n} + \beta_\mathrm{n})\mathrm{n} + \alpha_\mathrm{n}. \tag{4.5}$$

The general solution of the homogeneous differential Eq. (4.5) without the constant term α_n is as follows:

$$\mathrm{n} = \mathrm{A}\exp[-(\alpha_\mathrm{n} + \beta_\mathrm{n})\mathrm{t}]. \tag{4.6}$$

Let us find the integration constant A in formula (4.6) using the method of variation of this constant A, which gives

$$\mathrm{n(t)} = [(\alpha_\mathrm{n}/(\alpha_\mathrm{n} + \beta_\mathrm{n})] + \mathrm{A}_1\ \exp[-(\alpha_\mathrm{n} + \beta_\mathrm{n})\mathrm{t}]. \tag{4.7}$$

Then determine the constant A_1 from the natural initial condition, namely: we will assume that in initial time $t = 0$ there are no activated channel, i. e. $n = 0$ at $t = 0$. After substitutionof this initial condition in expression (4.7), we obtain: $A_1 = -\alpha_n/(\alpha_n + \beta_n)$.

Thus, the solution of the differential Eq. (6.4) of chemical kinetics, which determines the relative number $n(t)$ of activated channels is a function of the following form:

$$\mathrm{n(t)} = \alpha_\mathrm{n}/(\alpha_\mathrm{n} + \beta_\mathrm{n})\{1 - \exp[-(\alpha_\mathrm{n} + \beta_\mathrm{n})\mathrm{t}]\}. \tag{4.8}$$

To simplify the analytical form of formula (4.8), we will assume that at long times ($t \to \infty$), the relative number of activated channels becomes equal to N. Then, substituting condition $n(t \to \infty) = N$ in formula (4.8), we obtain such an expression for the multiplier $\alpha_n /(\alpha_n + \beta_n) = N$. As a result, the above solution (4.8) can be rewritten in the following equivalent form:

$$\mathrm{n(t)} = \mathrm{N}[1 - \exp(-\mathrm{t}/\tau_n)]. \tag{4.9}$$

The following designation is used in (4.9) for the relaxation time:

$$\tau_\mathrm{n} = 1/(\alpha_\mathrm{n} + \beta_\mathrm{n}), \tag{4.10}$$

Here, the relaxation time τ_n is such a time interval, during which the number of activated channels changes by e = 2.718 times.

As follows from the above consideration, ***the processes of activation of ion channels in axon membranes are described by exponential temporal kinetics***.

Conductivity of ion channels. Experimental studies by Hodgkin and Huxley, which used perfused axons with changes in their ion composition, have proved that the conductivity of the potassium channel appears only when four singly charged negative particles (ions) approach it simultaneously. The following phenomenological formula for the conductance g_K of the potassium channel was a reflection of this experimental fact:

$$g_K = (g_K)_{max} n^4, \quad (4.11)$$

where $(g_K)_{max}$ is the maximum value of the conductivity of the potassium channel, and n characterizes in this case the probability of approaching the membrane of one particle that activates the potassium channel. Degree "4" is connected with the assumption of independence of the events of the approach of each of the four particles to the given section of the membrane (theorem of multiplication of probabilities). The opening of the potassium channel can actually be interpreted, obviously, as a consequence of the removal of two Ca^{2+} ions from it.

Similarly, the following expression is used for the conductance of the sodium channel in the Hodgkin-Huxley model:

$$g_{Na} = (g_{Na})_{max} m^3 h, \quad (4.12)$$

i. e. the sodium channel opens if three activating negative particles (ions) approach it, and one blocking positively charged particle (ion) leaves. The differential kinetic equations for the functions $m(t)$ and $h(t)$, describing the probabilities of such independent processes, have the form completely analogous to the linear differential Eq. (4.4) for the function $n(t)$, which describes the processes of chemical kinetics for potassium channels:

$$dm/dt = \alpha_m(1 - m) - \beta_m m, \quad (4.13)$$

або

$$dh/dt = \alpha_h(1 - h) - \beta_h h, \quad (4.14)$$

The solution of the similar differential kinetic Eq. (4.4) was described in detail above. This makes it possible to write the solutions of similar differential Eqs. (4.13) and (4.14) in the same way:

$$m(t) = M[1 - \exp(-t/\tau_m)]. \quad (4.15)$$

$$h(t) = H[1 - \exp(-t/\tau_h)]. \quad (4.16)$$

Finally, the density J_M of the electric current through the biological membrane, which is described by the original Eq. (4.1) and takes into account changes in the conductivity of Na^+ and K^+ channels, has the following form:

$$J_m = C_m(d\Delta\varphi/dt) + (g_K)_{max} n^4(t)(\Delta\varphi - \Delta\varphi_K^0) + (g_{Na})_{max} m^3(t)(\Delta\varphi - \Delta\varphi_{Na}^0) + (g_{outflow})_{max}(\Delta\varphi - \Delta\varphi_{outflow}^0). \quad (4.17)$$

Note that the last term for outflow currents in the right-hand side of formula (4.17) does not require the use of chemical kinetics methods based on experimental data obtained by Hodgkin and Huxley.

The Hodgkin-Huxley differential equation. The next task requires the derivation of the Hodgkin-Huxley differential equation, which describes the process of nerve impulse propagation along the axon. For this purpose, let's analyze the change in the membrane current dI_m, which is caused by the change (decrease) in the current dI_a along the axis of the axon, i.e. $-dI_a = dI_m$. The current change through the membrane is written as

$$\mathrm{dI_m = J_m dS = J_m 2\pi r dx}, \tag{4.18}$$

since the area element of a cylindrical axon with radius r equals $dS = 2\pi r dx$. Hence, the current density J_M flowing through the lateral surface of the axon can be written as the derivative of the current I_a with respect to the coordinate along the axis of the axon

$$\mathrm{J_m = -(1/2\pi r)(dI_a/dx)}. \tag{4.19}$$

According to Ohm's law, the current strength in the area of the axoplasm is related to the voltage drop (decrease in potential difference) by the following expression:

$$J_{\mathrm{a}} = -\mathrm{d}\Delta\varphi/\mathrm{dR_a} \tag{4.20}$$

where $dR_a = \rho_a dx/\pi r^2$ is the resistance of the axoplasm in this area, and ρ_a is its specific resistance. Hence, we get the following formula for the current along the axon:

$$\mathrm{J_a} = -(\pi r^2/\rho_a)(d^2\Delta\varphi/dx^2). \tag{4.21}$$

Substituting formula (4.21) into the relationship (4.19), we get

$$J_m = (r/2\rho_a)((d^2\Delta\varphi/dx^2). \tag{4.22}$$

Since we have two expressions (4.17) and (4.22) for the same quantity—the electric current density J_{m} through the lateral surface of the axon, equating these expressions finally gives the following equation:

$$\begin{aligned}(r/2\rho_a)(d^2\Delta\varphi/dx^2) =& \mathrm{C_m}(\mathrm{d}\Delta\varphi/\mathrm{dt}) + (\mathrm{g_K})_{\mathrm{max}}\mathrm{n}^4(\mathrm{t})(\Delta\varphi - \Delta\varphi_K^0) \\ &+ (\mathrm{g_{Na}})_{\mathrm{max}}\mathrm{m}^3(\mathrm{t})(\Delta\varphi - \Delta\varphi_{Na}^0) \\ &+ (\mathrm{g_{outflow}})_{\mathrm{max}}(\Delta\varphi - \Delta\varphi_{outflow}^0).\end{aligned} \tag{4.23}$$

Equation (4.23) represents ***the Hodgkin-Huxley equation***, which is a linear differential equation of the second order in partial derivatives with variable coefficients. Indeed, it contains both the first partial derivative with respect to time $\partial \Delta\varphi / \partial t$ and the second partial derivative with respect to the coordinate $\partial^2 \Delta\varphi / \partial x^2$ of the membrane potential $\Delta\varphi$, and the function $\Delta\varphi$ itself and its two derivatives are included in Eq. (4.23) in first degree. Since the functions $n(t)$, $m(t)$ and $h(t)$ have an exponential dependence on time in accordance with formulas (4.9), (4.15) and (4.16), the Hodgkin- Huxley equation contains variable coefficients. Its solution determines the value of the membrane potential $\Delta\varphi(x,t)$ at some point of the axon at a given time. This equation is similar in structure to the so-called telegraph equation, which characterizes the propagation of an electric signal in conductors and which is widely used in electrical engineering, so the Hodgkin-Huxley equation is sometimes called the telegraph (or cable) equation.

Let's rewrite the left side of the Hodgkin-Huxley equation in another way, introducing the speed V of the propagation of the nerve impulse based on the wave equation, which has the following general form:

$$\partial^2 \Delta\varphi / \partial x^2 = (1/V^2)(\partial^2 \Delta\varphi / \partial t^2). \tag{4.24}$$

Then the Hodgkin-Huxley equation takes the following expression:

$$\begin{aligned}(r/2\rho_a V^2)(d^2\Delta\varphi/dx^2) =& \mathrm{C_m}(\mathrm{d}\Delta\varphi/\mathrm{dt}) + (\mathrm{g_K})_{\max}\mathrm{n}^4(\mathrm{t})(\Delta\varphi - \Delta\varphi_K^0) \\ &+ (\mathrm{g_{Na}})_{\max}\mathrm{m}^3(\mathrm{t})(\Delta\varphi - \Delta\varphi_{Na}^0) \\ &+ (\mathrm{g_{outflow}})_{\max}(\Delta\varphi - \Delta\varphi_{outflow}^0).\end{aligned} \tag{4.25}$$

The solution of Eq. (4.25) allows, in particular, to calculate the propagation speed of a nerve impulse, if the experimental values of the parameters included in the Hodgkin-Huxley telegraph equation are known. As is known, the speed of nerve impulse propagation depends on whether the axon has a myelin sheath or no such sheath. In the first case of myelinated axons, the propagation of the action potential (spike) occurs in jumps through Ranvier intercepts and is characterized by a much higher speed than in the second case of unmyelinated axons.

4.3.2 The Theory of Synaptic Transmission as a Phase Transition in the Mediator-Receptor System

Kinetic equations of the process of synaptic transmission in the cholinergic synapse. As was mentioned in Sects. 4.2.1 and 4.2.2, chemical reactions near the bifurcation threshold play a significant role in synaptic mechanisms of nerve impulse transmission. Earlier Figs. 4.3 and 4.4 showed a simplified scheme of the cholinergic synapse and described the sequence of events occurring in it. In particular, it was noted that the conduction of a nerve impulse through the synaptic cleft is a quantum and cooperative phenomenon, since the synchronous activation of a certain zone of receptors by acetylcholine can be considered as a phase transition [20, 24, 29], which is isomorphic to the phase transition of the "mixing-separation" type in a binary liquid mixture (see, e.g., [31–40]).

Consider a simplified model of synaptic transmission with two order parameters, which is described by the following scheme of biochemical reactions:

$$ACh + R \underset{K_{-1}}{\overset{K_1}{\leftrightarrow}} ACh - R^* \overset{K_3}{\rightarrow} P + R^* \overset{K_2}{\rightarrow} R. \tag{4.26}$$

Here, acetylcholine molecules are indicated by ACh (the concentration of these molecules $[A]$), by R the receptors in the initial (unactivated) state with an initial concentration of $[R]_0$, by P the product of the breakdown of acetylcholine under the action of the enzyme acetylcholinesterase $AChE$, and by K_i the rate coefficients of these reactions.

Let us write a system of two nonlinear differential equations that correspond to the scheme given in (4.26), namely:

$$\begin{aligned} dX/dt &= K_3 Y - K_2 X, \\ dY/dt &= K_1[A]([R]_0 - Y) - (K_1 + K_3)Y. \end{aligned} \tag{4.27}$$

The kinetic equations written in (4.27) describe the temporal evolution of two order parameters, which are the concentration of the receptor R^* in the activated state (variable X) and the concentration of acetylcholine-receptor complexes ACh -R^* (variable Y). It should be noted that similar systems of two nonlinear differential equations were first investigated in works [20, 24].

Later in this paragraph, we will dwell in more detail on the features of the properties of the acetylcholine-receptor system near the bifurcation state.

Correlation function of the acetylcholine-receptor system in the synaptic cleft. The anomalous growth of the correlation radius ξ predicted by the theory of scale invariance, strictly speaking, takes place for spatially unlimited environments. It is for such unbounded systems that the following expression for the pair correlation function (CF) G_2 is valid in the Ornstein–Zernike approximation:

$$G_2(r) = A\exp(-r/\xi)/r. \tag{4.28}$$

This formula represents the Green's function of the Helmholtz operator $\hat{L} = \nabla^2 - \kappa^2$ for an unbounded system with zero boundary conditions at infinity, where the value of κ^2 is proportional to the inverse isothermal susceptibility of the acetylcholine-receptor system. In real systems, to use the CF G_2 (r) from (4.28), it is enough that all linear dimensions of the system under study far exceed the correlation radius ξ of fluctuations of the order parameter.

For the acetylcholine-receptor system under study, a fundamentally important point is the fact that the synaptic cleft, in which the reactions responsible for the biochemical mechanisms of synaptic transmission take place, is a spatially limited object. For obvious reasons, it is clear that the correlation radius ξ in a finite-size system cannot exceed a linear size in the direction of its spatial limitation. These requirements are not satisfied by a CF $G_2\ (r)$ of the type (4.28). Therefore, the task of calculating the correlation function and the correlation radius of order parameter fluctuations for the system of acetylcholine-receptor complexes in a spatially limited synaptic cleft becomes urgent.

Let us choose the following geometry of the problem, for which the synaptic cleft is modeled by a plane-parallel layer bounded along the z axis. The coordinate $z = -L_0$ corresponds to the end plate of the presynaptic membrane of the axon, and the coordinate $z = L_0$ to the end plate of the postsynaptic membrane. Thus, the phase transition in the acetylcholine-receptor system takes place in the boundaries of the synaptic cleft $-\infty < x, y < \infty, \ -L_0 \leq z \leq L_0$.

The task of finding the CF $G_2\ (r)$ in such a layer is reduced to determining the Green's function of the Helmholtz differential operator from the following system of equations:

$$(\Delta - \kappa^2)G_2(r) = -C(r)/C_2, \tag{4.29}$$

$$\kappa^2 = (1 - C_0)/C_2, \tag{4.30}$$

$$C_0 = <\rho> \int C(r)d\vec{r},\ C_2 = (1/6) <\rho) \int C(r)r^2\,d\vec{r} \tag{4.31}$$

with zero boundary conditions on the boundaries $z = -L_0$ and $z = L_0$. The value κ, which is proportional to the inverse correlation radius of fluctuations of the order parameter ($\kappa \sim 1/\xi$), depends on the proximity of the studied system to the bifurcation state in terms of temperature and pressure (or temperature and density) depending on fixed external conditions.

The solution of the formulated problem of mathematical physics requires the application of the methods of Green's function theory and is performed as follows:

1. First, it is necessary to find an expression for the delta function $\delta(r)$, which is included in the right-hand part of (4.29), since the direct correlation function (DCF) $C(r) \sim \delta(r)$ in the zero approximation. For this purpose one should use the eigenfunctions of the Helmholtz operator $\hat{L}$ of the eigenvalue problem with zero boundary conditions for the desired function and its first derivative on the boundaries $z = \pm L_0$, provided that the eigenvalues λ_n of this operator satisfy the relation $\lambda_n = kx^2 + k_y{}^2 + k_z{}^2 + \kappa^2$, where k_x, k_y change continuously from -∞ to $+\infty$, and k_z takes a discrete series of values ($k_z = n^2\pi^2/4L_0{}^2$, $n = 0, 1, 2\ldots$). The expression for the delta function satisfying these requirements has the following form:

$$\delta(r) = (1/8\pi L_0)\sum_{n\geq 0}\iint [1-(-1)^n]\cos(\pi nx/2L_0)\exp[i(k_x x + k_y y)]dk_x dk_y. \tag{4.32}$$

2. On the basis of the obtained relation (4.32), a general analytical expression for the solution of Eq. (4.29) can be determined for the delta function:

$$G_2(r) = L^{-1}\delta(r) = (1/8\pi L_0)\sum_{n\geq 0}\iint [1-(-1)^n]\cos(n\pi x/2L_0)$$
$$\exp[i(k_x x + k_y y)]dk_x dk_y \times [k_x^2 + k_y^2 + (n\pi x/2L_0)^2 + \kappa^2]^{-1}. \tag{4.33}$$

Here, in formula (4.33), $\hat{L}^{-1}\delta(r)$ denotes the action of the inverse Helmholtz differential operator $\hat{L} = \nabla^2 - \kappa^2$ on the right-hand side of formula (4.32) for the delta function $\delta(r)$.

3. Further, it is convenient to switch to polar coordinates

$$k_x = k_{xy}\cos\varphi,\quad k_y = k_{xy}\sin\varphi,\quad \rho = (x^2+y^2)^{1/2}$$

and perform the integration in (4.33) taking into account the known relations for cylindrical functions:

$$\exp(ik\rho\cos\varphi) = J_0(k\rho) + 2iJ_1(k\rho) + 2i^2\cos 2\varphi J_2(k\rho) + \cdots,$$

$$\int_0^\infty \frac{kJ_0(k\rho)dk}{k^2+a^2} = K_0(\rho a).$$

Here $J_n(z)$ is a cylindrical Bessel function of order n, and $K_0(z)$ is a cylindrical McDonald function of zero order.

Then for the CF $G_2(r, z)$ of a plane-parallel layer we have

$$G_2(\rho, z) = (1/4\pi L_0)\sum_{n\geq 0}[1+(-1)^n]K_0\left[\rho\left(\kappa^2 + \frac{n^2\pi^2}{4L_0^2}\right)^{1/2}\right]\cos(n\pi z/2L_0). \tag{4.34}$$

4. Numerical evaluations showed that the main contribution to the CF $G_2(\rho,z)$ from formula (4.34) for a plane-parallel layer is given by the first term in the sum with the value $n = 1$, so that approximately we have

$$G_2(\rho, z) = (1/2\pi L_0)K_0\left[\rho(\kappa^2 + \pi^2/4L_0^2)\right]\cos(\pi z/2L_0). \tag{4.35}$$

An important result of the analysis of expressions (4.34) and (4.35) is not the exponential decay of CF $G_2(r)$, characteristic of bulk media, but the oscillating behavior of the correlation properties in the z direction for a bounded layer, i.e., along the perpendicular to the postsynaptic membrane of axon.

5. In the case of a large increase in the thickness of the layer $H = 2L_0$, when the layer's thickness H becomes much larger than the correlation radius $\xi (H \gg \xi)$ we obtain the following limiting transition from the CF (4.35) of a plane-parallel layer to the CF (4.24) of the Ornstein–Zernike theory for a spatially unlimited (bulk) medium:

$$
\begin{aligned}
G_2(\rho, z) &= (1/4\pi L_0)(2L_0/\pi)^2 \int_0^\infty K_0[\rho(\kappa^2 + \pi^2/4L_0^2)^{1/2}] \cos(k_z z) dk_z \\
&= (1/4\pi) \exp[-\kappa(\rho^2 + z^2)/(\rho^2 + z^2)^{1/2}.
\end{aligned}
$$

Correlation radius and activation zone of the acetylcholine-receptor system. Since the correlation function of concentration fluctuations for the binary acetylcholine-receptor system in a spatially limited synaptic cleft, defined by formula (4.35), does not have an exponential form, it is natural to determine the correlation radius ξ through the normalized second spatial moment of the CF $G_2(\rho,z)$ as follows:

$$
\xi = \sqrt{M_2}, \quad M_2 = \int G_2(r) r^2 d\vec{r} / \int G_2(r) d\vec{r}. \tag{4.36}
$$

To calculate the second spatial moment M_2, we will use polar coordinates. Then

$$
\mathrm{M}_2 = \frac{\int_0^\infty \rho^3 K_0(\rho\sqrt{\kappa^2 + \pi^2/4L_0^2})\, d\rho}{\int_0^\infty \rho K_0(\rho\sqrt{\kappa^2 + \pi^2/4L_0^2})\, d\rho}. \tag{4.37}
$$

Taking into account the following relation for the result of integrating the product of the power function and McDonald's cylindrical function depending on the imaginary argument (see integrals in the numerator and denominator of the previous formula), which reduced to the product of two gamma functions:

$$
\int_0^\infty x^\mu K_\nu(ax)\, dx = 2^{\mu-1} a^{\mu-1} \Gamma\left(\frac{1+\mu+\nu}{2}\right) \Gamma\left(\frac{1+\mu-\nu}{2}\right),
$$

we obtain the final formula for the of correlation radius ξ of concentration fluctuations (degree of completeness) of the acetylcholine-receptor system in the postsynaptic membrane

$$
\xi = \{\frac{4}{\kappa^2 + \pi^2/4L_0^2} + \frac{1}{4}(1 - 8/\pi^2)\mathrm{L}_0^2\}^{1/2}. \tag{4.38}
$$

Formula (4.38) for the correlation length ξ in (6) contains two different contributions, being an anisotropic quantity:

$$\xi = (\xi_{xy}^2 + \xi_z^2)^{1/2} \tag{4.39}$$

Here

$$\xi_{xy} = 2/(\kappa^2 + \pi^2/4L_0^2)^{1/2} \tag{4.40}$$

is the correlation length in the xy plane, and

$$\xi_z = (L_0/2)(1 - 8/\pi^2)^{1/2} \tag{4.41}$$

is the correlation length in the z direction, which are parallel and perpendicular to presynaptic and postsynaptic membranes, relatively.

Table 4.2 contains numerical values of the correlation length ξ and its components ξ_{xy}, ξ_z in the binary system "transmitter-receptor" inside the synaptic cleft with thickness $H = 2\ L_0$ within its interval $H = 10 \div 100$ nm.

As mentioned in 4.2.2, about several million receptor molecules are placed at the end plate of the postsynaptic membrane with a high packing density equals approximately $\rho_{surf.density} \approx 10^4$ molecular receptor per 1 μm^2. It allows to evaluate the number $N_{\text{ACh-R*}}$ of correlated ACh-R* complexes in the postsynaptic membrane. For this purpose, let us find the surface area $S_{\text{act}} = \pi d_{\text{act}}^2/4$ of the activation zone in the postsynaptic membrane of the chemical synapse with taking into account its maximum linear size—the diameter $_t\ d_{\text{act}} \approx 2(\xi_{xy})_{max} \approx 25.5$ nm. As a result, one has $S_{\text{act}} \approx \pi(\xi_{xy}^2)_{max} \approx 510\,\text{nm}^2 = 5.1.10^{-4}\ \mu m^2$. Then, the number N $_{\text{ACh-R*}}$ of correlated receptors or ACh-R* complexes in the postsynaptic membrane of the chemical synapse equals the product of $\rho_{surf.density}$ and S_{act}, i.e. one has $N_{\text{ACh-R*}} = \rho_{surf.density} \times S_{\text{act}} \approx 10^4$ receptors$/\mu m^2 \times 510 \cdot 10^{-6}\ \mu m^2 \approx 5.1$ receptors. This result definitely shows that there are about 5 correlated receptors in a chemical synapse.

In the case of the neuromuscular junction, one has a larger maximum value of the correlation length $(\xi_{\text{xy}})_{\text{max}} \approx 63.7$ nm than in the chemical sinapse. Therefore, the surface area S_{act} of the activation zone in the neuromuscular junction is 25 times more than in the chemical sinapse and equals $S_{\text{act}} \approx 127.5 \cdot 10^{-4}\ \mu m^2$. Therefore,

Table 4.2 Dependence of correlation length and its components on synaptic cleft's thickness

Thickness of synaptic cleft H, nm	ξ_{xy}, nm	ξ_z, nm	ξ, nm	Thickness of synaptic cleft H, nm	ξ_{xy}, nm	ξ_z, nm	ξ, nm
10	6.37	2.17	6.46	60	38.22	13.02	40.38
20	12.74	4.34	13.46	70	44.59	15.19	47.11
30	19.11	6.51	20.19	80	50.96	17.36	53.84
40	25.48	8.68	26.92	90	57.33	19.53	60.57
50	31.18	10.85	33.65	100	63.70	21.70	67.29

the number N $_{\text{ACh-R*}}$ of correlated complexes equals: $N_{\text{Ach-R}^*} = \rho_{surf.density} \times S_{\text{act}} \approx 10^4\,\text{receptors}/\mu m^2 \times 127.5 \cdot 10^{-4}\,\mu m^2 \approx 128\,\text{receptors}$. This result shows that approximately 128 receptors are mutually correlating in the neuromuscular junction.

4.3.3 A Strange Attractor in the Nonlinear Kinetic Model of Synaptic Transmission

It was already noted above that the problem of intercellular interaction and synaptic transmission has the same fundamental importance in living nature as the problem of intermolecular interaction in non-living nature. In particular, the intercellular interaction that occurs as a result of electrical processes of propagation of a nerve impulse and chemical processes of transmission of this electrical impulse through the synaptic cleft in chemical synapses is the basis of thinking processes.

As in the previous paragraph, here for certainty we will study the case in which the role of mediators is played by acetylcholine molecules (ACh), which interact with choline receptors (R). Consider the following sequence of processes that occur in the synaptic cleft:

1. Acetylcholine (ACh) molecules are synthesized by the additional special enzyme choline acetyltransferase and are formed in spherical vesicles in the presynaptic region. The membrane depolarizes during the passage of a nerve impulse when acetylcholine is released into the synaptic cleft. It is important to note here that during this process about 10^4 molecules of Ach are released and reach the postsynaptic membranes, where cholinergic receptors (R) are located.
2. Acetylcholine interacts with cholinergic receptors, causing a complex compound AChR, which leads to a dynamic equilibrium with free acetylcholine and receptors. As a result of the interaction of ACh and P, the permeability of membranes for Na^+ and K^+ ions increases, causing membrane depolarization. If the concentration of ACh is high enough, the depolarization process becomes suprathreshold, causing synaptic stimulation to spread along muscle fibers.
3. The postsynaptic membrane contains a high level of acetylcholinesterase enzyme AChE, which hydrolyzes acetylcholine. The process of removing acetylcholine from the area of receptor proliferation is very important for the renewal of the postsynaptic membrane. Under the influence of certain substances (for example, organophosphorus compounds), which inactivate AChE, the nature of neuromascular transmission changes. The intensity and duration of the ACh action increases. When AChE becomes completely inactivated, the transmission of nerve impulses is blocked as a result of the accumulation of ACh at the location of receptor.

Previous models of synaptic transmission discussed in this chapter took into account concentrations of acetylcholine and inactive receptors. The goal of this study is to create a more realistic model of synaptic transmission [24, 39]. Here we consider synaptic transmission in the cleft, where muscarinic cholinergic receptors predominate. It is well known that cholinergic receptors of this type are G-protein coupled. This means that they require the presence of G-protein to be activated. Let us assume that the concentration of G protein is constant and sufficient for activation. We will also assume that one acetylcholine molecule activates the receptor, and acetylcholinesterase is involved in the destruction of transmitter-receptor complexes. We will consider the function of acetylcholine release into the synaptic cleft to be short-term, approximating this process using the Dirac δ-function depending on time t. The main properties of this function, which belongs to the class of generalized functions, are characterized by the following relations:

$$(1)\quad \delta(t) = 0, \quad \text{if} \quad t \neq 0; \qquad (2) \quad \int_{-\infty}^{\infty} \delta(t)dt = 1. \tag{4.42}$$

Thus, we obtain the following system of three nonlinear differential equations that characterize the change in the concentration of ACh molecules (which is denoted by the variable x), the concentration of activated receptors (mediator-receptor complexes AChR (which is denoted by the variable y)) and the concentration of acetylcholinesterase AChE (which is denoted by the variable z):

$$\dot{x} = K\delta(t) - k_1 x(a - y) - k_2 xz - k_3 x + D_1 \Delta x, \tag{4.43}$$

$$\dot{y} = k_1 x(a - y) - k_4 xyz, \tag{4.44}$$

$$\dot{z} = k_5 z - k_2 xz - k_4 xyz - D_3 \Delta z. \tag{4.45}$$

The first equation describes the following processes: the first term on the right side of Eq. (4.43) describes the speed of acetylcholine delivery into the synaptic cleft, the second term is the rate of binding of acetylcholine to non-activated receptors (a is their total number on the postsynaptic membrane), the third term is the rate of cleavage of acetylcholine in the presence of acetylcholinesterase, the fourth and fifth terms describe the removal of acetylcholine from the synaptic cleft by reuptake and diffusion, respectively.

The second Eq. (4.44) describes the following processes: the first term determines the positive contribution to the rate of change in the concentration of mediator-receptor complexes due to the binding of acetylcholine to unactivated receptors, and the second term describes the rate of cleavage of mediator-receptor complexes in the presence of acetylcholinesterase.

In Eq. (4.45), the second, third, and fourth terms have the same meaning as similar terms in Eqs. (4.43) and (4.44). The first term, depending on the sign of the parameter k_5, which plays the role of a certain control parameter, will have a double meaning. At $k_5 < 0$ it is the rate of release of receptor complexes (normal functioning of the body). At $k_5 < 0$ it is the rate of binding of acetylcholinesterase enzyme molecules to organophosphorus compounds, which prevails over the rate of release of enzyme molecules. Note that the "+" sign before the diffusion term $D_1 \Delta x$ in the first kinetic equation reflects the fact that the mediator diffuses from the synaptic cleft, while the "−" sign before the diffusion term in the third kinetic equation $D_3 \Delta z$ means that acetylcholinesterase molecules can also move to the region with increased concentration of acetylcholine due to lateral diffusion.

We should also add the following explanation that the dots above the variables in the left parts of Eqs. (4.43)–(4.45) mean time derivatives, and the symbol Δ is used for the Laplace differential operator, which symbolizes the sum of three partial derivatives of the 2nd order with respect to the coordinates of concentrations x, y, z. Thus, these equations characterize a sufficiently complex object from a mathematical point of view, namely, a system of three differential equations in partial derivatives. At the same time, each of the differential equations contains four partial derivatives, and one of them is a derivative in time, and the other three are derivatives in spatial coordinates. To significantly simplify the search for a mathematical solution of such a system of differential equations, we will use an approximation that facilitates our task (without losing the fundamental features of its solution). Therefore, it is natural (from the point of view of preserving dimensionality) to introduce the following substitution: $D_1 \Delta x \approx D_1 \frac{x}{L^2}$, $D_3 \Delta z \approx D_3 \frac{z}{H^2}$. Here L and H are the characteristic linear dimensions of spatial inhomogeneity for the ACh concentration in (4.43), and the AChE concentration in (4.45).

Taking into account the above approximations, let us transform the system of partial differential Eqs. (4.43)–(4.45) into the following system of three ordinary differential equations for one variable depending on time t:

$$\frac{dx}{dt} = K\delta(t) - k_1 x(a - y) - k_2 xz - k_3 x + D_1 \frac{x}{H^2}, \tag{4.46}$$

$$\frac{dy}{dt} = k_1 x(a - y) - k_4 xyz, \tag{4.47}$$

$$\frac{dz}{dt} = k_5 z - k_2 xz - k_4 xyz - D_3 \frac{z}{L^2}. \tag{4.48}$$

It should be noted that the system of Eqs. (4.46)–(4.48) remains a nonlinear system of differential equations with respect to concentrations x, y, z. In system (4.46)–(4.48) H is the width of the synaptic cleft, and L is the characteristic geometric size of the postsynaptic zone, which should be taken as its radius.

Let us investigate the stationary solutions of the system of Eqs. (4.46)–(4.48) and the stability of these solutions according to the Lyapunov classification, which was diccusse in Sect. 4.2.4. Taking into account the fact that the first term in Eq. (4.46) is considered an instantaneous disturbance, then the corresponding system of stationary equations has the following form:

$$-k_1 x(a - y) - k_2 xz - k_3 x + D_1 \frac{x}{H^2} = 0, \tag{4.49}$$

$$k_1x(a - y) - k_4xyz = 0, \tag{4.50}$$

$$k_5z - k_2xz - k_4xyz - D_3\frac{z}{L^2} = 0. \tag{4.51}$$

Our further task is reduced to obtaining singular points of the system of algebraic Eqs. (4.49)–(4.51) and studying the following two problems: (a) finding the type of singular points in accordance with the Poincaré classification; (b) analyzing Lyapunov exponents in order to establish the possible chaotic behavior of the studied 3-dimensional model of synaptic transmission of a nerve impulse of the strange attractor type. To achieve this goal, it is necessary to consistently use the procedure that was described in Sect. 4.2.4.

The direct results of its use for finding three types of singular points and the corresponding roots of the characteristic equation are presented below and have the following physical and mathematical form, as well as neurophysiological meaning.

1. ***The first singular point*** takes such values in the 3-dimensional space of variables x, y, z of the nonlinear kinetic model of synaptic transmission or, what is the same, in the 3-dimensional space of ACh, AChR, and AChE concentrations:

$$\text{(a)}\quad x_0 = 0, y_0 = g, z_0 = 0, \tag{4.52}$$

where g is a certain concentration of activated receptors, which obviously satisfied the condition $g \leq a$. For a non-negative value of parameter $k_5 \geq 0$ (such a value corresponds to the physiological state of the studied model of synaptic transmission), the Lyapunov exponents λ_i, being solutions of the corresponding characteristic equation, turn out to be equal

$$\lambda_1 = k_1g + \frac{D_1}{H^2} - k_1a - k_3 < 0, \tag{4.53}$$

$$\lambda_2 = 0, \tag{4.54}$$

$$\lambda_3 = k_5 - \frac{D_3}{L^2} > 0. \tag{4.55}$$

It should be noted that the following experimental data [41] were used to establish the signs of inequalities (4.53) and (4.55):

$$\frac{D_1}{H^2} \approx 10^5, k_1a \approx 2\cdot 10^{12}, k_5 \approx 10^4, \frac{D_3}{L^2} \approx 10^{-2}. \tag{4.56}$$

2. ***The second singular point*** takes the following values

$$\text{(b)}\quad x_0 = m, y_0 = a, z_0 = 0, \tag{4.57}$$

where m is a certain concentration of free acetylcholine. It should be taken into account that over a fairly long period of time all free ACh will be split, returned back into the presynaptic membrane or diffused from the cleft. Therefore, the rate of reuptake will be zero. In this case, the Lyapunov exponents λ_i take the following values:

$$\lambda_1 = \frac{D_1}{H^2} - k_3, \tag{4.58}$$

$$\lambda_2 = -k_1 m < 0, \tag{4.59}$$

$$\lambda_3 = k_5 + \frac{D_3}{L^2} - k_2 m - k_4 m a. \tag{4.60}$$

Depending on the signs of λ_1 and λ_3 in (4.58) and (4.60), the second singular point can be both stable and unstable.

3. ***The third singular point*** is characterized by such concentrations

$$\text{(c)} \quad x_0 = 0, y_0 = g, z_0 = h, \tag{4.61}$$

where h is a certain concentration of acetylcholinesterase. For this singular point, we have the following values of the Lyapunov exponents:

$$\lambda_1 = k_1 g + \frac{D_1}{H^2} - k_1 a - k_2 h - k_3 < 0, \tag{4.62}$$

$$\lambda_2 = 0, \tag{4.63}$$

$$\lambda_3 = k_5 - \frac{D_3}{L^2} > 0, \tag{4.64}$$

To find out the signs of inequalities (4.62) and (4.64), the numerical values of the parameters in (4.56) were directly taken into account using experimental data [41]. Thus, based on the classification of Lyapunov exponents given in Sect. 4.2.4, the third singular point is classified as a ***strange attractor.***

In (4.52), (4.57), (4.61) the constants g, m, and h are certain concentrations of acetylcholine, activated receptors, and acetylcholinesterase, which can take different values. The only limitation on the concentration of mediator-receptor complexes is that $0 < g < a$. Thus, the general solution of the system of nonlinear differential Eqs. (4.49)–(4.51), which characterize, respectively, the change in concentrations x, y, z of ACh molecules, activated mediator-receptor complexes AChR and acetylcholinesterase AChE has the following form:

$$x = x_0 + K + F_1 e^{\lambda_1 t} + F_2 + F_3 e^{\lambda_3 t}; \tag{4.65}$$

$$y = y_0 + G_1 e^{\lambda_1 t} + G_2 e^{\lambda_2 t} + G_3 e^{\lambda_3 t}; \tag{4.66}$$

$$z = z_0 + H_1 e^{\lambda_1 t} + H_2 + H_3 e^{\lambda_3 t}. \tag{4.67}$$

where the coefficients F_i, G_i, and H_i can be found from the initial conditions.

It can be easily seen that the diffusion of the mediator has a greater influence than the lateral diffusion of cholinesterase on the process of synaptic transmission. The diffusion term changes the value of the solutions of the characteristic equations, but does not change the types of singular points, i. e. its contribution to the process of synaptic transmission is not so strong.

The main result of our study is as follows: ***the kinetic model of synaptic transmission with three order parameters exhibits behavior as a strange attractor under certain conditions that were outlined above.***

4.4 Conclusions

In this chapter, one of the most important problems of living nature is discussed and modeled, which is the problem of intercellular interaction and synaptic transmission of electrical signals along nerve fibers and through synaptic clefts.

In Sect. 4.2, at a fairly popular level, the following fundamental provisions of this problem are considered and analyzed: the electrical aspect of generation and propagation of action potential along nerve fibers in the Hodgkin-Huxley model (Sect. 4.2.1), the chemical aspect and main features of synaptic transmission of a nerve impulse (Sect. 4.2.2); the porocytosis hypothesis in the theory of synaptic transmission and its difference from the hypothesis of quantum-vesicular exocytosis (Sect. 4.2.3); synergetic approaches to synaptic transmission associated with the use of Poincaré classification of singular points, phase portraits and Lyapunov exponents (Sect. 4.2.4).

In Sect. 4.3, the main theoretical principles underlying the Hodgkin-Huxley model of the propagation of nerve impulses along axons are presented at a consistent physical and mathematical level (Sect. 4.3.1), as well as a new theoretical approach to the problem of synaptic transmission based on the hypothesis of isomorphism between (a) the guantum-like (cooperative) behavior of a neurotransmitter-receptor system in the synaptic cleft and (b) the phase transition in bounded binary mixture near its critical mixing point (Sect. 4.3.2).

In addition, our important findings, as a consequence of the above-mentioned hypothesis of isomorphism, are as follows: (1) establishing the number of correlated receptors on the postsynaptic membrane in a chemical synapse and in the neuromuscular junction (Sect. 4.3.3), and (2) studying a nonlinear kinetic model of synaptic transmission, which demonstrates chaotic behavior like a strange attractor.

References

1. J.V. Dyk (Ed.), *True Tales of Medical Physics: Insights into a Life-Saving Specialty*. Springer (2022)
2. M. Maqbool (Ed.), *An Introduction to Medical Physics*. Springer (2017)
3. D. Hamilton, *Diagnostic Nuclear Medicine: A Physics Perspectives*. Springer (2004)
4. V. Tsymbalyuk, V. Medvedev, *Man and his Brain*. (Vinnytsia, Nova Knyga, 2022)
5. *Medical and Biological Physics*, A.V. Chalyi (Ed.), 4th edition (Vinnytsia, Nova Knyga, 2020)
6. L.A. Bulavin (Ed.), L.G. Grechko, L.V. Lerman, A.V. Chalyi, *Medical Physics. Dynamic and Statistical Models* (Kyiv, Kyiv University Publishing Center, 2011).
7. B. Nolting, *Methods in Modern Biophysics* (Springer-Verlag, Berlin, 2006)
8. P. Nelson, *Biological physics. Energy, Information, Life* (W.H. Freeman and Company, New York, 2004)
9. R. Cotterill, *Biophysics. An Introduction* (Wiley, 2003)
10. J. Nolte, *The Human Brain* (Mosby, Missouri, 2002)
11. P.G. Kostyuk (Ed.), V.L. Zima, I.S. Magura, M.S. Miroshnichenko, M.F. Shuba, *Biophysics* (Kyiv, Oberegy, 2001)
12. H. Haken, *Principles of Brain Functioning: A Synergetic Approach to Brain Activity, Behavior and Cognition*. Springer Series in Synergetics (1995)
13. G.M. Shepard, *Neurobiology* (Oxford Univ. Press, New York, 1994)
14. P.G. Kostyuk, *Calcium Ions in Nerve Cell Function* (Oxford Univ. Press, New York, 1992)
15. P.G. Kostyuk, O.A. Kryshtal, *Mechanisms of Electrical Excitability of Nervous System* (Nauka, 1981)
16. D.H. Hubel, Ch.F. Stevens, E.R. Kandel, et al. *The Brain*, Scientific American, vol. 241, (1979)
17. Ch.F. Stevens. Neuron. In: *The Brain*, Scientific American, vol. 241, p. 54, (1979)
18. W. Betz, Centralblatt fur die Medizinischen Wissenschaften **12**, 595 (1874)
19. J. Eccles, A.L. Hodgkin, A.F. Huxley, *Nobel Lectures, Physiology or Medicine* 1963–1979 (Elsevier Publishing Company, 1972)
20. A.V. Chalyi, L.M. Chernenko, Phase transitions in finite-size systems and synaptic transmission, in D. Beysens, N. Boccara, G. Forgacs (Eds.) *Dynamic Phenomena at Interfaces, Surfaces and Membranes* (New York, Nova Sci. Publishers, 1993)
21. M.E. Kriebel, B. Keller, R.B. Silver, G.O. Fox, G.D. Pappas, Porocytosis: a new approach to synaptic function. Brain Res. Rev. **18**(1–2), 20–32 (2001). https://doi.org/10.1016/50165-0173(01)00066-2
22. M.E. Kriebel, B. Keller, R.B. Silver, G.D. Pappas, Porocytosis: A transient pore array secrets the neurotransmitter packet. *The Anatomical Record. Advance* https://doi.org/10.1016/50165-017 (in integrative anatomy and evolutionary biology, 282B:38–41 (2005); https://doi.org/10.1002/ar.b.20051
23. A. Dufour et al., Development of synaptic boutons in layer 4 of the barrel field of the rat somatosensory cortex: quantitative analysis. Cereb. Cortex **26**, 838–854 (2015)
24. A.V. Chalyi, A.N. Vasilev, E.V. Zaitseva, Synaptic transmission as a cooperative phenomenon in confined systems. Condens. Matter Phys. **20**(1), 13804 (2017)
25. A.V. Chalyi, Synergetic dialogue "physics-medicine": hexagones in living and inanimate nature. J. Molec. Liquids **320**, 114293 (2020). https://doi.org/10.1016/j.mol-liq.2020.114293
26. J. O'Keefe, M.-B. Moser, E.I. Moser, *The Nobel Prize in Physiology or Medicine*. Nobel Prize.org. Nobel Media AB 2020,<http:/nobelprize.org/prizes/medicine/2014/summery/ (2014)
27. A.V. Chalyi, Surface tension in bulk and bounded liquids. J. Mol. Liq. **288**, 110873 (2019)
28. E.N. Lorenz, Deterministic nonperiodic flow. J. Atmosph. Sci. **20**(2), 130–141 (1963)
29. L.D. Landau, E.M. Lifshitz, *Fluid Mechanics*, 2nd Edition (Pergamon Press, Oxford, 1987)
30. A.L. Hodgkin, A.F. Huxley, A quantitative description of membrane current and its application to conductionand exitationin nerve. J. Physiol. **117**(4), 500 (1952). https://doi.org/10.1113/jphysiolsp004764
31. H.E. Stanley, *Introduction to Phase Transitions and Critical Phenomena* (Glarendon Press, Oxford, 1971)

32. A.Z. Patashinskii, V.L. Pokrovskii, *The Fluctuation Theory of Phase Transitions* (Pergamon Press, Oxford, 1979)
33. M.A. Anisimov, *Critical Phenomena in Liquids and Liquid Crystals* (Gordon & Breach, 1991)
34. A.V. Chalyi, Critical phenomena in finite-size systems. J. Mol. Liquids **58**, 179–195 (1992)
35. K.A. Chalyy, K. Hamano, A.V. Chalyi, Correlating properties of a simple liquid at criticality in a reduced geometry. J. Mol. Liquids **92**(1–2), 153–164 (2001)
36. L.D. Landau, E.M. Lifshits, *Ststistical Physics* (Pergamon Press, Oxford, 2013)
37. A.V. Chalyi, L.A. Bulavin, V.F. Chekhun, K.A. Chalyy, L.M. Chernenko, A.N. Vasilev, E.V. Zaitseva, G.V. Khrapijchuk, A.V. Severin, M.V. Kovalenko, Universality classes and critical phenomenain confined liquid systems. Condens. Matter Phys. **16**, 23008 (2013)
38. A.V. Chalyi, Dynamic anomalies in confined supercooled water and bulk fluids, In: *Modern Problems of Molecular Physics*; (Eds) L. Bulavin, A. Chalyi (Springer International Publications: Cham, Switzerland, 1918)
39. A.V. Chalyi, E.V. Zaitseva, Strange attractor in kinetic model of synaptic transmission. J. Phys. Studies **11**, 322 (2007)
40. A.V. Chalyi, *Non-Equilibrium Processes in Physics and Biology* (Naukova Dumka, Kyiv, 1997)
41. N.R. Nigmatullin, V.A. Snetkov, E.E. Nikol'skii, L.G. Magazanik, *Neurophysiology*, 20(3), 289 (1988)

Chapter 5
Thermal Radiation of the Human Body and Its Medical Application

Abstract This chapter is devoted to the study of the peculiarities of thermal (temperature) radiation by objects of living (organic) and non-living (inorganic) nature. Thermal electromagnetic radiation is a manifestation of the presence of internal energy, which is associated with the thermal movement of atoms and molecules of all heated bodies. In Section 5A, the basic laws of thermal radiation, as well as the peculiarities of thermal radiation of the human body, will be considered at the popular level. Thermal radiation is the most powerful channel of radiation of the physical fields of the human body. It turns out that the average power of thermal radiation from the entire surface area of the human body in the infrared range of electromagnetic radiation is equal to a fairly significant value, which is approximately 100 W. The same section contains the results of our experimental studies of the process of temperature relaxation of thermal radiation, which will allow us to further determine the numerical values of the parameters of this process for the human body. It will be proven that the time dependence of the temperature of thermal radiation does not obey an exponential law, as for objects of an inorganic nature, but occurs in the form of damped temperature oscillations over time under the influence of thermoreceptors, which create a sensory system located in various parts and organs of the human body. Section 5B will provide the theoretical basis for the basic laws of thermal radiation obtained by using reliable mathematical methods and consistent approaches of theoretical physics and carrying out precise physical experiments. The theoretical processing of the obtained experimental results for the thermal radiation of the human skin made it possible to determine the following parameters of the temperature relaxation process: the period of decaying temperature oscillations, the decay coefficient and the relaxation time of the temperature oscillations of the thermal radiation of the human skin, as well as the frequency of the decaying and natural temperature oscillations. In addition to finding the specified parameters, one of the important theoretical results was the finding of an explicit form of the second-order differential equation, which describes the change in the temperature of thermal radiation of human skin. The use of thermal radiation of the human body as a diagnostic method of thermography has a unique value and a certain fundamental advantage in comparison with other diagnostic methods, such as X-ray computer tomography, magnetic resonance tomography, positron emission tomography, ultrasound studies. The main

A. V. Chalyi, *What is Medicine?*, https://doi.org/10.1007/978-3-031-64979-0_5

difference is explained by the fact that thermography uses the human body's own (and therefore non-invasive) thermal radiation, and not external (and therefore to a certain extent invasive) radiation, which can cause, under some conditions, quite significant dose loads on organs and tissues human body.

Keywords Thermal radiation · Basic laws of thermography · Absolute black body · Uniqueness of thermography diagnostic method · Experimental data for parameters of thermal radiation · Temperature relaxation of human skin

And if it is so, then what is beauty,

And why is it deified by people?

A vessel in which there is emptiness,

Or fire flickering in the vessel.

Nikolai Zabolotskii

5.1 Introduction

The subject matter of this chapter relates, on the one hand, to two branches of physics—optics and quantum mechanics, and, on the other hand, to several branches of medicine—primarily to physiology, medical diagnostics, and ophthalmology. In a broad sense, optics is the science of light and its interaction with matter [1–30]. At the same time, optics studies not only the visible part of the spectrum of electromagnetic waves with a wavelength of $\lambda = 0.40\,\mu\text{m} - 0.76\,\mu\text{m}$, i.e. light in the narrow sense of this term, but also shorter electromagnetic waves of ultraviolet radiation with a wavelength of $\lambda = 10\,\text{nm} - 400\,\text{nm}$ and longer electromagnetic waves of the infrared radiation with a wavelength from $\lambda = 0.76\,\mu\text{m}$ to $\lambda = 2000\,\mu\text{m}$. It is the last infrared range of electromagnetic waves, as will be shown below, that includes the thermal radiation of the human body. Sometimes even soft X-ray radiation with a wavelength from $\lambda = 0.2\,\text{nm}$ to $\lambda = 80\,\text{nm}$, as well as radio waves in the millimeter range, are considered optical radiation. Another interesting application of optical methods in neutron physics should be noted, which is related to neutron optics. As is known (see, for example, [31–33]), neutron optics combines the physical methods of slow, thermal and cold neutrons, which have a relatively low energy $E \leq 1\,\text{eV}$ and a relatively long de Broglie wavelength $\lambda \geq 3 \cdot 10^{-11}\,\text{m}$.

The various connections between natural phenomena create the inner beauty of science, the study of which is a great pleasure for every inquisitive person, especially for its creators. This is exactly what the poet-philosopher Nikolai Zabolotskii (1903–1958) wrote about in the words used in the epigraph for this chapter. In the 20 s of the last century, while studying at Moscow University at the Faculty of Medicine and

Faculty of History and Philology, and then at the Department of Language and Literature at the Petrograd Institute, Zabolotskii began to create his own poetic works filled with philosophical lyrics and natural philosophical ideas. He independently got acquainted with Albert Einstein's works on the theory of relativity, later became fascinated by the works of Konstantin Tsiolkovsky on space flights and the paintings of Brueghel, Chagall and Filonov. In 1938, Zabolotskii was repressed in connection with the accusation of anti-Soviet propaganda. Despite the physical torture, he pleaded not guilty to creating a counter-revolutionary organization, and most likely such courageous behavior during interrogations saved him from execution. Nikolai Zabolotskii was rehabilitated in 1963, 5 years after his death.

The development of science, like human life, has always gone through "trial and error". The famous Danish theoretical physicist Niels Bohr (1885–1962) once said that a specialist is one who knows the typical mistakes in his profession and knows how to avoid them. He did not say that a specialist is someone who does not make mistakes. Optics, like each of the branches of the physical and medical sciences, gives a lot of examples of human dramas and happy discoveries of the scientists who created them.

If any of the respected readers (in particular, respected medical students) may have doubts about the need to study optical phenomena, it is enough to mention a large number of physical optical instruments used in medicine, namely: microscopes, lasers, refractometers, nephelometers, concentration colorimeters, spectrophotometers, radiation thermometers (pyrometers). This is not a complete list of optical instruments used in medical clinics and hospitals, biomedical laboratories of scientific institutes and educational institutions. Without these devices, as well as without a deep understanding the rather complex processes occurring in living nature under the influence of optical radiation, the fruitful work of a modern doctor is impossible.

One of the creators of operational calculus[1] (not to be confused with surgery!) English physicist and mathematician Oliver Heaviside (1850–1925) said: "Am I going to refuse my lunch just because I do not fully understand the digestion process?". Let this phrase inspire dear readers to get acquainted with the contents of this chapter.

5.2 Section A

As noted above, the thermal radiation of the human body is the electromagnetic radiation arising from the presence of internal energy due to the thermal movement of atoms and molecules of this body (see, for example, [1–4]). This phenomenon played a decisive role in the creation of modern physics and, above all, quantum mechanics.

[1] There is such a part in mathematics (see, for example, [34,35]), with the help of which you can perform differentiation, integration operations and solve differential equations, transferring them from the "space of originals", where they are given, to the so-called "image space" where these operations and equations become algebraic and more easily solved.

The exact birth date of quantum mechanics is considered to be December 14, 1900, when Max Planck reported at a meeting of the German Physical Society that he had obtained a formula for the radiation spectrum of an absolutely black body using the hypothesis about its quantum nature. This formula solved not only the problem of the "ultraviolet Rayleigh-Jeans catastrophe" in the radiation spectrum of an absolutely black body at high frequencies, but also explained the regularities of this spectrum at low frequencies, discovered by Wien in 1893. In 1911, Wilhelm Wien was awarded the Nobel Prize in Physics "for his discoveries relating to the laws governing the radiation of heat" [5]. 7 years later, in 1918, Max Planck received the Nobel Prize in Physics "in recognition of the merits he made to the development of physics through the discovery of quanta of energy" [6]. The use of Planck's formula for quantum energy allowed Albert Einstein to give a theoretical explanation of the photoelectric effect in 1905 and to receive the Nobel Prize in Physics in 1921 "for his services in theoretical physics, especially for the discovery of the law of the photoelectric effect" [7]. The list of outstanding achievements awarded Nobel prizes in Physics during the next century due to the use of powerful methods of quantum mechanics, the origins of which are directly related to the laws of thermal radiation, would be too long and, at the same time, difficult to overestimate.

The thermal radiation of the human body has a continuous spectrum, the maximum of which depends on temperature. With an increase in temperature, the total energy of thermal radiation increases sharply, while the wavelength corresponding to the maximum thermal radiation of the human body shifts towards shorter wavelengths. The theoretical basis of thermography (thermal imaging) is formed by the laws of thermal radiation, which will be considered below, mainly at a qualitative level. A quantitative description of the laws of thermal radiation can be found, for example, in a textbook "Medical and Biological Physics" for students of the higher medical educational institutions [1]. Additional information about the use of modern physical methods in medicine (in particular, thermography or thermal imaging methods) is contained in [2–20].

5.2.1 *The Main Characteristics of Thermal Radiation*

First of all, we will introduce the exact definitions of such ***basic characteristics of the thermal radiative and absorptive capacity*** of heated bodies as the radiative flux F, the radiation emittance of the body (the integral energetic emittance) R, the spectral density of the radiating emittance of a body (or monochromatic radiation coefficient) r_λ, the absorption coefficient α, and the monochromatic absorption coefficient α_λ.

The radiation flux F is (a) the average radiation power over a period of time significantly exceeding the period of electromagnetic oscillations; (b) the energy of electromagnetic radiation per unit of time, measured in the watt (W).

The radiation emmitance of the body R is the flux of electromagnetic radiation coming out of a unit area S of the body surface and measured in the watt per m^2 (W/m^2).

The spectral density of the radiating emmitance of a body r_λ is the radiating ability related to a unit of length of the spectral interval of wavelengths λ and measured in the watt per m^3 (W/m^3). The dependence of the magnitude of the spectral density of the radiation emmitance on the wavelength λ is called ***the radiation spectrum of the body*** $r_\lambda = f(\lambda)$.

Since all the quantities introduced above characterized the radiative emmitance of the body, we will also introduce the characteristics of the absorbing properties of the body.

The absorption coefficient α is the value of the ratio of the radiation flux F_{abs} absorbed by the body to the radiation flux F_{inc} incident on the body. If the value of α refers to a single spectral interval of wavelengths λ, then it determines ***the monochromatic absorption coefficient*** α_λ.

Let us introduce the concept of ***«absolute black body»*** as a body, for which the monochromatic absorption coefficient is equal to unity in the entire spectral range and for any temperature, i.e. $\alpha_\lambda^{b.b.} = 1$. A cavity with a very small hole can serve as a model of an absolute black body (see Fig. 5.1). It is for such a model that the monochromatic absorption coefficient should be close to unity. A fairly good real example of an absolute black body is a platinum crucible at a temperature close to the melting point of platinum $T_{melt} = 2046.6$ K. In the theory of thermal radiation, the concept of ***«gray body»*** is also introduced, as such a body for which the absorption coefficient is less than unity. Thus, the human body can be considered gray in the infrared part of the spectrum, since in this spectral range the absorption coefficient is $\alpha_\lambda^{b.b.} \approx 0.9$.

Figure 5.2 shows the dependence of the monochromatic absorption coefficient on the wavelength λ for an absolute black body (upper solid straight line), gray bodies (dashed straight lines) and arbitrary bodies (lower curved line) at a given temperature. When the temperature of an arbitrary body changes, the nature of the

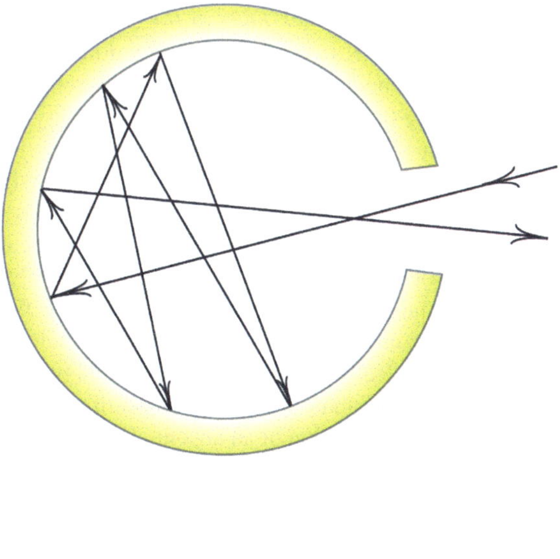

Fig. 5.1 Schematic analogue of an absolute black body

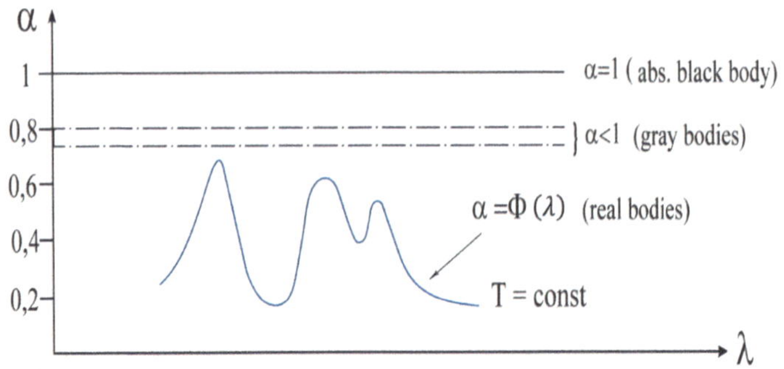

Fig. 5.2 Dependence of the monochromatic absorption coefficient on the wavelength λ

curve can change significantly: wavelengths that are absorbed at one temperature can be transmitted at another temperature, and vice versa.

5.2.2 *Basic Laws of Thermal Radiation*

Let us now turn to the consideration of thermal (temperature) radiation basic laws [1–4]:

1. ***The Kirckhoff's law***. This law was discovered by the German physicist Gustav Kirckhoff in 1859 and established a relationship between the radiating and absorbing abilities of a body. The Kirchhoff's law is formulated as follows: ***the ratio of the spectral density of the radiation emmitance r_λ of a body to the monochromatic absorption coefficient α_λ is the same for all bodies at a given temperature and is equal to the spectral density of the radiation emmitance ε_λ of an absolute black body at the same temperature*** (see the corresponding formula for the Kirckhoff's law in Section 5B). In other words, the ratio of the emissivity of bodies to their absorption capacity does not depend on the nature of the emitting body and is equal to the emissivity of a completely black body at a given temperature. On Fig. 5.3 presents the experimental curves of thermal radiation energy distribution of an absolutely black body (curve 1), a gray body (curve 2) and an arbitrary real body (curve 3). The curve of the spectral distribution of energy for a gray body can be obtained from the curve of the spectral distribution of an absolutely black body as a result of multiplying the ordinates of curve 1 at a fixed value of the wavelength by a factor less than unity, which is equal to the value of the monochromatic absorption coefficient $\alpha_\lambda(T)$ at a given temperature. Due to the selective absorption of some real bodies, curve 3 in Fig. 5.3 may have several maxima and minima, but this entire curve must always be located below curve 1 of the radiation of an absolutely black body, as follows from Kirchhoff's law, since for an arbitrary real body we have that $\alpha_\lambda(T) < 1$.

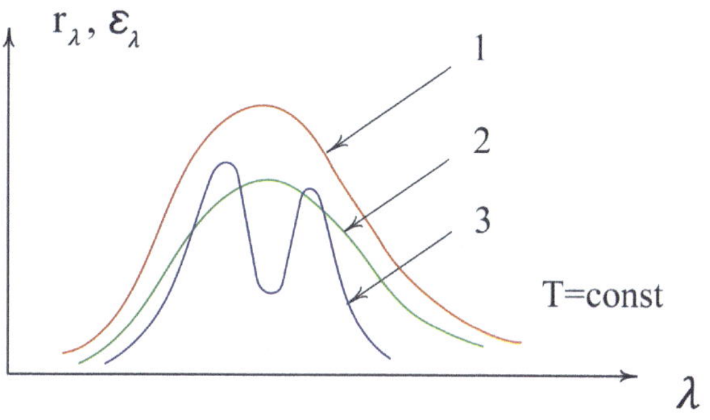

Fig. 5.3 The experimental distribution of energy in the spectra of thermal radiation of various bodies

2. ***The Planck's law of absolute black body radiation*** was discovered by the German physicist Max Planck in 1900 using the hypothesis, which was formulated by him about the change in the electromagnetic energy of radiation and absorption by minimal portions (quanta). This hypothesis marked the beginning of quantum mechanics, one of the most outstanding achievements of the natural sciences of the twentieth century, for which Max Planck was awarded the 1918 Nobel Prize in Physics. M. Planck's radiation law specifies the dependence of the spectral density of the radiation emmitance ε_λ of an absolute black body on the wavelength λ (or frequency ν) of electromagnetic radiation (see the corresponding formulas for the Planck's radiation law in section 5B). It should be noted that the radiation law of M. Planck is in full agreement with the experimental curves of the dependence of the emission spectra $r_\lambda = f(\lambda)$.
3. ***The Stefan-Boltzmann law*** is formulated as follows: ***the radiation emmitance R of an absolute black body is directly proportional to T^4, i.e. the fourth power of the absolute temperature T*** (see detals of the derivation of the corresponding formula for the Stefan-Bolltzmann law in section 5B). This law was established by two Austrian physicists Josef Stefan and Ludwig Boltzmann, first experimentally obtained by J. Stefan in 1879, and then theoretically derived in 1884 by the famous theoretical physicist L. Boltzmann, whose supervisor was J. Stefan. Thus, the integral radiation emmitance of an absolute black body increases quite sharply with an increase in absolute temperature. The Stefan-Boltzmann law can be applied to the thermal radiation of the human body, considering it a "gray body" (see below).
4. ***The Wien's displacement law.*** This law, established by the German physicist Wilhelm Wien in 1893, gives a relationship between the absolute temperature T and the wavelength λ_{max} corresponding to the maximum spectral density of the radiation emmitance in the spectrum of thermal radiation of an absolute black body. For the discovery of this law, Wilhelm Wien was awarded the 1911 Nobel Prize in Physics. The Wien's displacement law is formulated as follows: ***the wavelength λ_{max}, which accounts for the maximum spectral density of the radiation emmitance ε_λ of an absolute black body, is inversely proportional to its absolute temperature*** T. Thus, the wavelength λ_{max}, characterizing the maximum spectral density of the radiation emmitance of an absolute black body, shifts towards

shorter wavelengths with increasing temperature T. It is this fact, confirmed by the experimental curves of energy distribution in the spectra of thermal radiation (see Fig. 5.5), that gives reason to call Wien's law the "displacement law".

5.2.3 Special Properties and Experimental Studies of Thermal Radiation of the Human Body

Special properties of thermal radiation of the human body. Based on the Wien's displacement law, we may estimate the value of the wavelength λ_{max}, which accounts for the maximum value of the radiation emmitance of the human body. The surface layer of human skin has an approximate average temperature of $t \approx 32$ °C on the Celsius scale or

$$T \approx (273 + 32)\,\mathrm{K} \approx 305\,\mathrm{K}$$

on the Kelvin scale. Then, according to Wien's displacement law, we have such a numerical value of $\lambda_{Max} \approx 10$ μm. It follows that the wavelength of the human body falls on the infrared range of electromagnetic waves, which is not perceived by the human eye. The reason for this fact is that a person can see the world around him or her in a rather narrow range of wavelengths of the visible light from $\lambda = 0{,}40$ μm (blue light) to $\lambda = 0{,}76$ μm (red light). It should be noted that the Earth's atmosphere has a so-called "transparency window" for the wavelength interval, which also includes the wavelength of thermal infrared radiation of the human body $\lambda_{max} = 10$ μm. In other words, we can say that electromagnetic radiation of this wavelength is almost not absorbed by atoms and molecules of the Earth's atmosphere. If there was no such window of transparency in atmospheric air, then people could, under certain conditions, "roast" in their own electromagnetic radiation.

In addition to the calculation of the wavelength λ_{max}, which characterizes the maximum value of the radiation emmitance of the human body, the use of other laws of thermal radiation (more precisely, the Stefan-Boltzmann law) allows us to estimate the surface power of thermal radiation of the human body. Strictly speaking, the application of the Stefan-Boltzmann law to real bodies is, of course, a fairly significant approximation. The emissivity of real bodies (especially under the condition of their organic nature) should depend on many factors (the geometric shape of the body, the state of its surface, temperature changes, the course of various physiological processes, etc.), which are usually not taken into account in the Stefan-Boltzmann law.

At the same time, taking into account a certain limitation of this approach and considering the surface of the human body as a "gray" body, one may obtain the following estimate for the power of the thermal radiation of the human body: $P_{body} = 100$ W (see the corresponding numerical estimate in section 5B).

The obtained experimental data also show that the temperature relaxation of the thermal radiation of the human body in the infrared range of electromagnetic waves does not occur according to an exponential law with a gradual approach of the body temperature from above to the thermodynamically equilibrium and initial temperature T = 31 °C. Just such an exponential temperature change with time should be expected on the basis of the solution of the differential equation for the heat conduction processes (see, for example, monograph [4]). At the same time, in the biological systems including the human body, there are non-linear feedback connections due, as in this case, to the presence of thermal receptors conjugated with heat conduction processes.

Thus, as a result of such non-linear feedback effects, the process of temperature relaxation is characterized by a gradual approach of temperature to its equilibrium value in the form of free damped temperature oscillations in a certain area of the human body (see more details in the next Section B).

5.3 Section B

This Section is devoted to an in-depth study of the features of thermal radiation of objects of inanimate and living nature. The main purpose of this Section, which has the following structure, is the theoretical study of the features of the temperature relaxation process of thermal radiation of a locally heated part of the human body (see experimental results in Sect. 5.2.3).

In subsection 5B.1, the basic formulas and definitions of the main characteristics of thermal radiation will be introduced. In the next Sect. 5.3.2, exact mathematical formulas will be written that correspond to the formulations of the basic laws of thermal radiation. In accordance with the basic Wien's and Stefan-Boltzmann's laws of thermal radiation, which underlie the diagnostic method of thermography, the numerical estimations of the wavelength, which corresponds to the maximum value of the radiation-emittance spectrum, and the integral energetic emittance of the human body will be made in Sect. 5.3.3. The Sect. 5.3.4 will be devoted to the study of non-equilibrium thermodiffusion phenomena using the methods of thermodynamics of irreversible processes [1, 2]. In the last Sect. 5.3.5, the theoretical processing of the experimental results presented in Fig. 5.4. will be carried out. The immediate goal of these studies is to find the numerical values of the parameters and to obtain the differential equation, which characterize the process of themperature relaxation of the thermal radiation of human skin.

Temperature and diffusion relaxation, which characterizes the temporal dependence of thermodiffusion processes in objects of an inorganic nature, occurs, as a rule, according to an exponential law, which corresponds to the solution of the differential equations of thermal conductivity and diffusion in partial derivatives of the parabolic type [2–4]. Note that non-equilibrium phenomena in the presence of various thermodynamic forces (gradients of such physical quantities as not only concentration and temperature, but also gradients of pressure and electric field potential) for objects of inanimate and living nature in large volumes, when the linear size L_0 of the system

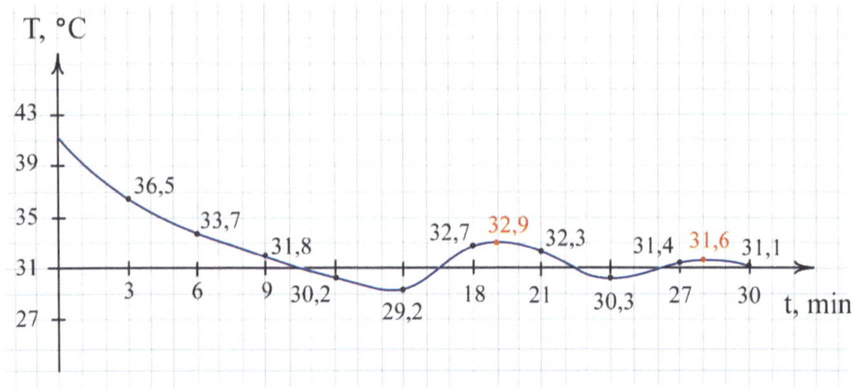

Fig. 5.4 The time dependence of the temperature relaxation of the human body

is significantly dominated by the correlation radius ξ of fluctuations of the order parameter.

5.3.1 Definitions and Formulas of the Main Characteristics of Thermal Radiation

First of all, we will introduce the exact formulas and definitions of such main characteristics of the thermal radiative and absorptive capacity of heated bodies as the radiative flux F, the radiation emittance (or the integral energetic emittance) R, the spectral density of the radiating emittance (or monochromatic radiation coefficient) r_λ, the absorption coefficient α, and the monochromatic absorption coefficient α_λ of the body.

The radiation flux F is the average radiation power over a period of time that significantly exceeds the period of electromagnetic oscillations. Since the power is characterized by the energy W of electromagnetic radiation per unit of time, the radiation flow is described by the following formula:

$$F = W/t \tag{5.1}$$

and is measured in watts $[F] = W$.

The flux of electromagnetic radiation emitted from a unit surface area characterizes ***the radiation emittance (the integral energetic emittance) of the body***

$$R = \frac{F}{S}, \quad [F] = W/m^2. \tag{5.2}$$

The radiation emittance related to the unit length of the spectral interval of wavelengths is called ***the spectral density of the radiating emittance of a body (or monochromatic radiation coefficient)*** r_λ:

$$r_\lambda = \frac{dR}{d\lambda}, \quad [r_\lambda] = [R]/[\lambda] = W/m^3. \tag{5.3}$$

The dependence of the value of the spectral density of the radiation emittance on the wavelength is called ***the radiation spectrum*** $r_\lambda = f(\lambda)$ of the body. It is obvious that the spectral density of the radiating emittance over the entire wavelength interval is determined by the following integral:

$$R = \int_0^\infty r_\lambda d\lambda, \tag{5.4}$$

which gives reason to call the value R an ***integral energetic emittance***.

In addition to the characteristics of the radiative capacity, we will also introduce the characteristics of the body's absorptive capacity.

The absorption coefficient α is the ratio of the radiation flux F_{abs} absorbed by the body to the radiation flux F_{inc} falling on the body:

$$\alpha = \frac{F_{abs}}{F_{inc}}. \tag{5.5}$$

If the value α refers to a unit spectral interval of wavelengths, then this value determines the monochromatic absorption coefficient

$$\alpha_\lambda = \frac{F_{abs}(\lambda)}{F_{inc}(\lambda)}. \tag{5.6}$$

We will also introduce the concept of "absolute black body", for which the monochromatic absorption coefficient is equal to unity in the entire spectral interval and for any temperature, i.e.

$$\alpha_\lambda^{bb} = 1. \tag{5.7}$$

As mentioned in section 5A, a platinum crucible with a small hole can serve as a model of an absolutely black body (see Fig. 5.1).

5.3.2 Basic Laws and Formulas of Thermal Radiation

The properties of thermal electromagnetic radiation caused by the thermal motion of atoms and molecules of heated bodies are used in medicine, in particular for diagnostic purposes. Inflammatory processes change the local body temperature. Therefore, recording the radiation of different areas of the body surface allows to determine their temperature remotely and contactlessly. This diagnostic method is called ***thermography***.

The theoretical basis of thermography is the laws of thermal radiation, which will be discussed below in more detail than was done in Chapter 5A. Thermal radiation is governed by the following basic laws: ***Kirchhoff's law***, which holds for any body, and the following three laws, which hold for an all-black body: ***Planck's radiation law***, ***Stefan-Boltzmann's law***, and ***Wien's displacement law***. Strictly speaking, the last three laws are fulfilled for an idealized absolutely black body, but under certain conditions, which will be discussed below, these laws can also be applied to bodies of organic nature, including the human body.

Kirchhoff's law. This law, which establishes a quantitative relationship between the emissive and absorptive capacities of a body, was established in 1859 by a German physicist Gustav Kirchhoff (1824–1887). This law states that ***the ratio of the spectral density of the radiation emittance (or monochromatic radiation coefficient)*** $r_\lambda(T)$ ***to the monochromatic absorption coefficient*** $\alpha_\lambda(T)$ ***is the same for all bodies at a given temperature and equals spectral density of the radiating emittance*** $\varepsilon_\lambda(T)$ ***of an absolutely black body at the same temperat***ure T. The following formula corresponds to this statement:

$$\left(\frac{r_\lambda(T)}{\alpha_\lambda(T)}\right)_1 = \left(\frac{r_\lambda(T)}{\alpha_\lambda(T)}\right)_2 = \cdots = \frac{\varepsilon_\lambda(T)}{1} = \varepsilon_\lambda(T), \tag{5.8}$$

where ε_λ is the spectral density of the radiating emittance of an absolutely black body, and the formula (5.7) for the monochromatic absorption coefficient of an absolutely black body is taken into account.

Gustav Kirchhoff (1824-1887)

The following consequences follow from formula (5.8) of Kirchhoff's law:

1. The spectral density of the radiating emittance $r_\lambda(T) = \alpha_\lambda(T) \cdot \varepsilon_\lambda(T)$. Since for real bodies $\alpha_\lambda(T) < 1$, then always $r_\lambda(T) < \varepsilon_\lambda(T)$), i.e., the radiating emittance of a real body is always lower than that of an absolutely black body.
2. If $\alpha_\lambda(T) = 0$, then $r_\lambda(T) = 0$, i.e., if the body does not absorb radiation, then it does not emit it in a certain range of wavelengths.

Figure 5.3 given in section 5A illustrates the experimental curves of the spectral distribution of the radiating emittance of an absolutely black body 1, a gray body 2, and an arbitrary body 3. Thus, an absolutely black body, for which the spectral distribution of the radiating emittance is characterized by curve 1, has the most intensive source of thermal radiation for a given temperature and fixed wavelength λ.

Planck's law of radiation. Experimental curves of the spectra of thermal radiation of bodies, which can be approximately considered completely black, demonstrate a shift of the maxima of the spectral density of the radiating emittance ε_λ towards short waves with increasing temperature (see Fig. 5.5).

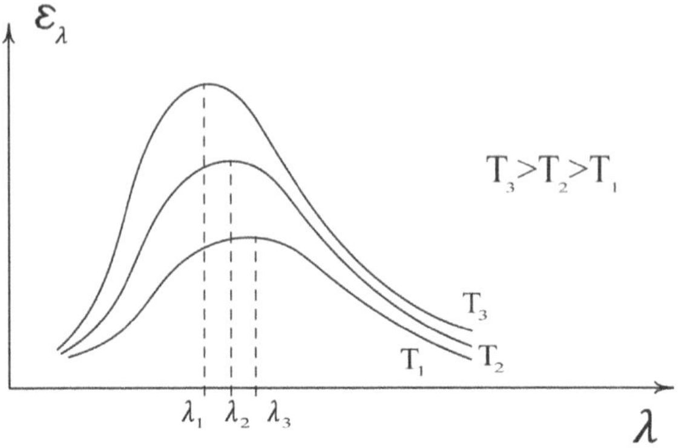

Fig. 5.5 Spectra of thermal radiation of an absolute black body at different temperatures

To determine the obvious type of functions $\varepsilon_\lambda(T)$ or $\varepsilon_\nu(T)$, completely different new ideas were needed regarding the mechanism of emission of electromagnetic waves by a completely black body and bodies that can be considered close to completely black in terms of their emissivity. Such a new idea was proposed in 1900 by a German theoretical physicist Max Planck (1858–1947), 1918 Nobel Prize winner in Physics. He formulated the hypothesis that the radiation and absorption of energy by atoms and molecules does not occur continuously, as was believed at that time in classical physics, but in separate portions (quanta). Planck's quantum hypothesis is characterized by the following famous formula:

$$E = h\nu = hc/\lambda, \tag{5.9}$$

marked the beginning of the development of quantum mechanics. This formula contains a new fundamental physical constant—the so-called Planck's constant $h = 6.626 \cdot 10^{-34}\,J \cdot s$. According to formula (5.9), the change in radiation or absorption energy occurs by a minimum energy portion—a quantum of energy $h\nu = hc/\lambda$ or by portions that are multiples of this minimum value, i.e., equal to $nh\nu = nhc/\lambda$, where $n = 1, 2, 3, \ldots$ is an integer number.

This hypothesis allowed Max Planck to obtain ***the law of radiation of an absolutely black body***, which is described by the following formula for the dependence of the spectral density of the radiating emittance of a body (or monochromatic radiation coefficient) ε_λ on the wavelength λ of thermal electromagnetic radiation and temperature T:

$$\varepsilon_\lambda(T) = \frac{2\pi hc^2}{\lambda^5} \cdot \frac{1}{e^{hc/kT\lambda} - 1}, \tag{5.10}$$

where c $= 2.998 \cdot 10^8$ m/s is the speed of light in a vacuum, $k = 1.381 \cdot 10^{-23}\,J/K$ is the Boltzmann constant.

Max Planck (1858-1945)

If in formula (5.10) a variable is replaced, namely, to which from wavelength λ to frequency $\nu = c/\lambda$, then it is possible to obtain the following Planck's formula for the frequency dependence of the spectral density of the radiating emittance, which is equivalent to expression (5.10):

$$\varepsilon_\nu = \frac{2\pi \nu^2}{c^2} \cdot \frac{h\nu}{e^{h\nu/kT} - 1} \tag{5.11}$$

The curves $\varepsilon_\lambda = f\ (\lambda,\ T)$ described by formula (5.10) for the Planck radiation law are fully consistent with the experimental curves shown in Fig. 5.5. In addition, it should be noted that from formula (5.11) for Planck's radiation law, two partial limiting cases directly follow, namely:

1. For relatively low frequencies $\nu \ll kT/h$, by expanding the exponential function $\exp(h\nu/kT) \approx 1 + h\nu/kT$ into a Taylor series, while preserving the linear term in the small exponent $h\nu/kT \ll 1$, we obtain the Rayleigh-Jeans equation for the frequency dependence of the spectral density of the radiating emittance

$$\varepsilon_\nu = (2\pi kT/c^2)\nu^2. \tag{5.12}$$

This law was independently discovered by John William Strett (Lord Rayleigh), winner of the 1904 Nobel Prize in Chemistry, and James Hopwood Jeans both on the basis of classical physics, using the notion of a uniform distribution of energy over the degrees of freedom, and on the basis of quantum theory as the above limiting case of Planck's radiation law.

John William Strett - Lord Rayleigh (1842-1919)

The Rayleigh-Jeans law is valid only for low frequencies. For high frequencies $\nu \gg kT/h$, the Rayleigh-Jeans law (5.12) gives an increase in the frequency dependence of the energy luminosity directly proportional to the square of the frequency: $\varepsilon_\nu \sim \nu^2$. This result of the Rayleigh-Jeans law contradicts experimental data (see Fig. 5.5) and is called the "ultraviolet catastrophe".

2. For relatively high frequencies $\nu \gg kT/h$, the exponential function in the denominator of the Planck formula for the frequency dependence of the energy luminosity becomes significantly greater than unity ($\exp(h\nu/kT) \gg 1$), which allows us to obtain the following expression from formula (5.10):

$$\varepsilon_\nu = \frac{2\pi h\nu^3}{c^2} \cdot \exp(-h\nu/kT). \tag{5.13}$$

Formula (5.13) is called Wien's formula, which was obtained by Wilhelm Wien in 1896 in a similar analytical form regarding the frequency dependence of the spectral density of the radiation emittance.

$$\varepsilon_\nu = C_1 \nu^3 \exp(-C_2 \nu / T), \tag{5.14}$$

where C_1 and C_2 are constant coefficients that naturally did not contain Planck's constant. Wien's formula was obtained using the ideas of classical physics 4 years before the appearance of Planck's quantum theory of thermal radiation. This formula described the frequency dependence of the spectral density of the radiation emittance of an absolutely black body in accordance with experimental data only at high frequencies.

Stefan-Boltzmann law. We will use Planck's radiation law (5.11) for the frequency dependence of the spectral density of the radiation emittance in order to obtain a formula for the integral energetic emittance of an absolutely black body. For this, we will use formula (5.4), which gives the relationship between the spectral density of radiation emittanceand the integral energetic emittance of an absolutely black body bodies, and calculate the following integral:

$$R = \int_0^\infty \varepsilon_\nu d\nu = \frac{2\pi h}{c^2} \int_0^\infty \frac{\nu^3 d\nu}{e^{h\nu/kT} - 1} \tag{5.15}$$

To carry out further calculations, it is convenient to replace the variable $\frac{h\nu}{kT} = x$, which allows to write the formula (5.15) in the form

$$\begin{aligned} R &= \frac{2\pi h}{c^2} \left(\frac{kT}{h} \right)^4 \int_0^\infty \frac{\left(\frac{h\nu}{kT}\right)^3 d\left(\frac{h\nu}{kT}\right)}{e^{h\nu/kT} - 1} \\ &= \frac{2\pi k^4 T^4}{h^3 c^2} \int_0^\infty \frac{x^3 dx}{e^x - 1} \end{aligned} \tag{5.16}$$

Calculation of the last integral in the right-hand side of expression (5.16) gives

$$\int_0^\infty \frac{x^3 dx}{e^x - 1} = \frac{\pi^4}{15}, \tag{5.17}$$

which after substituting (5.17) into (5.16) allows us to finally obtain the following formula:

$$R = \frac{2\pi^5 k^4}{15 h^3 c^2} T^4 = \sigma T^4 \tag{5.18}$$

The resulting expression (5.18) is ***the Stefan-Boltzmann law***. which is formulated as follows: ***the integral energetic emittance of an absolutely black body is directly proportional to the fourth power of its absolute temperature T.*** The coefficient of proportionality σ in the formula (5.18) of the Stefan-Boltzmann law $R = \sigma T^4$ is called the Stefan-Boltzmann constant, which is equal to $\sigma = \frac{2\pi^5 k^4}{15h^3c^2} = 5,67 \cdot 10^{-8}\,\mathrm{W/m^2K^4}$.

Josef Stefan (1835-1893)

As mentioned earlier, this law was first established experimentally by an Austrian physicist, mathematician, and poet Josef Stefan (1835–1893), and then derived theoretically by an Austrian theoretical physicist and philosopher Ludwig Boltzmann (1844–1906) (see portrait of Ludwig Boltzmann in Chap. 3). The Stefan-Boltzmann law can be applied to the thermal radiation of the human body using an approximation in which the human body is assumed to be a "gray body". In this case, instead of the Stefan-Boltzmann constant σ, the radiation coefficient δ is used, and the Stefan-Boltzmann law takes the form

$$R = \delta T^4, \tag{5.19}$$

where the radiation coefficient $\delta = \alpha\sigma$ is equal to the product of the Stefan-Boltzmann constant σ by the absorption coefficient α, which for gray bodies takes a numerical value less than one.

Wien's displacement law. This law, which was established in 1893 by a German physicist Wilhelm Wien (1864–1928), 1911 Nobel Prize winner in Physics, gives a relationship between temperature and the wavelength λ_{max} (or frequency ν_{max}) that corresponds to the maximum spectral density of the radiation emittance in the spectrum of thermal radiation of an absolutely black body.

Wilhelm Wien (1864-1928)

On the basis of formulas (5.10) and (5.11) of Planck's radiation law for the spectral density of the radiation emittance ε_λ or ε_ν, as well as the extremum conditions for the derivatives from ε_λ with respect to the wavelength the wavelength the wavelength λ (or from ε_ν with respect to the frequency ν)

$$\frac{d\varepsilon_\lambda}{d\lambda} = 0, \quad \frac{d\varepsilon_\nu}{d\nu} = 0, \tag{5.20}$$

it is possible to obtain Wien's law in the following 2 equivalent forms:

$$\lambda_{max} = \frac{b}{T}, \tag{5.21}$$

$$\nu_{max} = \frac{c}{b} T, \tag{5.22}$$

where the coefficient $b = 2{,}9 \cdot 10^{-3}$ m·K is called Wien's constant, and $c = 3 \cdot 10^{-8}$ m/s is the speed of light in a vacuum.

According to formulas (5.21) and (5.22), Wien's law has the following two equivalent formulations:

1. ***The wavelength at which the maximum spectral density of the energy luminosity of an absolutely black body falls is inversely proportional to its absolute temperature.***
2. ***The frequency at which the maximum spectral density of the energy luminosity of an absolutely black body falls is directly proportional to its absolute temperature.***

Thus, based on the experimental curves of energy distribution in the spectra of thermal radiation of an absolutely black body (see Fig. 5.5), the following conclusion can be drawn: ***with increasing temperature, the wavelength λ_{max} (or frequency ν_{max}), which characterizes the maximum spectral density of the radiation emittance, shifts toward shorter wavelengths (or higher frequencies).*** It is this fact that gives reason to call Wien's law the "***displacement law***".

5.3.3 Thermal Radiation of the Sun, Infrared and Ultraviolet Radiations

Solar thermal radiation. The thermal radiation of the Sun is a necessary condition for the existence of life on Earth. It is generally accepted that the source of the solar thermal energy is thermonuclear reactions, during which hydrogen is converted into helium. There are several cycles as a result of which such a transformation occurs. The simplest and most popular of these cycles is the so-called Bethe cycle or carbon cycle, discovered by the German theoretical physicist Hans Bethe in 1938. Without dwelling in detail on each of the six nuclear reactions of the Bethe cycle, it should be noted that the result of this cycle is the thermonuclear fusion of the alpha particle He with the participation of four protons, carbon, oxygen and nitrogen nuclei at intermediate stages.

Curve 1 in Fig. 5.6 sets the distribution of energy in the spectrum of the thermal radiation of the Sun at the upper boundary surface of the Earth's atmosphere. At the same time, for 1 square meter of this surface of the Earth's atmosphere, the power of the solar radiation flux is equal to 1350 watts, which is called the solar constant. The maximum spectral density ε_λ of the thermal radiation of the Sun falls at a wavelength $\lambda_{m1} = 470$ nm (see Fig. 5.6), which, in accordance with Wien's law (5.21), corresponds to the temperature of the radiative surface of the Sun, being equal to 6100 K. Comparison of curve 1 in Fig. 5.6 with the curves of the spectral density ε_λ of the thermal radiation of a black body presented in Fig. 5.5 shows their complete

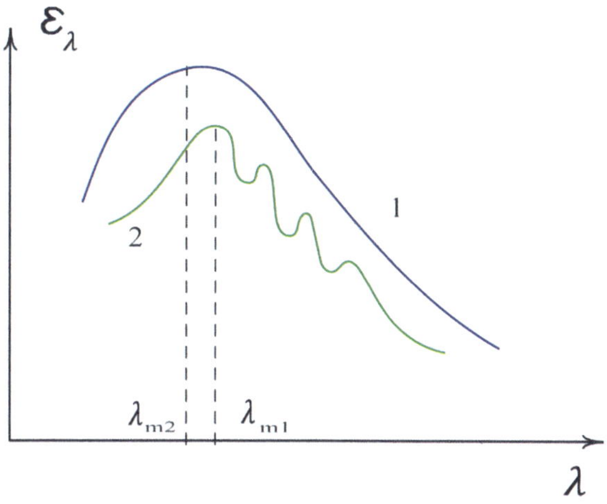

Fig. 5.6 The spectral density of solar thermal radiation: at the upper boundary of the Earth's atmosphere (curve 1); directly at the Earth's surface (curve 2)

similarity. This implies an important conclusion that the Sun, like a platinum crucible at a temperature of T = 2046.6 K, is also the standard of an absolutely black body.

Curve 2 in Fig. 5.6 describes the distribution of the spectral density ε_λ of the thermal radiation of the Sun directly near the surface of the Earth. This curve has a number of minima, which correspond to the absorption lines of solar radiation in the Earth's atmosphere. The maximum spectral density ε_λ of the thermal radiation of the Sun does not exceed 1120 W/m^2 near the surface of the Earth in the equator region and corresponds to the wavelength $\lambda_{m2} = 555$ nm (see Fig. 5.6). It should be noted the role of the specified value the wavelength for the process of biological evolution of the human visual organ, as a result of which the maximum spectral sensitivity of the human eye corresponds to this yellow-green part of solar radiation with the wavelength $\lambda_{m2} = 555$ nm.

Infrared thermal radiation. Electromagnetic radiation in the wavelength range from $\lambda = 0.76$ μm to $\lambda = 2000$ μm is called infrared (IR) radiation. Usually, the whole range of infrared radiation is divided into the following three zones: close IR zone from $\lambda = 0.76$ μm to $\lambda = 2.5$ μm, middle IR zone from $\lambda = 2.5$ μm to $\lambda = 50$ μm and distant IR zone from $\lambda = 50$ μm to $\lambda = 2000$ μm.

Using Wien's displacement law in the form

$$T = b/\lambda_{max}, \tag{5.23}$$

one may find the temperature interval, in which the sources of infrared thermal radiation should be located. For this purpose, let us substitute into formula (5.23) the extreme values of the IR wavelength interval. As a result, we obtain that the minimum value of the IR wavelength $\lambda = 0.76$ μm corresponds to the maximum temperature $T \approx 3810$ K, while the maximum value of the IR wavelength $\lambda = 2000$ μm corresponds to the minimum temperature $T \approx 1.5$ K. Thus, the results obtained mean that virtually all liquids and solids are sources of infrared thermal radiation.

The therapeutic effect of infrared radiation is mainly connected with its thermal influence. Infrared thermal radiation penetrates the human body to a depth of several centimeters, improving heat regulation, blood supply and other vital functions.

Ultraviolet thermal radiation. Electromagnetic radiation in the wavelength range from $\lambda = 10$ nm to $\lambda = 400$ nm is called ultraviolet (UV) radiation. Usually, the whole range of ultraviolet radiation is divided into the following four zones: UV zone A from $\lambda = 400$ nm to $\lambda = 315$ nm, UV zone B from $\lambda = 315$ nm to $\lambda =$ 280 nm, UV zone C from $\lambda = 280$ nm to $\lambda = 200$ nm and UV vacuum zone from λ $= 200$ nm to $\lambda = 10$ nm. The explanation for the name "vacuum" of the last zone is due to the fact that UV thermal radiation with $\lambda < 200$ nm is strongly absorbed by the molecules that make up the air. A direct consequence of this circumstance is the study of the properties of UV thermal radiation in vacuum for the wavelength range from 10 to 200 nm.

Using Wien's displacement law in the form (5.23), it can be determined that the sources of ultraviolet thermal radiation are bodies that are heated to a temperature from $T \approx 7250$ K (to obtain UV radiation with the maximum value of $\lambda = 400$ nm) to $T \approx 290000$ K (to obtain UV radiation with the minimum value of $\lambda = 10$ nm). This result practically means that in the immediate environment of us there can be no sources of significant UV thermal radiation. Of course, except for the Sun! Indeed, the Sun is a very powerful source of UV thermal radiation. It is interesting to note that 9% of all solar radiation reaching the upper boundary of the earth's atmosphere is ultraviolet thermal radiation.

The medical effect of UV radiation is mainly connected with its specific biophysical action due to photochemical and other reactions. In particular, in the UV zone A, processes of pigment formation occur, which determines the color of the human skin. In the UV zone B, processes occur that have an antirachitic effect at low doses of radiation, while at high doses of radiation in the same zone, a dangerous process of carcinogenesis can begin. Zone C is used to prevent viral diseases due to the bactericidal effect of ultraviolet radiation, which is inherent in electromagnetic waves in this range.

5.3.4 Non-equilibrium Thermodiffusion Processes in Systems of Inanimate and Living Nature, Exponential Temperature Relaxation of Thermal Radiation

Here, we consider non-equilibrium thermomechanical (thermodiffusion) processes in a one-component two-phase system. As is known (see, for example, monographs [2–4]), the conditions for equilibrium in such a system are the equality of pressures, temperatures, and chemical potentials of both phases. In the future, we will consider the case in which only the first condition of isobaricity is fulfilled, i.e. $p = const$, $T_1 \neq T_2$, $\mu_1 \neq \mu_2$. Then two flows J_i (i = 1,2) arise in such a system, namely: the diffusion flow of particles $J_1 = J_n$ and the flow of heat $J_2 = J_q$, which are directly connected with the presence of thermodynamic forces, which are determined by the following expressions $X_1 = -\nabla\mu/T$ and $X_2 = -\nabla T/T^2$. This choice of thermodynamic forces is related to gradients of chemical potential $\nabla\mu$ and temperature ∇T, which are assumed to be small enough, and to the following formula for entropy production:

$$\sigma = \partial S/\partial t = \sum_i J_i X_i. \tag{5.24}$$

It should be noted that expression (5.24) along with a number of other results is the basis of Onsager's theorem on reciprocity, or the principle of symmetry of kinetic coefficients. It was for this theorem that Lars Onsager (USA) received the 1968 Nobel Prize in Chemistry "for the discovery of reciprocity relations named after him, which are of fundamental importance for the thermodynamics of irreversible processes" [4].

Physical and mathematical details of the proof of Onsager's theorem can be found, for example, in [2, 4]) The starting relation for Onsager's theorem is the so-called linear law, which is a generalization of known empirical facts formulated in the form of Fick's, Fourier's, Ohm's laws, which define direct transport processes in which gradients of certain thermodynamic parameters (concentration, temperature, electric field potential, etc.) cause flows of the same parameters (diffusion flow, heat flow, electric charge flow or electric current density), etc. In addition, there are also indirect (cross-) transfer processes, where the gradient of one thermodynamic parameter causes a flow of another thermodynamic parameter. For example, a process called thermodiffusion or the Soret effect, when a temperature gradient causes a diffusion flow. The reverse cross process is called the Dufour effect, for which a concentration gradient causes a heat flow.

The linear law of thermodynamics of irreversible processes, which combines direct and cross transfer processes, states that each flow J_i is a linear combination of thermodynamic forces X_k, i.e.

$$J_i = \sum_k^N L_{ik} X_k, \tag{5.25}$$

where N is the total number of flows J_i and thermodynamic forces X_k, and the coefficients L_{ik} are called kinetic coefficients, or Onsager coefficients.

Onsager's theorem regarding the reciprocity relationship, or the principle of symmetry of kinetic coefficients, is formulated as follows: with a certain choice of thermodynamic forces and flows, the matrix of kinetic coefficients is symmetric, that is, the relation holds

$$L_{ik} = L_{ki}. \tag{5.26}$$

It should be noted that the linear law and, accordingly, the principle of symmetry of kinetic coefficients (Onsager's theorem) remain valid for relatively small deviations of the thermodynamic states of the system from the equilibrium position, when the gradients of the physical parameters are sufficiently small and quadratic and higher-order contributions to thermodynamic forces X_k can be neglected.

For the thermodiffusion phenomena considered above, the total number of flows and thermodynamic forces is 2, and in this case, there are only 3 independent kinetic coefficients, and two of them L_{11}, L_{22} describe direct transport processes corresponding to diffusion and heat conduction, and one kinetic coefficient $L_{12} = L_{21}$ corresponding to cross Soret processes and Dufour. Later, instead of these three independent Onsager kinetic coefficients, it makes sense to switch to empirical transfer coefficients, which are the diffusion coefficient D, the thermal conductivity coefficient λ, and the thermodiffusion ratio K_T. Note that the diffusion coefficient D arises as a proportionality coefficient in the direct transfer process between the diffusion flux J_n of the number of particles and the gradient of the density ρ of the number of particles. The corresponding linear law was established by Adolf Fick in 1855. The thermodiffusion ratio K_T, more precisely, K_T/T appears as the proportionality factor between the diffusion flux J_n and the temperature gradient ∇T in the second term of the following linear law:

$$J_n = -D[\nabla\rho + (K_T/T)\nabla T], \tag{5.27}$$

which describes the Soret thermodiffusion effect.

The second equation of the linear law for heat flow has the following form:

$$J_q = [(\partial\mu/\partial\rho)_T\delta^*/D]J_n - \lambda\nabla T, \tag{5.28}$$

where δ^* due to the indicated kinetic coefficient responsible for the Dufour effect, while λ due to the indicated thermal conductivity coefficient.

Two Eqs. (5.27) and (5.28) contain four unknown quantities—diffusion flow J_n, heat flow J_q, density $\rho\{r, t)$ and temperature $T(r, t)$, as well as three parameters—diffusion coefficient D, thermal conductivity coefficient λ and thermal diffusion ratio K_T. The kinetic coefficient δ^* in formula (5.28) is not an independent parameter due to the Onsager reciprocity ratio (5.26) between the kinetic coefficients in the Soret and Dufour effects, its explicit form through other kinetic coefficients is given in [10].

For the completeness of the system of equations that describe thermomechanical non-equilibrium phenomena, let us add the equation of discontinuity

$$\partial\rho/\partial t = -div\, J_n \tag{5.29}$$

and the entropy balance equation, which is written in isobaric conditions $p = const \neq f(t)$ as follows:

$$\rho C_p \partial T/\partial t = -div\, J_q + \mu\, div\, J_n. \tag{5.30}$$

The resulting system of Eqs. (5.27)–(5.30) is complete for describing thermo-mechanical non-equilibrium phenomena. Consider the partial case when there is no diffusion flow in the isobaric system, i.e., $J_n = 0$. Then from Eq. (5.30) we have $J_q = -\lambda \nabla T$, which gives the following result after substituting this expression into the entropy balance Eq. (5.30):

$$\partial T/\partial t = \chi \Delta T + \nabla T \nabla \lambda / \rho C_p. \tag{5.31}$$

Here, the coefficient of the thermal diffusivity

$$\chi = \lambda/\rho C_p, \tag{5.32}$$

is introduced, where λ is the coefficient of the thermal conductivity and C_p is the isobaric heat capacity.

If the coefficient of thermal conductivity λ does not depend on the spatial coordinates, i.e. $\lambda = const \neq f(r)$ and, therefore, $\nabla\lambda = 0$, then Eq. (5.31) turns into the following differential equation of thermal conductivity:

$$\partial T/\partial t = \chi \Delta T. \tag{5.33}$$

In the case when the initial value of the temperature is given $T(r, t = 0) = T_0(r)$, then the solution of the corresponding Cauchy problem for the differential equation (5.33) of the parabolic type in partial derivatives has the following form:

$$T(r,t) = \int \frac{T_0(r')}{(4\pi\chi t)^{3/2}} exp[-(r-r')^2/4\chi t]dr'. \tag{5.34}$$

Let us find the temperature relaxation time τ_{relax} using the differential equation of thermal conductivity (5.33) for the one-dimensional case when the following approximation is fulfilled:

$$\Delta T = \partial^2 T/\partial x^2 \approx T/L^2, \tag{5.35}$$

where L is the characteristic linear size of the region of smooth temperature inhomogeneity. We will focus on the process of temperature relaxation of the body, which is heated to a temperature T_0 at the initial moment of time $t = 0$ that exceeds the temperature T_{inv} of the external environment. In this case, the temperature relaxation of such a body will correspond to a negative value of the derivative $\partial T/\partial t$, since over time the temperature $T(t)$ of the body should decrease, approaching the external temperature T_{inv}. Taking into account the above, the differential Eq. (5.33), which describes the relaxation of the thermal radiation of such a body under condition (5.35), takes the following form:

$$\partial T/\partial t = -\chi T/L^2. \tag{5.36}$$

The corresponding solution of Eq. (5.36), which satisfies the initial condition $T(t = 0) = T_0$, corresponds to the following exponential dependence of the temperature change T over time t:

$$T(t) = T_0 \exp(-\chi t/L^2). \tag{5.37}$$

Thus, the temperature relaxation time τ_{relax} is described by the formula

$$\tau_{relax} = L^2/\chi = \rho C_P L^2/\lambda, \tag{5.38}$$

from which it follows that the temperature relaxation time τ_{relax} increases with an increase in the density ρ (concentration) of the number of particles and isobaric heat capacity C_p, as well as with a decrease in the coefficients of thermal conductivity λ and thermal diffusivity χ.

5.3.5 *Numerical Estimates of the Wavelength and Integral Energetic Emittance of Human Body*

For the numerical estimation of the wavelength λ_{max}, which corresponds to the maximum value of the radiation emittance in the spectrum of the human body, we will use Wien's displacement law. As mentioned above, the surface layer of human skin has an approximate average temperature of $t = 32$ °C on the Celsius scale or T = 305 K on the Kelvin scale. Then we have for λ_{max} such a numerical value

$$\lambda_{max} = \frac{2.9 \cdot 10^{-3}}{305} \mathrm{m} \approx 10^{-5}\,\mathrm{m} = 10\,\mu\mathrm{m}. \tag{5.39}$$

It follows from formula (5.39), the wavelength λ_{max} of the human body falls on the infrared range of electromagnetic waves, which is not perceived by the human eye, since a person sees the surrounding world in a rather narrow interval of wavelengths of the visible range of electromagnetic radiation from $\lambda = 0,40\,\mu\mathrm{m}$ (blue light) to $\lambda = 0,76\,\mu\mathrm{m}$ (red light). It should be noted that the Earth's atmosphere has a so-called "window of transparency" has a so-called "window of transparency" for the wavelength of thermal infrared radiation of the human body $\lambda_{max} = 10\,\mu\mathrm{m}$, that is, this wavelength is almost not absorbed by atoms and molecules that are part of atmospheric air.

In addition to the calculated wavelength λ_{max}, which characterizes the maximum value of the emissivity of the human body, the use of other laws of thermal radiation (more precisely, the Stefan-Boltzmann law) allows one to estimate the surface power of thermal radiation of the human body. Strictly speaking, the application of the Stefan-Boltzmann law to real bodies is, of course, a rather significant approximation. The emissivity of real bodies (especially given their organic nature) must depend on many factors (the geometric shape of the body, the condition of its surface, changes in temperature, the course of various physiological processes, etc.), which are not taken into account in the Stefan-Boltzmann law.

At the same time, taking into account certain limitations of this approach and considering the surface of the human body as a "gray" body, one can use the following estimation formula for the integral energetic emittance of the human body:

$$R = \delta\,(T^4_{\text{body}} - T^4_{inv}), \tag{5.40}$$

where $T_{\text{body}} = 305$ K is the surface temperature of the human body, $T_{\text{inv}} = 293$ K is the temperature of the surrounding environment, and the coefficient $\delta = \alpha\sigma$ (it is also called the emissivity coefficient) is equal to the product of the Stefan-Boltzmann constant σ and the absorption coefficient α, which for the human body, as mentioned above, has a value of 0.9 in the infrared part of the spectrum. Thus, the use of the Stefan-Boltzmann law in the form (5.40) proves a rapid increase of heat radiation power with the rise of body temperature.

Both basic laws of thermal radiation—Stefan-Boltzmann law and Wien law underlie the method of ***thermography***, i.e., diagnostic method that uses heat radiation of human body for determination of local temperature. The considerable advantage of thermography as compared with other methods (for example, methods of radio-diagnosis) is in that at thermographic examination we use intrinsic heat radiation of human body and not external radiation which is often accompanied by heavy dose burden. Determination of temperature of body surface in thermography may be carried out in two ways. The first approach lies in application of liquid–crystal indicators, which optical properties are very sensitive to any temperature fluctuations. Placing such indicators on patient's body we may visually determine local deviation of temperature values. The second approach lies in visualisation of temperature field of body surface using devices called ***thermovision cameras (infrared scanners)***. The main principle of this approach lies in correlation between level of signal registered by the device and radiation property of body surface, which, in its turn, depends on temperature according to Stefan-Boltzmann law.

As a result of substitution of the given numerical values of the emissivity coefficient $\delta = \alpha\sigma$ and temperatures T_{body} and T_{inv} to the formula (5.40), we obtain the following estimate of the surface radiation emittance (surface power) R of the infrared thermal radiation of the human body: $R = 64.5\,\text{W/m}^2$.

Taking into account that the average value of the entire surface of the human body $S \approx 1.5m^2$, we have the numerical estimate for the power P_{body} of the thermal radiation of the human body in the infrared range of electromagnetic waves: $P_{body} = R \cdot S \approx 100$ Watt. It should be noted that thermal radiation in the infrared spectrum of electromagnetic waves is the most significant source (channel) of the physical fields of the human body.

5.3.6 Parameters of the Temperature Relaxation Process of Thermal Radiation of the Human Body

This paragraph will analyze the experimental results of temperature relaxation of thermal radiation of human body (strictly saying, human skin) presented in Fig. 5.4.[2] The main result of this analysis will be the finding of numerical values of the parameters that characterize the process of temperature relaxation of thermal radiation, namely: the damping decrement δ^* and the logarithmic decrement λ^* of damping temperature oscillations; the period T^* of damping temperature oscillations; the damping coefficient β of temperature oscillations; the relaxation time τ_{relax} of damping temperature oscillations; the frequency ω of damping oscillations and the frequency ω_0 of own oscillations of the thermal radiation temperature. In addition, the obtained numerical values of the specified parameters will provide an opportunity to write down in explicit form a differential equation of the 2nd order with constant coefficients, which characterizes the experimentally investigated process of temperature relaxation of thermal radiation of the human body.

The damping decrement δ^*. Numerical results of experimental data processing, which are presented in Fig. 5.4. allow finding the value of the damping (attenuation) decrement δ^* according to the following formula:

$$\delta^* = \frac{A(t_1)}{A(t_2)} = \frac{A_0 \exp(-\beta t_1)}{A_0 \exp[-\beta(t_1 + T^*)]} = \exp(\beta T^*), \tag{5.41}$$

where $A(t_1)$ and $A(t_2)$ are two successive amplitudes of decaying oscillations, which persist in time for the period T^*, i.e., $t_2 - t_1 = T^*$. Based on the experimental values of successive amplitudes $A(t_1) = 1.900$ and $A(t_2) = 0.600$, which are presented in Fig. 5.4, we get the numerical value of the damping decrement $\delta^* = 3.167$.

The logarithmic decrement λ^* ***of damping temperature oscillations***. The numerical value of the logarithmic decrement λ^* of the damping temperature oscillations is given using the following formula:

$$\lambda^* = \ln \delta^* = \beta T^*, \tag{5.42}$$

which is directly obtained from formula (5.41). Hence, we have that the numerical value of the logarithmic decrement of the damping temperature oscillations is equal to $\lambda^* = 1.153$.

The period T^* ***of damping temperature oscillations.*** As can be seen from the graph of damping oscillations in Fig. 5.4, successive amplitudes $A(t_1)$ and $A(t_2)$ are separated

[2] The Assistant Professor, PhD Natalia Grytsenko (Dept. Medical and Biological Physics and Informatics of the Bogomolets National Medical University, Kyiv, Ukraine), as well as students Victoria Khmil and Yevhen Kornytskyi (the Faculty of Medicine of the Bogomolets National Medical University, Kyiv, Ukraine) also took part in this experiment.

from each other in time by a distance $t_2 - t_1 = 9$ min. Thus, the period of decaying oscillations is equal to $T^* = 9$ min.

The damping coefficient β of temperature oscillations. By substituting the numerical values of the logarithmic decrement λ^* of damping temperature oscillations and the period T^* of damping oscillations into formula (5.42), we find the following numerical value of the damping coefficient β of temperature oscillations of the thermal radiation of the human body: $\beta = \lambda^*/T^* = 0.128\,\text{min}^{-1}$.

The relaxation time τ_{relax} of damping temperature oscillations. The inverse value of the damping coefficient β of temperature oscillations determines the relaxation time τ_{relax} of the damping temperature oscillations of the thermal radiation of the human body: $\tau_{relax} = 1/\beta = 7.813\,\text{min}$. The obtained value of the relaxation time τ_{relax} means that approximately every 470 s the temperature deviation of the thermal radiation will decrease by $\text{e} = 2.718$ times.

The frequency ω of damping temperature oscillations. The frequency ω of damping temperature oscillations of the thermal radiation of the human body is calculated from the known period T^* of damping temperature oscillations according to the formula $\omega = 2\pi/T^* = 0.698\,\text{rad/min}$.

The frequency ω_0 of own temperature oscillations. The frequency ω_0 of own temperature oscillations is related to the frequency ω of damping temperature oscillations and the damping coefficient β of temperature oscillations according to the following formula:

$$\omega_0 = (\omega^2 + \beta^2)^{1/2}. \tag{5.43}$$

which gives the numerical value of the frequency ω_0 of own temperature oscillations of the thermal radiation of the human body: $\omega_0 = 0.710\,\text{rad/min}$.

The precise form of the differential equation of damping temperature oscillations. A differential equation describing damping temperature oscillations has the following general form:

$$\frac{d^2T}{dt^2} + 2\beta\frac{dT}{dt} + \varpi_0^2 T = 0. \tag{5.44}$$

By substituting the values of the found parameters $\beta = 0.128\,\text{min}^{-1}$ and $\omega_0 = 0.710\,\text{rad/min}$ into (5.44), we obtain the following precise form of the differential equation of the 2nd order with constant coefficients:

$$\frac{d^2T}{dt^2} + 0.256\frac{dT}{dt} + 0.504T = 0, \tag{5.45}$$

which characterizes the experimentally investigated process of temperature relaxation of thermal radiation of the human body in accordance with the experimental data presented in Fig. 5.4.

5.4 Conclusions

Thermal radiation, which was discussed in this chapter, is electromagnetic radiation arising from the presence of internal energy due to the thermal movement of atoms and molecules of all heated bodies at positive absolute temperatures. In section 5A at the popular scientific level and in section 5B at the advanced scientific level, the main physical characteristics and fundamental laws of thermal radiation were considered. The study of thermal radiation played an outstanding role in the development of physical science, laying the foundations for quantum mechanics (see Sects. 5.2.2 and 5.3.2). In Sect. 5.3.3, some important medical applications of the thermal radiation of the Sun, the spectrum of which being close to the spectral features of an absolutely black body, were discussed.

The experimental data obtained in Sect. 5.2.3 demonstrated that the temperature relaxation of the thermal radiation of the human skin in the infrared range of electromagnetic waves was characterized by damping temperature oscillations with parameters given in Table 5.1.

The physical properties of thermal radiation find their use in medicine, in particular for diagnostic purposes. Inflammatory processes change the local body temperature. Therefore, remote registration of electromagnetic radiation signals from different parts of the human body allows non-contact determination of their temperature. Summing up, it should be emphasized the important fact that the use of thermal radiation of the human body as a diagnostic method of thermography or thermal imaging has a fundamental advantage compared to other diagnostic methods.

Indeed, a method of thermography uses the body's own (and therefore—non-invasive) thermal radiation, while such diagnostic methods as X-ray computer tomography (CT), magnetic resonance imaging (MRI), positron emission tomography

Table 5.1 Parameters of temperature relaxation of thermal radiation of the human skin

NNo	Parameter	Formula and numerical value
1	The damping decrement δ^*	$\delta^* = \exp(\beta T^*) = 3.167$
2	The logarithmic decrement λ^*	$\lambda^* = \ln \delta^* = \beta T^* = 1.153$
3	The period T^* of damping oscillations	$T^* = 9$ min
4	The damping coefficient β	$\beta = \lambda^*/T^* = 0.128\,\text{min}^{-1}$
5	The relaxation time τ_{relax}	$\tau_{relax} = 1/\beta = 7.813$
6	The frequency ω of damping oscillations	$\omega = 2\pi/T^* = 0.698\,\text{rad/min}$
7	The frequency ω_0 of own oscillations	$\omega_0 = (\omega^2 + \beta^2)^{1/2} = 0.710\,\text{rad/min}$

(PET) and even ultrasound examinations use external (and therefore—invasive to a different extent) radiation that can cause, under certain conditions, quite significant dose loads on the organs and tissues of the human body.

References

1. *Medical and Biological Physics*/Textbook for students of higher medical institutions. Edited by Prof. A.V. Chalyi, 2nd edition (Knyga plus, Kyiv, 2005)
2. S.R. de Groot, P. Mazur, *Non-Equilibrium Thermodynamics* (Dover Publications, New York, 2013)
3. L.A. Bulavin, L.G. Grechko, L.B. Lerman, A.V. Chalyi, *Medical Physics. Dynamic and Statistical Models* (Publising Center "Kyiv University", Kyiv, 2011)
4. AV. Chalyi, *Non-Equilibrium Processes in Physics and Biology* (Naukova Dumka, Kyiv, 1997)
5. *Wilhelm Wien*, MLA style: The Nobel Prize in Physics 1911. NobelPrize.org. Nobel Prize Outreach AB (2023). https://www.nobelprize.org/prizes/physics/1911/summary/
6. *Max Planck*. MLA style: The Nobel Prize in Physics 1918. NobelPrize.org. Nobel Prize Outreach AB (2023). https://www.nobelprize.org/prizes/physics/1918/summary/
7. *Albert Einshtein*. MLA style: The Nobel Prize in Physics 1921. NobelPrize.org. Nobel Prize Outreach AB (2023). https://www.nobelprize.org/prizes/physics/1921/summary/
8. A.V. Chalyi, S.P. Samko, Membrane cell potentials with thermodiffusion distribution of ion concentration. Phys. Alive **10**(2), 66 (2002)
9. A.V. Chalyi, S.P. Samko, Electrical membrane rest potentials with the temperature gradients taking into account. Ukr. J. Phys. **48**(2), 365 (2003)
10. A.V. Chalyi, S.P. Samko, Influence of an external electrical field on membrane potentials. Ukr. J. Phys. **48**(9), 966 (2003)
11. A.V. Chalyi, A.N. Vasil'ev. Correlation properties, critical parameters and critical light \scattering in finite-size systems. J. Mol. Liquids **84** (2), 203 (2000)
12. K.A. Chalyy, K. Hamano, A.V. Chalyi, Correlating properties of a simple lquids at criticality in a reduced geometry. J. Mol. Liquids **92**(1–2), 153 (2001)
13. K.A. Chalyy, L.A. Bulavin, A.V. Chalyi, Dynamic scaling and central component width of critical opalescence spectrum in liquids with restricted geometry. J. Phys. Studies **9**(1), 66 (2005)
14. A.V. Chalyi, E.V. Zaitseva, Strange attractor in kinetic model of synaptic transmission. J. Phys. Studies **11**(3), 153 (2007)
15. A.V. Chalyi, A.N. Vasilev, E.V. Zaitseva, Synaptic transmissionas as a cooperative phenomenon in confined systems. Cond. Matter Phys. **20**(1), 13804 (2017)
16. D.A. Gavryushenko, K.V. Cherevko, V.M. Sysoev, The influence of the chemical reactions on the diffusion phenomena in the cylincrical systems bounded with the membranes. J. Mol. Liquids **127**(1–3), 71 (2006)
17. K.V. Cherevko, D.A. Gavryushenko, O.V. Korobko, V.M. Sysoev, Entropy production in the diffusion of a Margules solution in a flat-parallel pore. Ukr. J. Phys. **58**(10), 988 (2013)
18. K.V. Cherevko, D.A. Gavryushenko, L.A. Bulavin, Entropy production in a model biological system with facilitated diffusion. Ukr. J. Phys. **66**(8), 714 (2021)
19. A.V. Chalyi, G.V. Hrapijchuk, L.M. Chernenko, K.A. Chalyy, E.V. Zaitseva, Barodiffusion phenomena in nanoscale liquid mixturein the critical region. Ukr. J. Phys. **55**, 1113 (2010)
20. A.V. Chalyi, E.V. Zaitseva, K.A. Chalyy, G.V. Hrapijchuk, Dimensional crossover and thermodynamic properties of nanoscale condensed matter. Ukr. J. Phys. **60**, 885 (2015)
21. J. Eccles, A.L. Hodgkin, A.F. Huxley, *MLA style: The Nobel Prize in Physiology or Medicine 1963*. NobelPrize.org.Nobel Prize Outreach AB 2023. https://www.nobelprize.org/prizes/medicine/1963/summary/

22. D. Hubel, C. Stevens, E. Kandel, et al. The Brain, Scientific American **10** (1979)
23. P.G. Kostyuk, O.A. Kryshtal, *Mechanisms of electrical excitability of nervous system* (Nauka, 1981)
24. A.V. Chalyi, L.M. Chernenko, Phase transitions in finite-size systems and synaptic transmission, in *Dynamic Phenomena at Interfaces, Surfaces and Membranes* Edited by D. Beysens, N. Boccara, G. Forgacs (Nova Sci. Publ., 1993)
25. H. Haken, *Principles of Brain Functioning: A Synergetic Approach to Brain Activity, Behavior and Cognition* (Springer Series in Sinergetics, 1995)
26. H. Haken, *Brain dynamics: An introduction to models and simulations* (Springer Series in Sinergetics, 2002)
27. J. Nolte,*The Human Brain.* (Mosby 2002)
28. V.Tsymbalyuk, V. Medvedev, *Man and His Brain* (Nova Knyga, 2022)
29. *Медична біологія* / Навчальний посібник для практичних занять. За редакцією О.В. Романенка.- Здоров'я, Київ (2005)
30. *Lars Onsager*. MLA style: The Nobel Prize in Chemistry 1968/ NobelPrize/org. Nobel Prize Outreach AB 2023. https://www.nobelprize.org/prizes/chemistry/1968/summary/
31. L.A. Bulavin, K.A. Chalyy, *Neutron Optics of Mesoscale Liquids* (Naukova Dumka, Kyiv, 2006)
32. *Modern Problems of Molecular Liquids*. Edited by L.A. Bulavin and A.V. Chalyi (Springer, Cham 2018)
33. A.V. Chalyi, L.A. Bulavin, K.A. Chalyy, On the neutron optics of liquids. J. Mol. Liquids **383**, 121979 (2023)
34. I.V. Lindell, Heaviside operational rules applicable to electromagnetic problems. Progress Electromag. Res. **26**, 293–331 (2000)
35. Schwartz Laurent, *Mathematics for the Physical Sciences*. Dover Books on Mathematics, New York: Dover Publications, pp. 215–241 (2008)

Chapter 6
Physical Principles of Radiation Medicine

Abstract This chapter deals with physical, biophysical and some medical problems related to the action of ionizing radiation on humans and other biological objects. Ionizing radiation is a general term that includes *X*-rays, all types of radioactive radiation, i.e. alpha, beta, and gamma radiation, as well as streams of particles (protons, neutrons, etc.). Radiation medicine (in a broad sense) studies the physical, chemical and biological mechanisms of the impact of ionizing radiation on the human body, methods of treatment of radiation damages and prevention of medical consequences of ionizing radiation. The main issues covered below are as following. In Sect. 6.2A—general information (at a qualitative level) about *X*-rays (Sect. 6.2.1), radioactive radiation (Sect. 6.2.2), ionizing radiation and its interaction with biological tissues (Sect. 6.2.3), physical, chemical and biological stages of radiation damage (Sect. 6.2.4). In Sect. 6.3B—a more detailed analysis (at the quantitative level) of deceleration (bremsstrahlung) and characteristic *X*-ray radiation (Sect. 6.3.1), theoretical problems of radioactive radiation (Sect. 6.3.2), theoretical foundations of ionizing radiation dosimetry (Sect. 6.3.3), theoretical basis of the interaction of ionizing radiation with biolodical tissues (Sect. 6.3.4). It is difficult to overestimate the importance of knowledge in this field for a modern doctor, especially in connection with the problems facing humanity today.

Keywords Ionizing radiation · X-ray radiation · Deceleration (bremsstrahlung) and characteristic spectra · Radioactive radiation · Dosimetry of ionizing radiation · Processes of interaction of ionizing radiation with biological tissues

> *This threat was created by science, but*
>
> *the real key to solving the problem before*
>
> *us is in the minds and hearts of people.*
>
> Albert Einstein

A. V. Chalyi, *What is Medicine?*, https://doi.org/10.1007/978-3-031-64979-0_6

6.1 Introduction

The words given in the epigraph were said by Albert Einstein about 70 years ago, assessing the threat of nuclear war and feeling a certain guilt before humanity, because he had inadvertently contributed to the creation of the atomic bomb with his discoveries. These words remain relevant even now at a time when discoveries are made that are not only outstanding in terms of their scientific significance, but may also pose a threat to the lives of individuals and the entire biosphere of the Earth.

Doctors have a special role in overcoming dangerous trends in the development of modern civilization. It is to them, first of all, that the words of another well-known physicist of the twentieth century, Viktor Weiskopf, should be attributed: "Human being relies simultaneously on two whales, which are knowledge and feelings. But feelings alone without knowledge are not effective, while knowledge alone without feelings is not humane."

Some physical and chemical as well as medical and biological problems will be discussed in this chapter, being associated with the effect of various components of ionizing radiation on humans and other living objects (see Fig. 6.1).

When writing this chapter, the author did not set himself the task of providing a detailed coverage of the entire topic devoted to the physical aspects of radiation medicine, since there are quite a few excellent monographs and textbooks on this topic (see, for exaple, [1–8], as well as our explanations contained in the introduction to Chap. 1). Therefore, here we will touch only on those individual topics that are reflected in the discipline "Medical and Biological Physics" for students of the Bogomolets National Medical University in Kyiv, being the most famous medical University in Ukraine. Of course, it should be noted that the choice of problems discussed in this Chapter was also dictated by the scientific publications of the author and his colleagues [9–11].

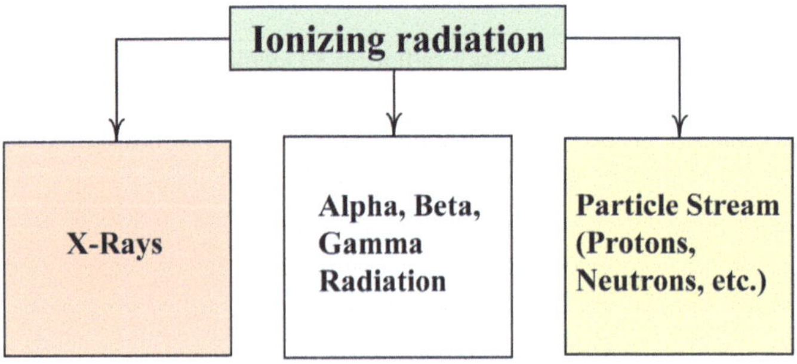

Fig. 6.1 Main components of ionizing radiation

6.2 Section A

6.2.1 X-ray Radiation

The history of the discovery of X-rays by Wilhelm Röntgen, the works of Ivan Puluj. Wilhelm Conrad Röntgen's discovery of X-rays in 1895 was indeed a groundbreaking moment in the history of science and medicine. Röntgen was experimenting with cathode rays when he noticed that a fluorescent screen in his lab began to glow even when it was not directly in the path of the cathode rays. He realized that some unknown radiation was passing through the vacuum and causing the screen to fluoresce. Further experiments revealed that these rays could penetrate materials like cardboard and even human tissue but were blocked by denser materials like bone and metal. Röntgen named these mysterious rays "X-rays," with the 'X' symbolizing their unknown nature. His discovery revolutionized various fields, particularly medicine. X-rays quickly became indispensable in medical diagnostics, allowing doctors to see inside the human body without invasive procedures. This breakthrough dramatically improved the diagnosis and treatment of injuries and illnesses. For his discovery, Röntgen was awarded the first Nobel Prize in Physics in 1901. The significance of his work continues to be felt today, with X-rays remaining a fundamental tool in medical imaging technology.

In connection with this outstanding event, it should be noted the significant contribution of the Ukrainian and Austro-Hungarian electrotechnical physicist Ivan Puluj, who constructed his own cathode-ray tube in the 80 s of the nineteenth century, which essentially became the first source of X-rays [12]. This statement is contained in research conducted by physicist Prof. Vasyl Shenderovsky [13] (see also the monograph by Roman Goida and Roman Plyatsko [14]). Let us dwell briefly on the biography of Ivan Puluj (1845–1918), a native of Hrymailiv (now this small town is located near the Ukrainian regional center of Ternopil, and in the middle of the nineteenth century it was part of the Austrian Empire). Puluj entered the theological faculty of the University of Vienna after high school. At the same time, he attends lectures on mathematics, physics, and astronomy, which were given at the Faculty of Philosophy. And he is so fascinated by them that after completing the theology course, he renounces the rank of a priest and is engaged in physical and mathematical sciences. He became a professor of the Higher Technical School in Prague in 1884, its rector in 1899–1900, and in 1902 he became the Dean of the first electrical engineering faculty of the Higher Technical School in Europe. Ivan Puluj received international recognition for the development of electric lighting lamps and cathode ray tubes, he was the first to research "cold light" lamps. Ivan Puluj, as already noted above, independently produced a cathode ray tube, which was used to obtain the first clinical X-rays images in America in February 1896 [15]. However, Ivan Puluj did not pay due attention to his invention in the sense that he did not protect it with a patent, being absorbed in other types of his activities. In particular, he made the first translation of the Bible into Ukrainian together with the famous Ukrainian writers Panteleimon Kulish and Ivan Nechuy-Levytskyi. It should also be noted that both scientists Röntgen and Puluj worked in the laboratory of Professor August Kundt at the University of Strasbourg. Unfortunately, we still do not know, and we will probably never know, with which tube Röntgen discovered X-rays on November 8, 1895. Moreover, before his death, Röntgen ordered to burn his entire archive, including laboratory records.

Indeed, after Wilhelm Conrad Röntgen's discovery of X-rays in 1895, scientists around the world quickly began to investigate and disseminate information about these new rays and their remarkable properties. One significant milestone in understanding the nature of X-rays came in 1912 when the German physicist Max von Laue conducted experiments on the diffraction of X-rays by crystal lattices. By directing X-rays at crystalline substances, Laue observed that the rays were diffracted, producing a pattern of spots on a photographic plate. This diffraction pattern provided strong evidence that X-rays behaved as waves, rather than particles, and obeyed the laws of wave optics. This experimental confirmation of the wave nature of X-rays was a crucial step in our understanding of their electromagnetic properties. For his groundbreaking work on X-ray diffraction, Max von Laue was awarded the Nobel Prize in Physics in 1914. His discoveries laid the foundation for the development of X-ray crystallography, a technique used to determine the atomic and molecular structure of crystals, which has had profound implications in various scientific disciplines, including chemistry, biology, and materials science.

The description provided outlines the basic functioning of an X-ray tube, a crucial component in generating X-rays for various applications, including medical imaging and industrial testing. The X-ray tube is a sealed glass cylinder from which air is evacuated to create a vacuum. This vacuum environment is necessary to prevent air molecules from interfering with the generation of X-rays. The X-ray tube contains two electrodes: the cathode and the anode. The cathode typically consists of a spiral of tungsten wire, which is heated by an electric current to emit electrons through thermionic emission. The anode, also known as the anticathode, is a metal rod with a plate made of a refractory metal fixed on its beveled end. This plate serves as the target for the electrons emitted by the cathode. When a high voltage is applied between the cathode and the anode (anticathode), typically several tens of thousands of volts, it creates an electric field that accelerates the emitted electrons towards the anticathode. The fast-moving electrons emitted by the cathode collide with the target material (anticathode) on the anode. These collisions result in the deceleration of the electrons, during which a portion of their kinetic energy is converted into X-ray radiation. This process is known as bremsstrahlung, or "braking radiation," where the sudden deceleration of charged particles (electrons) produces electromagnetic radiation (X-rays). The X-rays generated by the electron-anticathode collisions are emitted in various directions. The area of the anticathode where most of the electrons collide and X-rays are produced is called the focus of the tube. The width and intensity of the X-ray beam depend on the size of this focus area. Thus, the X-ray tube operates based on the principle of converting the kinetic energy of accelerated electrons into X-ray radiation through interactions with a target material, ultimately producing the focused X-ray beam used for imaging purposes.

The interaction of fast electrons with the atoms of metallic anode involves several processes, each contributing to the generation of X-ray radiation: (a) the fast electrons that are emmited by a heated cathode (previously called cathode rays) can knock out outer-shell electrons, a process known as ionization. This accounts for the majority (up to 96%) of the energy loss of fast electrons. The energy lost in ionization is converted into heat, leading to the heating of the anticathode material. To prevent the anticathode from melting, materials with high melting points, such as tungsten or molybdenum, are commonly used. Additionally, cooling mechanisms, such as circulating water or oil, are employed to dissipate heat from the anticathode; (b) because of deceleration of fast electrons within a metallic anode, they emit electromagnetic radiation due to their negative acceleration. This process, known as bremsstrahlung (German for "braking or deceleration"), results in the conversion of a portion of the kinetic energy of fast electrons into X-ray radiation. The energy of the emitted X-rays depends on the degree of deceleration, which is influenced by the interaction between the electrons and the atomic structure of the material; (c) high-energy electrons can also interact with inner-shell (K, L, M) electrons of the atoms in the anode (anticathode) material, causing them to be ejected. Electrons from higher energy levels then fill the vacancies in the inner shells, resulting in the emission of characteristic X-rays with discrete energy spectrum. The energy of these characteristic X-rays depends on the specific energy levels of the electrons within the atoms of the material.

Hence, the interaction of fast electrons with the anticathode material in an X-ray tube leads to the production of both continuous and characteristic X-ray spectra, which are essential for various applications in medical imaging, material analysis, and other fields.

Bremsstraglung (deceleration) X-ray radiation with continuous spectrum. This radiation has a continuous spectrum, which is generated as a result of the deceleration of fast electrons within the substance of the anticathode in an X-ray tube. When electrons are accelerated towards the anode (anticathode) and collide with its material, they undergo sharp deceleration, emitting electromagnetic radiation in the form of *X*-rays. The degree of deceleration varies for different electrons, leading to different fractions of their kinetic energy being converted into the energy of X-ray quanta. As a consequence, the continuous *X*-ray spectrum covers a range of wavelengths, starting from λ_{min}, which is determined by a specific formula (6.1) from Sect. 6.3.1. This means that there is a lower limit to the wavelengths of *X*-ray radiation in the continuous spectrum, dictated by the operating voltage U_a of the *X*-ray tube. Shorter wavelengths cannot be produced at a given voltage. An important implication of this phenomenon is that measurements of the short-wavelength limit of the *X*-ray continuous spectrum can yield a highly accurate value for Planck constant *h*, contributing to the precision of fundamental constants in physics.

The X-rays flux emitted by the *X*-ray tube is influenced by several factors, including the strength of the cathode filament current *I*, the square of the potential difference between the cathode and the anode U_a, and the atomic number Z of the material constituting the anode. Higher values of these parameters result in increased X-ray flux. Furthermore, the power of X-ray radiation, which indicates the penetrating ability of the X-rays, increases with increasing frequency (or decreasing wavelength). This characteristic is directly related to the voltage applied to the tube, with higher voltages producing harder X-rays, as described by a specific formula (6.2).

Characteristic X-ray radiation with line spectrum. Part of the fast electrons bombarding the anode material penetrates the atoms, exciting them in the process. These electronic transitions occur within the inner electron shells of the atom, namely the K, L, and M shells, which are closer to the nucleus. Unlike visible light, which is emitted during electronic transitions between outer shells of an atom, X-ray quanta arise from transitions within the inner shells. Consequently, the energy of X-ray quanta is greater than that of visible light quanta. The characteristic X-ray radiation manifests as a line spectrum, with distinct lines corresponding to specific energy levels involved in the electronic transitions. Each line in the spectrum corresponds to a particular transition between electron shells. Characteristic X-ray radiation is named as such because it uniquely characterizes the substance of the anode material, irrespective of its chemical state (free or bound). These lines are observed against the background of the continuous spectrum of deceleration X-rays, which serves as a pedestal for the characteristic radiation. The characteristic spectrum comprises various series of lines, with each series corresponding to transitions ending at different electron shells. For example, the K-series arises from transitions ending at the K-shell, with lines such as K_{α}, K_{β}, and K_{γ} corresponding to specific transitions from higher shells (L, M, N, respectively) to the K-shell. Similarly, the L-series and M-series originate from transitions ending at the L-shell and M-shell, respectively. In summary, the characteristic X-ray spectrum, with its distinct line structure, provides valuable information about the composition of the anode material and plays a crucial role in various applications of X-ray technology, including material analysis and elemental identification.

6.2.2 *Radioactive Radiation*

Discovery of radioactivity, types of radioactivity, its properties. The discovery of radioactivity by Antoine Henri Becquerel in 1896 marked a significant milestone in the understanding of atomic physics and led to groundbreaking research by scientists such as Pierre Curie, Marie Sklodowska-Curie, and Ernst Rutherford. Here's an overview of the discovery, types, and properties of radioactivity.

(a) In 1896, Antoine Henri Becquerel (1852–1908) discovered that uranium salts emit invisible rays with high penetrating power. These rays were initially named "Becquerel rays." Becquerel's colleagues, Pierre Curie (1859–1906) and Marie Sklodowska-Curie (1867–1934) further investigated these rays and discovered that other elements, such as polonium and radium, also emitted similar radiation.

Alpha (α) Decay: In alpha decay, a radioactive nucleus emits an alpha particle, consisting of two protons and two neutrons (essentially a helium nucleus). This process reduces the atomic number of the nucleus by 2 and the mass number by 4.

Beta (β) Decay Beta decay involves the emission of a beta particle, which can be either a negatively charged electron (β_-) or a positively charged positron (β_+). In β_- decay, a neutron in the nucleus converts into a proton, emitting an electron and an antineutrino. In β_+ decay, a proton in the nucleus converts into a neutron, emitting a positron and a neutrino.

Gamma () Decay: Gamma decay involves the emission of gamma rays, which are high-energy photons emitted from the nucleus following an alpha or beta decay. Gamma decay typically occurs after alpha or beta decay to stabilize the nucleus further. Gamma rays have no mass or charge but carry significant energy.

Main Properties of Radioactive Decay: Radioactive decay is spontaneous and occurs randomly at the level of individual nuclei. The rate of radioactive decay is characterized by the half-life time (period), which is the time taken for half of the radioactive nuclei in a sample to decay into other nuclei. Radioactive decay is independent of external conditions such as temperature, pressure, or illumination. The emission of alpha, beta, and gamma radiation can be affected by external factors such as magnetic fields, which can cause the radiation to bend or deflect. The discovery of radioactivity revolutionized our understanding of atomic structure and led to advancements in various fields, including medicine, energy production, and materials science. Marie Sklodowska-Curie's pioneering work in radioactivity earned her two Nobel Prizes in Physics and Chemistry, making her the first woman and scientist to win Nobel Prizes in two different scientific fields.

Beta Decay and the Law of Energy Concervation. Understanding the specifics of the explanation for beta decay has been particularly challenging for scientists. Physicists initially struggled to explain the continuous spectrum of beta radiation within the framework of the law of energy conservation. The understanding of beta radiation and its continuous spectrum was significantly advanced by Fermi's theory, which introduced the concept of neutrinos and antineutrinos to explain the observed phenomena in beta decay.

Enrico Fermi provided an explanation for the departure of electrons and positrons from the nucleus during beta decay in 1934. He proposed that the transformation of a neutron into a proton (or vice versa) in the nucleus is accompanied by the emission of electronic or positron beta radiation, respectively. The existence of the positron was theoretically predicted by Wolfgang Pauli and Paul Dirac in 1930 and experimentally confirmed by Carl Anderson in 1932 during the study of cosmic rays.

To explain the continuous spectrum of beta decay, Fermi's theory resolved this issue by proposing the existence of neutrinos and antineutrinos, which carry away variable amounts of energy, resulting in a continuous spectrum of emitted beta particles. The introduction of neutrinos and antineutrinos allowed for the conservation of energy while accounting for the variability in energy observed in beta decay processes. Neutrinos were hypothesized by Pauli in 1930, but Fermi's theory of beta decay provided a concrete context for their existence and introduced the term "neutrino" to describe these elusive particles.

It should be noted that the beta radiation is associated, as a rule, with electrons, since we live in the world of particles, not antiparticles. As is known, in the process of interaction of an electron–positron pair (in the general case—matter and antimatter), the phenomenon of annihilation occurs, as a result of which the electron–positron pair turns into two gamma quanta (see more details about this annihilation process in Chap. 7, where the theoretical foundations of the diagnostic method of positron emission tomography (PET) are considered.

Summarizing what has been said about the various types of radioactivity, we now have the basis for the following formulation of the concept of radioactivity. Radioactivity encompasses the spontaneous transformation of atomic nuclei, leading to the emission of various particles and radiation. This process includes:

(a) **Spontaneous Transformation of Nuclei**

Alpha Decay(α decay): In alpha decay, a nucleus emits an alpha particle, which consists of two protons and two neutrons. This emission results in the transformation of the original nucleus into the nucleus of a different chemical element.

Beta Decay (β-decay): Beta decay involves the emission of beta particles, which can be either electrons (β_{-}) or positrons (β_{+}). Beta decay leads to the transformation of a neutron into a proton (β_{-} decay) or a proton into a neutron (β_{+} decay) within the nucleus, changing the identity of the nucleus.

(b) **Spontaneous Radiation from the Nucleus**

Gamma Decay(γ-decay) Gamma decay involves the emission of high-energy gamma rays from the nucleus. Gamma radiation accompanies alpha and beta decay processes, often serving to stabilize the nucleus further by releasing excess energy.

The basic law of radioactive decay. Natural radioactivity, which was discussed above, involves the decay of naturally occurring radioactive isotopes found in minerals or elements. This process occurs spontaneously without external influence. On the other hand, artificial radioactivity refers to the induced or artificial creation of radioactive isotopes through various nuclear reactions. These reactions can occur in nuclear reactors, particle accelerators, or through other means of nuclear engineering. The discovery of artificial radioactivity by Irene and Frederic Joliot-Curie in 1934 marked a significant milestone in nuclear physics and paved the way for numerous applications in medicine, industry, and research. Despite the difference in origin, both natural and artificial radioactivity are governed by the same laws of radioactive decay, including the exponential decay law and the concept of half-life period. These laws provide a fundamental understanding of the behavior of radioactive isotopes and are essential for various practical applications, such as radiometric dating, medical imaging, and nuclear power generation.

The basic law of radioactive decay states that ***the relative decrease in the number of radioactive nuclei per unit of time is a constant value, known as the constant***

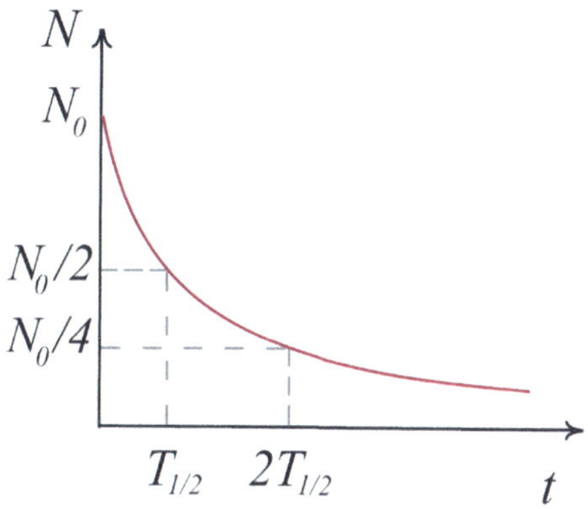

Fig. 6.2 Graph of the law of radioactive decay in integral form, which describes the change in the number of radioactive nuclei over time

of radioactive decay (see formulas of the law of radioactive decay in so-called differential and integral forms in Sect. 9.3.2).

Half-life period $T_{1/2}$. The concept of half-life period $T_{1/2}$. is introduced as ***the time during which half of the original number of radioactive nuclei decay***. It determines the time scale for the decrease in the number N of radioactive nuclei. Figure 6.2 illustates the exponentsal decrease in the number N over time t. It is conventionally accepted that radionuclides (isotopes of radioactive elements) practically cease to have a significant effect after a period of approximately 10 half-lives. This estimate is based on the fact that after ten half-life periods, the number of remaining radioactive nuclei will have decreased by a factor of $2^{10} = 1024$. Thus, the substance is considered effectively non-radioactive for practical purposes after this time period.

Table 6.1 presented below, contains information on the $T_{1/2}$ half-life values for a number of radioactive elements (radionuclides), as well as on the types of radioactivity inherent in these radionuclides. As can be seen from this table, the half-lives $T_{1/2}$ of the presented radioactive elements can vary within very wide limits.

The results presented in Table 6.1 show that the uranium isotope ${}^{238}_{92}U$, which is the most widespread in the natural environment, has its half-life period $T_{1/2} = 4.5 \cdot 10^9$ years. This value for the half-life period of this uranium isotope ${}^{238}_{92}U$ roughly coincides with the age of the planet Earth. In other words, the following statement is correct: for the entire time of the Earth's existence, the number of radioactive nuclei of the uranium isotope ${}^{238}_{92}U$ has decreased only two times.

The provided information highlights the specific properties and behaviors of certain radioactive isotopes, particularly iodine-131, strontium-90, and cesium-137, in biological systems and the environment:

1. ***Half-Life Periods***: Iodine-131, half-life period $T_{1/2} = 8$ days; Strontium-90, half-life period $T_{1/2} = 28.8$ years; Cesium-137 half-life period $T_{1/2} = 30.2$ years.
2. ***Behavior in Biological Systems***: Strontium-90 behaves similarly to calcium (Ca) in biological processes. This is significant because calcium plays a crucial role in bone formation and structure. Consequently, Strontium-90 can replace

Table 6.1 Main physical properties of radionuclides

Name of radioactive element	Symbol of element	Number of protons (Z)	Number of nuclons (Z + N)	Half-life period $T_{1/2}$ in years (yr), days (d) and seconds (s)	Type of radioactive radiation
Uranium	U	92	238	$4.5 \cdot 10^9$ yr	α, γ
Uranium	U	92	235	$7 \cdot 10^8$ yr	α
Protactinium	Pa	91	231	$3.4 \cdot 10^4$ yr	α, γ
Thorium	Th	90	232	$1.4 \cdot 10^{10}$ yr	α, γ
Actinium	Ac	89	227	21.7 yr	β
Radium	Ra	88	226	1860 yr	α, γ
Radon	Rn	86	222	3.8 d	α
Polonium	Po	84	214	$1.6 \cdot 10^{-4}$ s	α
Polonium	Po	84	210	138 d	α
Radiolead	Pb	82	210	22.2 yr	β
Cesium	Cs	5	137	30.2 yr	β
Iodine	J	53	131	8 d	γ
Strontium	Sr	38	90	28.8 yr	β

calcium and become incorporated into bone tissues. Cesium-137 behaves similarly to potassium (K). As potassium is essential for various physiological processes, Cesium-137 can be taken up by organisms and incorporated into tissues, particularly muscle tissues.

3. ***Distribution in the Environment***: Following radioactive contamination of the biosphere, the highest concentrations of Strontium-90 are typically found in organs and products rich in calcium. This includes bone tissues, where Strontium-90 can accumulate due to its similarity to calcium. Cesium-137 tends to accumulate in biological objects rich in potassium. This includes muscle tissues, where Strontium-90 can replace potassium and become concentrated.

Understanding the behavior and distribution of these radioactive isotopes in biological systems and the environment is crucial for assessing the potential risks associated with radioactive contamination and implementing appropriate measures to mitigate these risks.

6.2.3 Interaction of Ionizing Radiation with Biological Tissues

A. ***The concept of ionizing radiation*** encompasses various forms of radiation and streams of elementary particles that possess sufficient energy to detach an electron from an atom or molecule, resulting in the formation of positively charged ions.

This process is known as ionization and can occur in both inorganic and organic substances, including biological tissues.

To characterize the effects of ionizing radiation on substances, particularly biological tissues, ***three doses of ionizing radiation*** are commonly used:

(a) **Exposure Dose (D_0)**: This dose measures the amount of ionizing radiation delivered to a specific area or volume of material. It quantifies the amount of charge imparted to the material per unit mass. The exposure dose is expressed in units such as coulombs per kilogram (C/kg) or roentgens (R). More precisely, the exposure dose D_0 determines the amount of charge generated as a result of ionization in a unit mass of dry air when exposed to X-ray and gamma radiation (refer to formula (6.29) for calculation of ionizing radiation exposure dose). This measure characterizes the degree of air ionization caused by X-rays and gamma rays, which can be reliably measured through experimental methods. It's important to note that ionization energy is significantly different from thermal energy. For example, a dose of gamma radiation potent enough to be fatal to an individual only marginally increases body temperature by a fraction of a degree.

(b) **Absorbed Dose (D_a)**: The absorbed dose represents the amount of ionizing radiation energy absorbed per unit mass of the irradiated material. It accounts for the energy deposited within the material and is crucial for assessing the potential biological effects of radiation exposure. The absorbed dose is typically measured in units such as gray (Gy) or rad. Speaking in more details, the absorbed dose D_a of ionizing radiation represents the portion of energy from the radiation that is absorbed per unit mass of the substance. It's crucial to understand that not all radiation incident upon a substance contributes to its effect; only the fraction that is absorbed during interactions with the atoms and molecules of the substance is relevant. The absorbed dose D_a is quantified by a specific formula provided in Sect. 6.3.3, which accurately calculates the amount of energy absorbed per unit mass of the substance. This parameter is essential for assessing the potential biological effects and understanding the impact of ionizing radiation on various materials, including biological tissues and other substances

(c) **Equivalent Dose (Deq) or Biological Dose (Db)**: This dose takes into account not only the absorbed dose but also the biological effectiveness of the type of radiation involved. Different types of radiation (e.g., X-rays, alpha particles) have varying biological effects even when they deliver the same absorbed dose. The equivalent dose is expressed in units such as sieverts (Sv) or rems. More accutately, the equivalent dose D_{eq} is a crucial concept in assessing the biological effects of ionizing radiation. It takes into account the varying effectiveness of different types of radiation on biological objects, even when they deliver the same absorbed dose D_a. The biological effect of ionizing radiation can differ significantly for different types of radiation. For instance, the effectiveness (radiosensitivity) of biological objects can vary by factors such as the type of radiation. For example, the transition from gamma radiation to fast neutrons can increase radiosensitivity by up to 10 times, while the transition from gamma radiation to alpha radiation can increase radiosensitivity by up to 20 times. To assess the biological effect of each type of ionizing radiation accurately, the absorbed dose D_a is multiplied by a factor k, known as the quality factor or relative biological effectiveness (RBE) of radiation. The product of k and D_a characterizes the equivalent dose D_{eq}, also known as the biological dose D_b. In summary, the equivalent dose Deq accounts for the varying effectiveness of different types of radiation on biological tissues, providing a more accurate representation of the biological effects of ionizing radiation.

In addition to these doses, three corresponding dose powers (dose rates) of ionizing radiation are also considered.

(a) **Power of the Exposure Dose** (P_0): The power of the exposure dose P_0 is the exposure dose related to the unit of time, i.e., it represents the charge generated per unit of time in the unit mass of dry air under the influence of X-ray and gamma radiation (see the corresponding formulas for the power of the exposure dose in Sect. 6.3.3). This represents the rate at which exposure dose is delivered to a material or tissue over time. It is measured in units such as coulombs per kilogram per second (C/kg × s) or roentgens per second (R/s).

(b) **Power of the Absorbed Dose (P_a)**: This measures the rate at which absorbed dose is delivered to a material or tissue over time. It is expressed in units such as gray per second (Gy/s) or rad per second. Thus, the power of the absorbed dose P_a represents the rate at which energy from ionizing radiation is absorbed per unit mass of the substance over time. It provides insight into the dynamic process of energy deposition within the material and is critical for evaluating the temporal aspects of radiation exposure. When the effect of ionizing radiation on a substance varies over time, the power of the absorbed dose is characterized by the rate of change of the absorbed dose Da with respect to time t. In mathematical terms, this is represented by the derivative dD_a/dt, where dDa is a small change in the absorbed dose and dt is a small time interval during which the absorbed dose does not change significantly. In other words, the power of the absorbed dose P_a is determined by the rate at which the absorbed dose is changing over time. This rate of change indicates how rapidly the absorbed dose is increasing or decreasing with respect to time, providing valuable information about the dynamics of the radiation exposure process.

(c) **Power of the Equivalent Dose (P_{eq}) or Biological Dose (P_b)**: This represents the rate at which equivalent dose is delivered to a material or tissue over time. It is measured in units such as sieverts per second (Sv/s) or rems per second (rem/s). As in two previous cases, to estimate the rate of change in the equivalent dose over time, which represents the velocity of change in the biological dose, we introduce the concept of the power of the equivalent (biological) dose P_{eq}. This power is determined by the ratio of the change in the equivalent (biological) dose to the time interval over which this change occurs and provides a valuable information, reflecting the dynamics of the biological effects of ionizing radiation on living organisms.

These doses and dose powers (dose rates) are essential for assessing radiation exposure levels, understanding biological effects, and ensuring radiation safety in various applications, including medical diagnosis and treatment, industrial processes, and environmental monitoring.

B. ***Interaction of X-rays radiation with biological tissues.*** When X-rays enter a substance (for example, the human body), some of the X-ray radiation is absorbed and scattered in the substance, while the rest passes through it. There are three main primary processes of *X*-ray interaction with biological tissues: ***coherent (elastic)scattering, photoeffect and incoherent (inelastic) scattering (or Compton effect).***

(a) ***Coherent (elastic) scattering*** occurs at low energy values of an *X*-ray quantum, when as a result of its interaction with an atom, the energy of the quantum does not change, but only the direction of its propagation changes (see Fig. 6.4 in Sect. 6.3.4). Coherent X-ray scattering is the basis of X-ray structural analysis, which is based on *X*-ray diffraction and allows determining the spatial structure of crystal lattices of solid solutions, alloys, pure liquids and their mixtures, as well as other systems, including biological objects. The spatial structure of the whale hemoglobin protein molecule was determined for the first time using *X*-ray structural analysis. It was the use of coherent *X*-ray scattering that allowed Rosalind Franklin in 1952 to make one of the most outstanding discoveries of the twentieth century—the helical structure of DNA.

To calculate the quantitative parameters of the spatial structure of substances, the basic formula (6.47) of *X*-ray structural analysis (see Fig. 6.5) is used for its derivation in Sect. 6.3.4). Formula (6.47) is called the Bragg formula (in the Ukrainian physics literature—the Bragg-Wulf formula) [15–17]. This formula was obtained in 1913 independently by two scientists: the English physicist William Lawrence Bragg (1890–1971), as well as the crystallographer and physicist Georgy (Yuriy) Wulf (1963–1925), who was born in the Ukrainian city of Nizhyn [17, 18]. It should be noted that William Lawrence Bragg, together with his father, the English physicist William Henri Bragg (1862–1942), were awarded the Nobel Prize in Physics in 1915 for services to the study of crystal structure using *X*-rays.

(b) ***The photoelectric effect*** occurs when the energy of the *X*-ray quantum turns out to be sufficient to detach an electron from an atom, that is, to perform the work of ionizing the atom, as well as to give the free electron a certain amount of kinetic energy (see Fig. 6.6 and Einstein's formula (6.48) in Sect. 6.3.4). Albert Einstein (1879–1955) obtained the mentioned formula for the photoeffect in 1905 and received the 1921 Nobel Prize in Physics for it. *T*he absorption of X-rays by matter depends on the energy of the *X*-ray quanta. As the energy of the X-ray quanta increases (corresponding to decreasing wavelength or increasing frequency), the absorption coefficient typically decreases. There are specific frequencies at which the absorption coefficient increases greatly. This phenomenon occurs due to additional absorption processes caused by the ejection of electrons from the corresponding inner shell of the atom, known as the photoelectric effect. The frequencies at which this great increase in absorption occurs allow us to determine the energy levels of electrons in different electron shells of atoms. When X-rays interact with matter through the photoelectric effect, they can cause intense ionization of atoms and molecules. This ionization process disrupts the stability of the atoms and molecules, leading to various chemical and biological effects depending on the type of material and the dose of X-ray radiation.

(c) ***Incoherent inelastic scattering*** (***the Compton effect***) takes place when an X-ray quantum (primary photon) impinging on an atom has enough energy to not only cause the photoeffect, but also transfer the rest of its energy to an additional (secondary) quantum. In other words, during incoherent scattering, the following processes take place: (a) work is done to ionize the atom and (b) kinetic energy is given to the detached electron—in full accordance with the two terms on the right side of Einstein's equation, and (c) another term appears (secondary) quantum (photon) with energy lower than that of the primary photon striking an atom. Electrons that break away from an atom as a result of the photoeffect are called Compton electrons. Together with the Compton electrons, secondary photons with lower energy than primary photons also arise. The direction of motion of these photons is also different from that of primary photons. Of course, in the process of incoherent scattering (Compton effect), the basic laws of conservation of energy and momentum of all interacting elementary particles and atoms (molecules) must be observed. Figure 6.7 and Eq. (6.49) in Sect. 6.3.4 illustrate all processes and conservation laws corresponding to incoherent scattering (Compton effect) during the interaction of high-energy quanta (photons) of X-ray radiation with matter.

C. ***Interaction of radioactive radiation and particle flows with biological tissues.*** The interaction of charged α- and β-particles of radioactive radiation with matter is reduced mainly to the excitation and ionization of atoms, that is, the transition of the electrons of the atom to higher energy levels and the separation of electrons from the atom. At the same time, if the energy of α- and β-particles is high enough, they can knock out electrons from deep energy levels, which leads to the appearance of characteristic X-ray radiation.

As for γ-radiation, the primary mechanisms of its interaction with matter are mostly similar to the action of X-ray radiation, namely the photoelectric effect and the Compton effect. Note that coherent scattering of γ-radiation is not included in the list of primary mechanisms of interaction of gamma rays with matter due to the low energy of quanta during coherent scattering.

It should be noted that another primary mechanism of interaction of γ-rays with matter becomes characteristic of high-energy γ-radiation, which is caused by the process of formation of electron–positron pairs (see in more detail in Sect. 6.3.4). It is indeed important to note that ionizing power and penetrating power exhibit distinct characteristics across different types of radiation as follows:

1. ***The ionizing power*** refers to the ability of radiation to produce ionization in a material. ***In the series of γ-, β-, α-radiation, the ionizing power increases***. Gamma (γ) radiation, being electromagnetic radiation, has relatively lower ionizing power compared to charged particles. Beta (β) radiation, consisting of high-speed electrons or positrons, has greater ionizing power than gamma radiation. Alpha (α) radiation, composed of helium nuclei, has the highest ionizing power among these types of radiation due to the heavy and charged nature of alpha particles.
2. ***The penetrating power*** refers to the ability of radiation to pass through matter. ***In the reverse series of α-, β-, γ-radiation, the penetrating power increases***. Alpha (α) radiation has the lowest penetrating power among these types. It can be stopped by a sheet of paper or a few centimeters of air. Beta (β) radiation, being more penetrating than alpha radiation, can penetrate through materials such as clothing or aluminum. Gamma (γ) radiation, with the highest penetrating power, can penetrate through most materials, requiring dense materials like lead or concrete for effective shielding.

This inverse relationship between ionizing power and penetrating power (see Fig. 6.3) is a crucial consideration in radiation protection and the design of shielding materials for various applications.

Ionizing radiation encompasses protons and neutrons as well. The velocity and energy of protons generated during nuclear reactions, along with their ionizing and absorption capabilities, closely resemble those of alpha particles. Neutrons emitted during nuclear reactions typically possess an initial velocity on the order of 3×10^7 m/s and an energy ranging up to 4–5 MeV. Due to their lack of charge, their primary ionization potential is low, resulting in relatively high permeability. The ionizing effect of neutrons on matter primarily arises from secondary processes. Upon interaction with atomic nuclei, neutrons can undergo elastic scattering, inelastic scattering, and capture by nuclei (radiative capture). In inelastic interactions, particularly

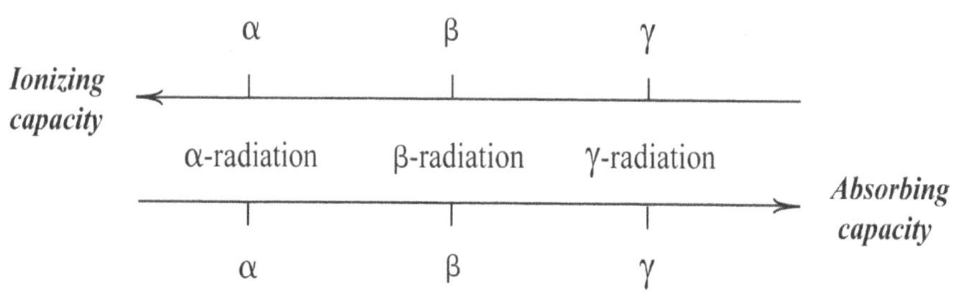

Fig. 6.3 Relative ionizing and absorbing capacity of radioactive radiation

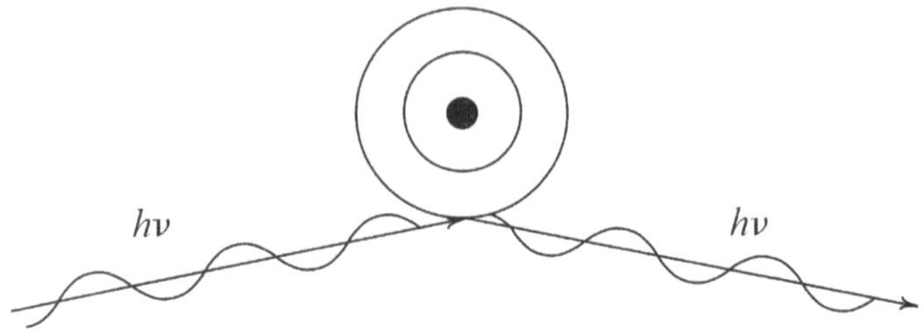

Fig. 6.4 Schematic representation of the process of coherent (or elastic) scattering of an X-ray quantum during its interaction with an atom of a substance, provided that the energy of the quantum hν, which is significantly less than the ionization work A_i, does not change in magnitude, but only the direction of propagation of the quantum changes before and after interaction with the atom

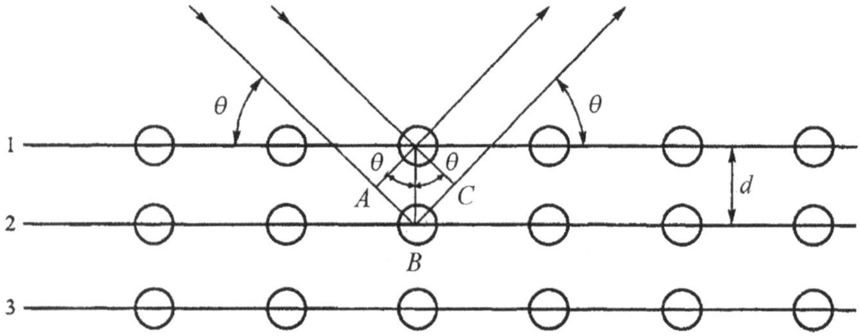

Fig. 6.5 Schematic representation of the crystal lattice for obtaining the basic Bragg-Wulf formula of X-ray structural analysis

Fig. 6.6 Schematic representation of the external photoeffect process

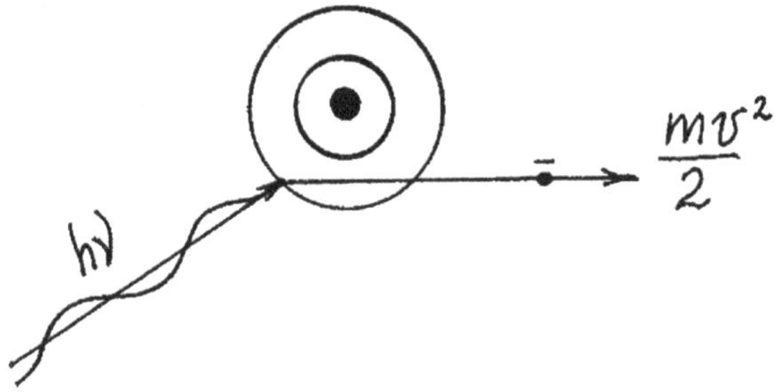

Fig. 6.7 Schematic representation of the process of incoherent scattering (the Compton effect)

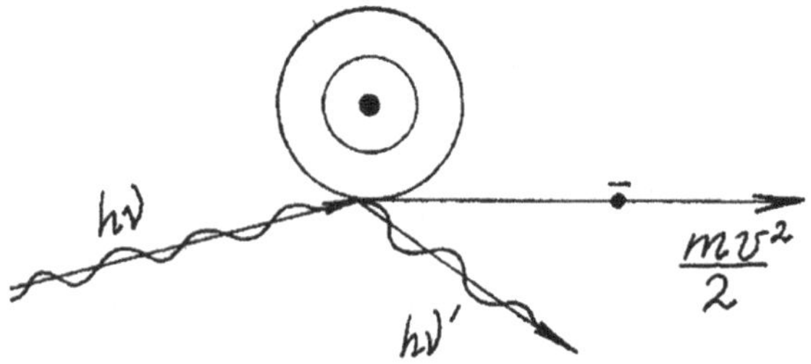

with light-element nuclei, the neutron transfers a portion of its kinetic energy to the nucleus. This transferred energy prompts secondary ionization, which can be significant owing to the presence of an electric charge within the nucleus. Since human body tissues contain abundant hydrogen, fast neutrons readily dissipate their energy within them, leading to substantial ionization effects.

6.2.4 *Physical, Chemical and Biological Stages of Radiation Damage*

In radiation medicine and radiation protection (see the list of references to this chapter), several fundamental physicochemical mechanisms underlie the damage caused by the action of ionizing radiation on living organisms. These mechanisms include (see e.g. [4]):

During the primary stage of radiation interaction, particularly with ionizing radiation, water molecules play a pivotal role due to their abundance in biological tissues, constituting approximately three-quarters of the total body weight. In this primary stage, which can be called ***physical stage***, ionizing radiation can directly ionize water molecules. As a result, a water molecule (H_2O) can undergo ionization to form a positively charged water molecule ($(H_2O)^+$) and a free electron (e^-), represented by the reaction:

$$H_2O \rightarrow (H_2O)^+ + e^-.$$

This process occurs very rapidly, typically lasting for a very short time on the order of 10^{-16} s.

During the subsequent ***physico-chemical stage*** following the initial ionization of water molecules, several processes or reactions occur, each contributing to the complex cascade of events initiated by radiation exposure. These reactions typically occur over a longer timescale, approximately on the order of 1 microsecond (1 μs). The key processes during this stage include:

(a) Separation of a proton from a positive ion. The positively charged water ion ($(H_2O)^+$) loses a proton, forming a hydrogen ion (H^+) and a hydroxide ion (OH^-):
$(H_2O)^+ \rightarrow H^+ + OH^-$

(b) Attachment of an electron to a neutral water molecule. An electron attaches to a neutral water molecule, forming a negatively charged ion $(H_2O)^-$:
$H_2O + e^- \rightarrow (H_2O)^-$

(c) Transformation of the negative ion into a hydrogen atom and a hydroxide ion. The negatively charged water ion $(H_2O)^-$ undergoes transformation into a hydrogen atom (H) and a hydroxide ion (OH^-):
$(H_2O)^- \rightarrow H + (OH)^-$

These processes collectively contribute to the dynamic chemical alterations occurring in the irradiated biological tissue, leading to further ionization and the formation of reactive chemical species. The products of described reactions include both charged ions, such as $(OH)^-$ and H^+, as well as neutral particles like H and OH. It's worth noting that the H^+ and $(OH)^-$ ions are abundant due to the dissociation of water molecules in ordinary water and therefore do not significantly impact subsequent processes. On the other hand, the neutral formations, hydrogen molecules (H) and hydroxyl groups (OH), are free radicals, possessing unpaired electrons, which render them highly chemically reactive. As such, these neutral particles play a crucial role in subsequent chemical reactions and biological effects induced by radiation exposure. The formation of hydrogen peroxide (H_2O_2) is a notable aspect of the chemical reactions occurring during this physico-chemical stage:

$$OH + OH \rightarrow H_2O_2,$$

Hydrogen peroxide is indeed a potent oxidizing agent and can induce oxidative stress in biological systems. Its formation adds to the complexity of the chemical environment resulting from radiation exposure and contributes to the subsequent biochemical processes and cellular damage.

Following the physico-chemical stage, the subsequent phase, often referred to as ***the chemical stage***, ensues. During this stage, which typically spans a few seconds, the reaction products generated in the preceding stages commence interactions with organic molecules within cells. These reactive species can target various cellular components, including molecular complexes within chromosomes and other vital cellular structures. The chemical alterations induced during this stage contribute to the overall biological effects of radiation exposure, leading to potential cellular damage, genetic mutations, and other physiological consequences.

Exactly, the initial stages of radiation exposure—physical, physico-chemical, and chemical—primarily lay the groundwork for subsequent biological effects. While these stages themselves may not inflict significant tissue disturbances, they set the stage for secondary reactions within complex organic molecules. Free radicals and other reactive biochemical products generated during these stages act as catalysts for subsequent biological processes, which unfold in a chain reaction manner. These secondary reactions, escalating like an avalanche, can precipitate various biological effects, ultimately culminating in conditions such as radiation sickness. This process underscores the complex interplay between radiation and biological systems, highlighting the importance of understanding the intricate mechanisms underlying radiation-induced damage.

Indeed, all these secondary effects of radiation exposure manifest in ***the final biological stage***, which can span from mere minutes to several decades. ***Radiation-induced damage at this biological stage is typically categorized into two main classes***:

1. ***Somatic effects***, which impact only the irradiated organism, such as an individual human being. Somatic effects may manifest as acute radiation sickness or chronic health issues, depending on the severity and duration of exposure.
2. ***Hereditary effects***, which are transmitted to reproductive organs and can be passed on to subsequent generations. Hereditary effects involve mutations in germ cells, which can lead to genetic disorders or abnormalities in offspring. Understanding the distinction between somatic and hereditary effects is crucial in assessing the long-term consequences of radiation exposure and formulating strategies for radiation protection and risk management.

During the biological stage, the effects of the physico-chemical mechanisms of radiation action on living organisms manifest as remote consequences. These effects often involve the slowing down or cessation of cell division processes, leading to various changes in subsequent generations of cells. Radiation-induced alterations can result in premature cell death and the decline of cellular populations, ultimately affecting the overall health and viability of the organism. The biological ramifications of radiation exposure can be far-reaching, impacting cellular function, tissue integrity, and organismal health. Understanding these effects is essential for comprehensively evaluating the risks associated with radiation exposure and developing strategies to mitigate its adverse consequences on living organisms.

6.3 Section B

6.3.1 Deceleration and Characteristic X-ray Radiation

Features of the deceleration spectrum of X-rays. The electrons are accelerated in the electric field, if a voltage U_a is applied between cathode and anode. The energy of such electrons is equal to eU_a, where e is the charge of the electron. When the energy of the electron is completely transformed into the energy of the X-ray quantum, we get $h\nu_{max} = eU_a$. Here, h is a Planck constant and ν_{max} is the highest frequency of the X-ray deceleration spectrum at a given voltage U_a. Taking into account the formula $\nu_{max} = c/\lambda_{min}$, where c is the speed of light in a vacuum, λ_{min} is the limiting wavelength of X-ray radiation corresponding to ν_{max}, we obtain

$$\lambda_{min} = \mathrm{hc}/\mathrm{eU_a} = 1.24[\mathrm{nm}]/\mathrm{U_a}[\mathrm{kV}] \tag{6.1}$$

The short-wavelength limit λ_{min} of the continuous deceleration X-ray spectrum is a crucial parameter. According to expression (6.1), this limit is determined by the given voltage U_a, and wavelengths shorter than λ_{min} cannot occur under this condition. It is worth highlighting that the value of a Planck constant h, derived from experimental measurements of λ_{min} using formula (6.1), is exceptionally precise and dependable. This underscores the significance of such experimental data in establishing fundamental constants of nature.

The flow Φ of X-rays is determined by the following formula:

$$\Phi = \mathrm{kIU_a^2Z}\,, \tag{6.2}$$

where the coefficient is $k = 10^{-9}$ W/AV^2. As is seen from formula (6.2), the X-rays flow grows proportionally to the strength of the current I in the tube, the square of the voltage on the tube U_a and depends on the atomic number Z of the substance of the anode (anticathode).

Moseley's law for characteristic X-ray radiation. Henry Moseley (1887–1915), an English physicist, made significant contributions to the understanding of atomic structure. One of his notable achievements was the establishment of a fundamental relationship known as Moseley's law. This law describes the dependence of the frequencies (or wavelengths) of characteristic X-ray spectral lines on the atomic number Z of elements.

Moseley's law can be summarized as follows:

$$\nu = R(Z - \sigma)^2\left(\frac{1}{1^2} - \frac{1}{2^2}\right) \tag{6.3}$$

$$\nu = R(Z - \sigma)^2\left(\frac{1}{2^2} - \frac{1}{3^2}\right) \tag{6.4}$$

for L-series. Here $R = 3.3 \cdot 10^{15}$ s^{-1} is the Rydberg constant, σ is a so-called shielding constant, being equal $\sigma = 1$ for the K_α lines, and $\sigma = 0.75$ for the L-series.

Moseley's law for characteristic X-ray spectral lines laid the groundwork for understanding the inner structure of atoms and contributed to the validity of the nuclear model of the atom and the Mendeleev's periodic table of chemical elements.

6.3.2 Radioactive Radiation: Theoretical Problems

Beta radiation and the law of conservation of energy. According to the theory proposed by Enrico Fermi, beta decay involves the spontaneous transformation of nuclear particles, specifically nucleons (this term was proposed by Werner Heisenberg to describe protons and neutrons within the nucleus). Beta decay occurs through the following nuclear reactions:

$${}_0^1n \rightarrow {}_1^1p + {}_{-1}^{\ 0}\beta + \tilde{\nu} \tag{6.5}$$

$${}_1^1p \rightarrow {}_0^1n + {}_{+1}^{\ 0}\beta + \nu$$

for electron beta (β^-) and positnon beta (β^+) decays, respectively.

The first nuclear reaction in (6.5) for electron beta decay involves the spontaneous conversion of a neutron (n) into a proton (p), emitting an electron ($^{0}_{-1}\beta$) and an antineutrino ($\tilde{\nu}$). Correspondingly, the second nuclear reaction in (9.5) for positron beta decay involves the spontaneous reverse conversion of a proton (p) into a neutron (n), emitting a positron ($^{0}_{+1}\beta$) and a neutrino (ν).

Electronic K-capture is another process classified as beta decay, during which nucleons within the nucleus undergo mutual transformation. In this process, an electron located in the K-shell closest to the nucleus is captured by the nucleus. This results in the conversion of a proton into a neutron, accompanied by the emission of a neutrino. The reaction for K-capture is as follows:

$$^{1}_{1}p + ^{0}_{-1}\beta \rightarrow ^{1}_{0} n + \tilde{\nu} \tag{6.6}$$

An example of such a transformation is the conversion of Beryllium-7 into Lithium-7

$$^{7}_{4}Be + ^{0}_{-1}\beta \rightarrow ^{7}_{3} Li + \tilde{\nu} \tag{6.7}$$

During K-capture, the number of protons (Z), which determines the atomic number of a chemical element, decreases by one. As a result, the nucleus of the element transitions to the position in Mendeleev's periodic system occupied by the element with one fewer proton. This transformation is similar to what occurs during positron beta decay.

The formula for the frequency (wavelength) of gamma quanta. The formula for the frequency (or wavelength) of gamma (γ) quanta is derived from the energy difference between the stationary levels of the atomic nucleus, where energy transitions of nucleons occur. This relationship is described by the Bohr-Sommerfeld formula:

$$hf = \Delta E, \tag{6.8}$$

where h is Planck constant, f is the frequency of the gamma quantum (here we use the frequency notation f instead of v, so as not to confuse the frequency notation with the neutrino symbol), ΔE is the energy difference of the stationary levels of the atomic nucleus

The speed of light c can be represented as $c = \lambda f$, where λ is the wavelength. So, we can rearrange the Eq. (6.8) to express the wavelength as $\lambda = c/f$. Combining this with the Bohr-Sommerfeld formula, we find the following formula for the wavelength λ of gamma quanta, which is completely equivalent to the relation (6.8):

$$\lambda = hc/\Delta E \tag{6.9}$$

Thus, the wavelength λ of gamma rays can be calculated from the energy difference ΔE using the formula (6.9). For example, if $\Delta E = 0.5$ MeV, the wavelength of the gamma rays would be approximately equal to $\lambda \approx 2.5 \times 10^{-12}$ m.

Formulas for the basic law of radioactive decay. The law of radioactive decay can be expressed in two forms—differential and integral. The differential form describes the rate (velocity) of decay of radioactive nuclei over a small time interval. It is given by the equation:

$$-dN/Ndt = \lambda, \tag{6.10}$$

where dN is a small change in the number of radioactive nuclei, N is the number of radioactive nuclei at time t, dt is a small interval of time, and λ is the radioactive decay constant. The negative sign in the left side of the Eq. (6.10) indicates that the number of radioactive nuclei decreases over time. The constant λ represents the probability per unit time that a nucleus will decay and is determined by the properties of the specific radioactive isotope.

The formula (6.10) represents ***the differential form of the law of radioactive decay***, which can be formulated as follows: **the relative reduction in the number of radioactive nuclei over a single time interval (−dN/Ndt) is a constant value, which is determined by the radioactive decay constant (λ).**

To obtain ***the law of radioactive decay in integral form***, we start with the linear differential Eq. (6.10) and separate both variables N and t in it as follows:

$$dN/N = -\lambda dt \tag{6.11}$$

Next, we integrate both sides of the Eq. (6.11). Integrating the left side gives us the natural logarithm of N, and integrating the right side gives us $-\lambda t + C$, where C is the constant of integration. So, we have

$$\ln N = -\lambda t + C. \tag{6.12}$$

To determine the value of C, we can use the initial condition that at $t = 0$, the initial number of radioactive nuclei was N_0. Substituting $t = 0$ and $N = N_0$ into the equation, we get $C = \ln N_0$.

Now, substituting this value of $C = \ln N_0$ back into the Eq. (6.12), we have:

$$\ln N = -\lambda t + \ln N_0 \tag{6.13}$$

Using the properties of logarithms, we can rewrite the Eq. (6.13) as

$$\ln(N/N_0) = -\lambda t \tag{6.14}$$

Taking the exponent, i.e., performing the inverse of logarithm operation, on both sides of (6.14), we get:

$$N/N_0 = e^{-\lambda t} \tag{6.15}$$

Multiplying both sides of (6.15) by N_0, we finally obtain:

$$N = N_0 e^{-\lambda t} \quad (6.16)$$

This is ***the law of radioactive decay in integral form***, denoting N as the number of radioactive nuclei at time $t > 0$, N_0 as the initial number of radioactive nuclei at time $t = 0$, N as the number of radioactive nuclei at time $t > 0$, and λ as the radioactive decay constant specific to a particular radio-isotope.

The graph of the law of radioactive decay in integral form (6.16), which describes the exponential change in the number of N radioactive nuclei with time t, was given in Fig. 6.2 in Section A. Based on formula (6.16), it follows that the greater the radioactive decay constant λ, the faster the curve $N(t)$ falls.

Formulas for half-life period $T_{1/2}$. As mentioned earlier, the half-life time (period) $T_{1/2}$ characterizes the period of time during which half of the initial number of radioactive nuclei decays. To find the relationship between the half-life time $T_{1/2}$ and the radioactive decay constant λ, we start with the law of radioactive decay in integral form (6.16).

At time $t = T_{1/2}$, the number of radioactive nuclei N decreases by half, and therefore $N = N_0/2$. Substituting these values of N and t into the Eq. (6.16), we have:

$$N_0/2 = N_0 \exp(-\lambda T_{1/2}) \quad (6.17)$$

Now, we can solve this Eq. (6.17) with respect to the half-life period $T_{1/2}$. Dividing both sides of (6.17) by N_0, we get the following relation:

$$\exp(\lambda T_{1/2}) = 2 \quad (6.18)$$

Taking the natural logarithm of both sides of the Eq. (6.18) and solving it to find $T_{1/2}$, we have the following formula, connecting the half-life time $T_{1/2}$ and the radioactive decay constant λ:

$$T_{1/2} = \ln 2/\lambda = 0.693/\lambda. \quad (6.19)$$

The formula (6.19) between $T_{1/2}$ and λ confirms the statement drawn from the law of radioactive decay in integral form (6.16): ***the greater the radioactive decay constant λ, the shorter the half-life period $T_{1/2}$***. In other words, ***a larger λ implies a faster decrease in the number of radioactive nuclei over time.***

As previously noted, radionuclides (radioactive isotopes) essentially cease to have an impact on the environment after approximately 10 half-life periods. Substituting the expression (6.19) for the radioactive decay constant λ in terms of the half-life period $T_{1/2}$ into the law of radioactive decay in integral form (6.16), we obtain

$$N = N_0 \exp(-t \ln 2/T_{1/2}). \quad (6.20)$$

After 10 half-life periods, the number of radioactive nuclei N will be equal to

$$N = N_0 \exp[-\lambda(10T_{1/2})]. \quad (6.21)$$

Substituting the expression (6.19) for λ into the Eq. (6.21), we finally have the following result:

$$N = N_0 \exp(-10 \ln 2) = \frac{N_0}{2^{10}} = \frac{N_0}{1024} \tag{6.22}$$

The obtained result in (6.22) means that after 10 half-life periods, the number of radioactive nuclei will become approximately a thousand times smaller. It follows from this fact that since the half-life time (period) of isotopes $^{90}_{38}Sr$ and $^{137}_{55}Cs$ is about 30 years (see Table 6.1), the impact of these danger radionuclides may be felt for about three centuries.

Activity of radioactive isotopes. The concept of ***activity A of a radioactive isotope characterizes the velocity (rate) of radioactive decay or the number of decays of radioactive nuclei per unit of time***. The total amount of energy released as a result of radioactive decay per unit time will be determined by the activity itself.

Taking into account (a) the relationship (6.19) between the half-life period $T_{1/2}$ and the radioactive decay constant λ and (b) the law of radioactive decay (6.16), the formula for activity A has the following form:

$$A = -dN/dt = \lambda N = 0.693N/T_{1/2}, \tag{6.23}$$

As follows from the formula (6.23), the activity A of a radioactive isotope is directly proportional to the number N of radioactive nuclei present and inversely proportional to the half-life period $T_{1/2}$ of the isotope. This means that ***a larger number of radioactive nuclei and a shorter half-life period will result in a higher activity, indicating a faster rate of decay.***

Thus, with the same value of the number of radioactive nuclei N and the activity of radionuclide A, ***the danger of radioactive damage will increase with a decrease in the half-life period*** $T_{1/2}$. One of the most dangerous radionuclides, as can be seen from Table 6.1, are the α-active isotope of polonium $_{84}Po^{214}$ with an extremely short half-life $T_{1/2} = 1.6 \cdot 10^{-4}$ s, as well as radioactive isotopes of radon $_{86}Rn^{222}$ and iodine $_{53}J^{131}$, which have rather short half-lives periods $T_{1/2}$ according to 3.8 days and 8 days. Thus, it is known that radioactive iodine accumulates mainly in the human thyroid gland in the absence of a non-radioactive isotope of iodine. That is why specialists in the field of radiation oncology recommend using this non-radioactive isotope of iodine in case of danger of radioactive damage due to the action of γ-active isotope of iodine.

As for the units of measurement of activity A, its basic units are Becquerel (Bq) and Curie (Ci). Becquerel characterizes the activity of a radioactive isotope, which corresponds to one decay per second is the activity of a radionuclide, for which one radioactive decay occurs in one second (1 Bq = 1 decay/s). The second unit of activity in 1 Curie (Ci) significantly greater activity, which corresponds 37 billion decays in one second (1 Ci = $3.7 \cdot 10^{10}$ decays/s = $3.7 \cdot 10^{10}$ Bq).

Specific features of radioactive decay processes. A crucial role in understanding radioactive decay belongs to ***the law of conservation of electric charge***. According to this law, the total electric charge before and after a decay process remains constant. In the context of nuclear transformations, this means that ***the sum of the charge numbers of the particles and nuclei formed during radioactive decay equals the charge number of the original nucleus.***

For instance, uranium isotopes typically have a charge number Z equal to 92, while helium isotopes and alpha particles have Z equal to 2. Therefore, when a uranium isotope ${}_{92}U^{238}$ undergoes radioactive decay, the resulting nuclei of Thorium (${}_{90}Th^{234}$) and alfa particles (${}_{2}He^{4}$) must have charge numbers that sum up to 92, ensuring the conservation of electric charge throughout the process of the following nuclear reaction:

$$ {}_{92}U^{238} \rightarrow_{90} Th^{234} +_{2} He^{4} \tag{6.24}$$

This principle underscores the fundamental nature of charge conservation in nuclear processes.

The law of conservation of mass is a fundamental principle in physics, including in the context of radioactive decay. When a nucleus undergoes radioactive decay, the total mass before and after the decay must remain constant. This principle is described by the equation $M_n = \sum M_i + E/c.^2$. Here, M_n is the mass of the original nucleus, $\sum M_i$ is the sum of the masses of the formed nuclei and particles, E is the energy released during radioactive decay, and c is the speed of light in a vacuum. For most radioactive decays, the energy E released is relatively small compared to the rest mass energy mc^2 of the particles involved. This means that the change in mass due to the energy released is negligible in comparison to the total mass of the particles. As a result, the law of conservation of mass can be simplified in terms of mass numbers: ***the sum of the mass numbers of the nuclei and particles formed during radioactive decay is equal to the mass number of the original nucleus***.

During beta decay of chemical elements, the corresponding equations of neuclear reactions must describe the processes of the emission of electrons and antineutrinos (in electron beta decay), as well as positrons and neutrinos (in positron beta decay). These equations can be summarized as follows:

For electron β- decay

$$ {}_{Z}X^{M} \rightarrow_{z+1} Y^{M} + {}_{-1}e^{0} + \tilde{\nu} \tag{6.25}$$

For positron β- decay

$$ {}_{Z}X^{M} \rightarrow_{z-1} Y^{M} + {}_{+1}e^{0} + \nu \tag{6.26}$$

In these equations, ${}_{z}X^{M}$ represents the atomic number of the parent nucleus (before decay), ${}_{z+1}Y^{M}$ or ${}_{z-1}Y^{M}$ represents the atomic number of the daughter nucleus (after decay), ${}_{-1}e^{0}$ is an electron, $\nu_e{}^{-}$ is an electron antineutrino (for β^- decay), ${}_{+1}e^{0}$ is a positron, and ν_e is an electron neutrino (for β^+ decay). As a result of electronic or positronic β-decay, nuclei of elements are formed, which are located in Mendeleev's periodic table at the next or previous places in relation to the original element.

Formulas (6.24), (6.25) and (6.26) are sometimes called ***displacement rules*** during radioactive decay.

The distinction between soft and hard beta radiation based on their energy levels is crucial for understanding their behavior and effects. Soft beta radiation typically ranges from 0.05 MeV up to the maximum energy $E_{\max}$, while hard beta radiation extends beyond $E_{\max}$ to several MeV. In cases of hard beta radiation, where energies approach or exceed several MeV, relativistic effects must be considered due to the high speeds of the beta particles, which can approach the speed of light. These effects are accounted for using formulas derived from Einstein's special theory of relativity.

In this regard, below in Sect. 10, the fundamental problem of the development of the concept of mass in physics from the 17th to the twenty-first centuries, which is directly related to one of the precise methods of medical diagnostics—positron emission tomography (PET), will be considered. In medical diagnostics, techniques like positron emission tomography (PET) utilize the behavior of particles such as positrons, which are emitted in certain types of radioactive decay, including beta-plus decay. Understanding the relativistic effects on particle behavior is crucial for accurately interpreting PET scan results and obtaining precise diagnostic information. This highlights the intersection of fundamental physics principles with real-world applications in healthcare, demonstrating the importance of theoretical understanding in practical fields.

The experimental discovery of neutrinos was received in Raymond Davis's pioneering experiments in the Homestake Mine in South Dakota, which detected neutrinos produced by nuclear reactions in the Sun. It was done in1956, almost a quarter of century after the theoretical discovery of neutrinos by two outstanding theoretical physicists Wolfgang Pauli (1900–1958) and Enrico Fermi (1901–1954), both Nobel Prize winners in Physics, respectively, in 1945 and 1938. In fact, Raymond Davis has realized Bruno Pontecorvo's theoretical proposal to detect solar neutrinos using chlorine, according to the following nuclear reaction:

$$ {}_{17}Cl^{37} + \nu \rightarrow {}_{18}Ar^{37} + {}^{0}_{-1}\beta \tag{6.27}$$

Such an idea played a crucial role in guiding Davis's experimental efforts. In these experiments, neutrinos interacted with chlorine atoms in a large tank of cleaning fluid, transforming them into argon atoms, which were then detected and counted. This successful collaboration between theory and experiment led to a deeper understanding of solar processes and neutrino physics, ultimately earning Raymond Davis the Nobel Prize in Physics in 2002. The confirmation of neutrinos and antineutrinos not only confirmed key aspects of the Standard Model of particle physics but also opened up new avenues of research in astrophysics, cosmology, and particle physics.

It should be noted that the antineutrino was discovered by Frederick Reines, winner of the 1995 Nobel Prize in Physics, in the reaction of the conversion of a proton into a neutron

$$ {}^{1}_{1}p + \tilde{\nu} \rightarrow {}^{1}_{0}n + {}^{0}_{+1}\beta \tag{6.28}$$

Neutrinos and antineutrinos participating in nuclear reactions (6.7)–(6.8) and (6.25)–(6.26) are called electron neutrinos and antineutrinos (sometimes denoted by ν_e and $\tilde{\nu}_e$). There are other known meson neutrinos and antineutrinos (ν_μ and $\tilde{\nu}_\mu$), as well as tau neutrinos and antineutrinos (ν_τ and $\tilde{\nu}_\tau$).

Fundamentally important from the point of view of the problem of the evolution of the Universe is the question of the zero or non-zero rest mass of neutrinos. If the neutrino has a non-zero rest mass, then according to modern cosmology—the science of the evolution of the Universe—the galaxies, which nowadays are scattered due to the accelerated expansion of the Universe, after some time (of course, very long on the scale of the life span of an individual) will begin to converge. This means that the Universe will gradually shrink and such a process, according to the general laws of thermodynamics, will be accompanied by an increase in the average density and temperature of matter in the Universe. The 2015 Nobel Prize in Physics was awarded to Arthur McDonald and Takaaki Kajita for the discovery of neutrino oscillations, which proved the presence of the neutrino mass.

Estimating the age of organic matter by measuring radiocarbon concentrations. This method was proposed by the American physical chemist Willard Libby (1908–1980), who received the Nobel Prize in Chemistry in 1960 for the development of the radiocarbon dating method of organic substances. The theoretical basis of this method is as follows. Under the influence of neutrons ${}_0n^1$ in cosmic radiation, the constant formation a carbon radioisotope (or radiocarbon) ${}_6C^{14}$ from nitrogen ${}_7N^{14}$ occurs in the earth's atmosphere due to such a nuclear reaction:

$$ {}_7N^{14} + {}_0n^1 \rightarrow {}_6C^{14} + {}_1p^1 $$

Radiocarbon ${}_6C^{14}$ is β-active with a half-life period $T_{1/2} \approx 5730 \pm 40$ years.

The process of radiocarbon dating, based on the radioactive decay of carbon-14 (${}_6C^{14}$) in organic materials, is described by the following main points:

Formation and decay of radiocarbon: Radiocarbon is continuously formed in the atmosphere through the reaction of cosmic rays with nitrogen. This process results in the formation of dN^+ radioactive nuclei per unit volume. At the same time, radiocarbon undergoes radioactive decay dN^- at a rate described by the decay constant λ and the number of radiocarbon nuclei N present. The rate of increase in radiocarbon due to cosmic ray interactions is equal to the rate of decrease due to radioactive decay, i.e. $dN^+ = dN^-$.

Steady-state concentration: Due to the balance between formation and decay, the average concentration of radiocarbon remains constant in different parts of the globe. This concentration corresponds to an activity where 14 decays occur per minute for each gram of radiocarbon.

Incorporation into living organisms: Radiocarbon is absorbed by plants during photosynthesis and becomes part of the organic material. As long as the organism is alive, the concentration of radiocarbon remains constant due to continuous exchange with the environment.

Dating process: When an organism dies, it no longer takes in new radiocarbon, and the concentration of radiocarbon begins to decrease solely due to radioactive decay. By measuring the remaining concentration of radiocarbon in the remains of organic materials, scientists can determine the approximate age of the material based on the known decay rate of $_6C^{14}$. Thus, radiocarbon dating is a valuable tool in archaeology, anthropology, and other fields for determining the age of organic materials up to around 50,000 years old.

6.3.3 Theoretical Foundations of Ionizing Radiation Dosimetry

(a) ***Formulas for the exposure dose, its power, units of measurement.***

The corresponding formula for the exposure dose D_0, which determines (see Sect. 6.2.3) the amount of charge q resulting from the ionization of a unit mass m of dry air under the action of X-ray and gamma radiation, is written as follows:

$$D_0 = q/m \tag{6.29}$$

In accordance with formula (9.29), ***the unit of exposure dose in the SI system is the exposure dose of X-ray and γ-radiation, at which a charge of 1 C is formed as a result of complete ionization of 1 kg of dry air, i.e. $[D_0] = 1$ C/kg.***

Another common and extrasystemic unit of exposure dose is ***Roentgen*** (**R**). The connection between exposure dose units C/kg and R is given by the following relation:

$$1\ \mathrm{R} = 2.58 \times 10^{-4}\ \mathrm{C/kg}. \tag{6.30}$$

The physical meaning of the exposure dose unit of Roentgen gives such a definition:***1 Roentgen is such an exposure dose of X-ray and γ-radiation, which causes the appearance of an electric charge of* $2.58{\cdot}10^{-4}$ *Coulomb due to the ionization of 1 kg of dry air.*** Since each positive ion formed when one electron is removed from a neutral atom carries a charge of $1.6{\cdot}10^{-19}$ C, it is easy to calculate the following number N of pairs of positive and negative ions, occurring in 1 kg of air under the influence of exposure dose $D_0 = 1$ R:

$$\mathrm{N} = 2.58 \times 10^{-4}/1.6 \times 10^{-19} \approx 1.6 \times 10^{15}\ \text{ion pairs/kg}. \tag{6.31}$$

It is possible to give other equivalent (9.30) definitions of the exposure dose in 1 Roentgen:

(1) Roentgen is the exposure dose at which, as a result of complete ionization under the action of X-ray and γ-radiation, about 2 billion pairs of ions (more precisely, $2.07{\cdot}10^9$ ion pairs) are formed in 1 cm^3 of dry air. Note that 1 cm^3 of

dry air at temperature $T = 273$ K and pressure $P = 10^5$ Pa, as follows from the Claiperon-Mendeleev equation, has a mass of $m = 0.001293$ g $\approx 1.3{\cdot}10^{-6}$ kg;

(2) Roentgen is the exposure dose at which 8.69 mJ of X-ray and γ-radiation energy is absorbed in 1 kg of dry air. Indeed, the average energy that causes the appearance of a "positive ion + electron" pair in air is equal to $5.43{\cdot}10^{-18}$ J $\approx$ 34 eV (this number takes into account the relative concentrations and ionization potentials of atoms and molecules, which determine the composition of atmospheric air). Therefore, the total energy required for the formation of $1.6{\cdot}10^{15}$ ion pairs/kg is equal to

$$W = 5.43 \times 10^{-18} \times 1.6 \times 10^{15} = 8.69 \times 10^{-3}\ \text{J/kg} \approx 5.4 \times 10^{16}\ \text{eV/kg}. \quad (6.32)$$

A certain disadvantage of Roentgen as a unit of exposure dose is that it characterizes 1) only X-ray and γ-radiation, 2) ionization effect only in air. It is generally accepted that for the tissues of the human body, the exposure dose $D_0 = 1$ R causes the absorption of approximately 10^{-2} J per 1 kg of biomass (more precisely, 1 R = $9.6{\cdot}10^{-3}$ J/kg).

The power P_0 of the exposure dose is such an exposure dose D_0 that is related to a unit of time t, i.e.

$$P_0 = D_0/t\,, \quad (6.33)$$

or, taking into account the definition (6.29) of the exposure dose D_0, we have

$$P_0 = q/m \times t = I/m\,, \quad (6.34)$$

where $I = q/t$ is the electric current.

The following additional definitions of the power P_0 of the exposure dose can be given:

(1) ***the power P_0 of the exposure dose is the charge q that occurs in 1 s in 1 kg of dry air under the action of X-ray and γ-radiation,***
(2) ***the power P_0 of the exposure dose is the current I that occurs in 1 kg of dry air as a result of its ionization by X-rays and γ-rays.***

In accordance with formulas (6.33) and (6.34) the power P_0 of the exposure dose is measured in the following units:

$$[P_0] = \text{C/kg} \times \text{s} = \text{A/kg} \quad (6.35)$$

(in the SI system),

$$[P_0] = \text{R/s, R/min, R/hour} \quad (6.36)$$

(off-system units), i.e. Ampere per kilogram or Roentgen per second (minute, hour).

(b) **Formulas for Absorbed Dose, Its Power, Units of Measurement.**

The main formula for the value of ***the absorbed dose D_a, which is characterized by the energy E of ionizing radiation of all types, being absorbed by a unit of mass m*** and is determined by the formula

$$D_a = E/m \tag{6.37}$$

According to this definition and formula (6.35), the unit of measurement of the absorbed dose D_a in the SI system is $[D_a] = 1$ J/kg, i.e. the absorbed dose of ionizing radiation at which 1 Joule (J) of energy of any type is absorbed in 1 kg of the mass of the irradiated substance. Another equivalent name for this unit is ***Gray (Gy)***, i.e. 1 Gy = 1 J/kg = m^2/s^2. This unit is named after Louis Gray (1905–1965), the English physicist, one of the founders of radiobiology.

An extrasystemic unit of the absorbed dose of ionizing radiation is also used, which is called ***rad*** (it is an abbreviation of the term "***radiation absorbed dose***"). The unit of measurement of the absorbed dose in 1 rad is the amount of the absorbed dose of any type of ionizing radiation, at which the energy of ionizing radiation equal to 0.01 J is absorbed in 1 kg of substance. Thus, we have 1 rad = 0.01 Gy, i.e. a unit of absorbed dose of 1 Gy turns out to be 100 times greater than a unit of absorbed dose of 1 rad.

Between the absorbed dose D_a and the exposure dose D_0 there is the following directly proportional relationship:

$$D_a = f\, D_0, \tag{6.38}$$

where f is the transition coefficient that depends on the atomic number and density of the irradiated substance, as well as on the energy of ionizing radiation. The numerical value of the coefficient f is usually found experimentally.

For a fairly wide class of soft-matter substances in human bodies (the term "soft matter" was introduced by a famous French physicist Pierre Gilles de Gennes (1932–2007), the 1991 Nobel Prize laureate in Physics), it can be assumed that the exposure dose D_0 measured in Roentgens roughly corresponds to the absorbed dose measured in rads. But there are exceptions to this rule: for example, for bone tissue, the coefficient f varies from 2 to 5 (smaller values correspond to higher energies of γ-photons).

Below we will present two examples of determining the transition coefficient f. In both examples, the $D_0 = 1$ in terms of Roentgens (R).

Example 1. For dry air, 1 Roentgen (R) of exposure dose corresponds to the absorption of 8.69 mJ of X-ray and gamma radiation energy per kilogram of dry air. Therefore, the absorbed dose D_a in dry air is calculated as:

$$D_a = [8.69 \times 10^{-3}(\text{J/kg}) / 10^{-2}(\text{J/kg})]\text{rad} = 0.869\text{rad}$$

This means that D_a (in rad) is equal to 0.869 times D_0 (in R) for air. Therefore, in this case, the transition coefficient $f = 0.869$.

Example 2. For water, biological fluids, and soft tissues of the human body, 1 R of exposure dose corresponds to the absorption of 9.6×10^{-3} J/kg of gamma radiation energy. Therefore, the absorbed dose *Da* in these environments is calculated as:

$$D_a = [9.6 \times 10^{-3}(\text{J/kg}) / 10^{-2}(\text{J/kg})]\text{rad} = 0.96\text{rad}$$

This means that D_a (in rad) is equal to 0.96 times D_0 (in R) for water and biological fluids. Therefore, in this case, the transition coefficient *f* is approximately 0.96, which is approximately 1.

Thus, the transition coefficient *f* represents the conversion factor between exposure dose and absorbed dose in a particular medium, and it depends on the characteristics of that medium.

The concept of power P_a of the absorbed dose. As mentioned in 9.3.2, the effect of ionizing radiation on matter, especially on the tissues of human and animal bodies, depends significantly not only on the total dose of absorbed radiation D_a, but also on the speed with which this dose changed, i.e. on the derivative dD_a/dt from the absorbed dose over time. It is clear that this derivative determines the power P_a of the absorbed dose. With a fairly uniform effect of ionizing radiation on a substance, the power of the absorbed dose P_a is determined by the ratio of the value of the absorbed dose Da to the time interval *t* during which the ionizing radiation was active, i.e.

$$P_a = D_a/t = E/m \times t. \tag{6.39}$$

In accordance with formulas (6.35) and (6.37), the units of absorbed dose power Pa are the

$$[P_a] = \text{J /kg} \times \text{s} = \text{Gy /s} = \text{Watt /kg} \tag{6.40}$$

(in the SI system),

$$[P_a] = \text{rad /s , rad /min , rad /hour} \tag{6.41}$$

(off-system units).

(c) **Formulas for the Equivalent (Biological) Dose, Its Power, Units of Measurement.**

Conducted studies of the action of ionizing radiation have shown that biological effects significantly depend on the type of radiation at the same amount of absorbed dose, as well as on the type of biological objects in which absorption of ionizing radiation occurs at a fixed value of the exposure dose. Therefore, as mentioned above in Sect. 9.2.3, ***for the qualitative and quantitative assessment of the effect of***

ionizing radiation on objects of a biological nature, the equivalent dose D_{eq} (or, as it is also called, ***the biological dose D_b) is introduced***.

The corresponding ***formulas for this equivalent (biological) dose of D_{eq} take into account the dependence of the biological effect of ionizing radiation on the following two factors***:

(1) ***the type of radiation at the same amount of the absorbed dose D_a*** by entering the multiplier k into the right-hand side of the following formula for the equivalent (biological) dose

$$D_{eq} = kD_a. \tag{6.42}$$

This multiplier k is called ***the quality factor*** or ***relative biological effectiveness (RBE)*** of radiation. Formula (6.40) thus provides a relationship between the equivalent dose D_{eq} (or biological dose D_b) and the absorbed dose D_a and also takes into account the dependence of the biological effect on the type of ionizing radiation.

Table 6.2 contains data on the dependence of the coefficient of biological efficiency k for the most common types of ionizing radiation. It should be noted that this coefficient depends on many parameters (mass, charge, energy, etc.) of different types of radiation, which determine ionization and other physical and chemical mechanisms of radiation effects.

(2) ***the type of biological tissue in which absorption of ionizing radiation occurs at a fixed value of the exposure dose D_0***, in accordance with the following formula:

$$D_{eq} = k\,f\,D_0. \tag{6.43}$$

Indeed, formula (6.41) contains both the biological efficiency coefficient k and the transition coefficient f (see formula (6.36) for the direct-proportional relationship between absorbed D_a and exposure D_0 doses, as well as examples of calculating the transition coefficient f for different types of biological tissues).

Table 6.2 Dependence of the coefficient k of relative biological effectiveness on the type of ionizing radiation

Type of ionizing radiation	Coefficient of relative biological effectiveness
X-ray radiation	1
Gamma-radiation	1
Electrons (beta—radiation)	1
Thermal neutrons	2–3
Slow neutrons	5
Fast-moving neutrons	10
Protons	10
Alpha—radiation	20

The basic unit of equivalent dose in the SI system is 1 ***Sievert (Sv),*** which is the equivalent dose that corresponds to the absorbed dose in 1 Gray (Gy) when X-ray, γ- and β-radiation affects a biological object. In the general case, which includes other types of ionizing radiation, based on formula (6.40), we have the following relationship between the equivalent dose D_{eq} measured in Sv and the absorbed dose D_a in Gy:

$$D_{eq}(\mathrm{Sv}) = kD_a(\mathrm{Gy}). \tag{6.44}$$

Another (off-system) unit of equivalent dose is ***rem***. This abbreviation arose from the term "***roentgen equivalent man***". Since $k = 1$ for the first three types of radiation (Table 6.2), then 1 rem is the equivalent dose that corresponds to the absorbed dose in 1 rad of X-ray, γ- and β-radiation. For other types of radiation by analogy with (6.40) and (6.41), one has

$$\mathrm{D_{eq}\ (rem) = k\ D_a\ (rad).} \tag{6.45}$$

$$\mathrm{D_{eq}\ (rem) = k\ f\ D_0(R).} \tag{6.46}$$

The last relationship (6.44), which connects the biological dose D_{eq}, measured in rem, and the exposure dose D_0, measured in roentgen (R), gave rise to the term "roentgen equivalent man". Since 1 Gy = 100 rad, the same relation remains between Sievert and rem, i.e. 1 Sv = 100 rem.

The power of the equivalent dose $\mathbf{P_{eq}}$ is determined by the value of the equivalent dose $\mathbf{D_{eq}}$, referred to a unit of time, i.e.

$$\mathrm{P_{eq} = D_{eq}\ /t\ .} \tag{6.47}$$

Accordingly, the units of equivalent dose power P_{eq} are as follows:

$[\mathrm{P_{eq}}]$ = Sv /s, and $[P_{eq}]$ = rem/s, rem/min, rem/hour,

i.e., Sievert per second, or rem per second (minute, hour).

An important and difficult problem of ionizing radiation dosimetry is the calculation (reconstruction) of doses, if the radiation flow through a certain surface is known. So, for example, the absorbed dose of γ-radiation (in rad) can be calculated using the formula

$$\mathrm{D_a} = 1.6 \times 10^{-5}\mathrm{N}\ h\nu(\mu_{\mathrm{C}}+\mu_{\mathrm{ph}}+\mu_{\mathrm{e\,-\,p}}), \tag{6.48}$$

where N is the number of γ-quanta with energy $h\nu$ falling perpendicular to 1 cm^2 of the surface. The expression in round brackets characterizes the linear attenuation coefficient per unit mass of the substance, which corresponds to the known primary mechanisms of action of γ-radiation—the Compton effect (μ_C), the photoeffect (μ_{ph}) and the formation of electron–positron pairs ($\mu_{e\text{-}p}$) (see below in the next Sect. 6.3.4). The numerical coefficient allows converting the absorbed dose D_a (MeV/kg) into D_a

(rad). The dose of β-radiation can be found if the number of β-particles falling on the substance and the average value of the ionization losses during the passage of β-rays through the substance are known.

Table 6.3 contains consolidated information regarding a number of important characteristics of ionizing radiation, their systemic and other units of measurement and the relationships between these units.

At the end of this paragraph, the equivalent (biological) dose D_{eq} is calculated based on given data of the absorbed dose or the power of the absorbed dose.

Example 1. The average power of the exposure dose due to gamma radiation is approximately $(P_0)_\gamma = 15\mu$R/hour. Therefore, we can find the total exposure dose D_0 of gamma radiation for a year:

$$D_0 = (P_0)_? \times t\,(\text{hours/year}) = 15 \times 10^{-6}\,(\text{R}) \times 24 \times 365 \approx 0.13\ \text{R}.$$

Since the energy absorption of gamma radiation occurs in biological fluids and soft tissues, where the transition coefficient $f = 1$, and the quality factor $Q = k = 1$, the equivalent dose D_{eq} is given by the following result:

$$D_{eq} = k \times f \times D_0(\text{R}) = 0.13\ \text{rem}$$

Table 6.3 Activity and doses of ionizing radiation, its powers and units

Quantity	Units in SI	Off-system units	Notes
Activity	Becquerel (Bq)	Curie (Ci)	1 Bq = 1 decay/s 1 Ci = $3.7 \cdot 10^{10}$ decays/s = $3.7 \cdot 10^{10}$ Bq
Exposure dose	Coulomb/kilogram (C/kg)	Roentgen (R)	1 R = $2.58 \cdot 10^{-4}$ C/kg
Power of exposure dose	Ampere/kilogram (A/kg)	Roentgen/s, min, hour (R/s, min, hour)	1 R/s = $2.58 \cdot 10^{-4}$ A/kg
Absorbed dose	Joule/kilogram (J/kg) = Gray (Gy) = m^2/s^2	Rad (radiation Absorbed dose)	1 Gy = 100 rad 1 rad = 10^{-2} Gy
Power of absorbed dose	Gray (Gy)/s = Watt (W)/kilogram (kg)	Rad/s, min, hour	1 Gy/s = 100 rad/s 1 rad/s = 10^{-2} Gy/s
Equivalent (biological) dose	Sivert (Sv)	Rem (roentgen Equivalent man)	1 Sv = 100 rem 1 rem = 10^{-2} Sv
Power of equivalent (biological) dose	Sivert (sv)/s, min, hour	Rem/s, min, hour	1 Sv/s = 100 rem/s 1 rem/s = 10^{-2} Sv/s

Example 2. A nuclear power plant worker received the following absorbed doses D_a of ionizing radiation during a year:

$(D_a)\gamma = 0.0015$ Gy $= 0.15$rad via gamma radiation channel.

$(D_a)\beta = 0.0005$ Gy $= 0.05$rad via beta radiation channel,

$(D_a)n = 0.0005$ Gy $= 0.05$rad via slow neutrons channel,

$(D_a)\alpha = 0.0001$ Gy $= 0.01$rad via alpha radiation channel.

Using the quality factors k (or Q) for different types of radiation given in Table 9.2, the total equivalent (biological) dose D_{eq} (or D_b) received by the worker in a year is calculated as follows:

$$\begin{aligned} \text{Deq} &= \text{k}\gamma \times (D_a)\gamma + \text{k}\beta \times (D_a)\beta + \text{kn} \times (D_a)\text{n} + \text{k}\alpha \times (D_a)\alpha \\ &= 1 \times 0.15 + 1 \times 0.05 + 5 \times 0.05 + 20 \times 0.01\text{rem} \\ &= 0.50\text{rem} \end{aligned}$$

These examples illustrate the calculation of equivalent doses based on known absorbed doses or absorbed dose rates for different types of ionizing radiation.

6.3.4 Theoretical Basis of the Interaction of Ionizing Radiation with Biological Tissues

(a) ***Theoretical foundations of the interaction of X-ray radiation with biological tissues.*** The physical mechanisms of interaction of *X*-ray radiation with matter (in particular, with biological tissues) are determined primarily by the energy of *X*-ray quanta $E = h\nu$ (that is, by their frequency ν or wavelength $\lambda = c/\nu$). Radiation with a wavelength of $\lambda < 0.01$ nm $= 10^{-11}$ m $= 0.1$ Angstrom (Å) has greater energy and penetrates deeper into matter, it is called ***hard*** *X*-ray radiation. Radiation with $\lambda = 10$–100 Å $= 1$–10 nm is called ***intermediate***, and radiation with $\lambda > 100$ Å $= 10$ nm is called ***soft***, since it has a relatively low penetrating power. Let us consider the quantitative aspects of the three main primary physical mechanisms of coherent (electic) scattering, the photoeffect and incoherent (inelectic) scattering or the Compton effect, which characterize the interaction of *X*-ray radiation with biological tissues.

Coherent (elastic) scattering is observed under the condition that the energy $h\nu$ of an *X*-ray quantum (photon) is less than the energy required for the ionization of an atom, which is equal to the work A_i on the separation of an electron from an atom, that is, when the inequality $h\nu < A_i$ is fulfilled. As noted in Sect. 9.2.4, in this case, the energy of the *X*-ray quantum (photon) remains unchanged, but the direction of its propagation changes during such an elestic interaction with atoms (molecules) of matter. Figure 6.4 schematically shows the process of coherent (elastic) scattering, where wavy lines with arrows show the directions of propagation of low-energy X-ray quanta before and after their interaction with an atom, the nucleus and electron orbits of which are marked by a black dot and concentric circles.

Let us derive the basic Bragg-Wulf formula for X-ray diffraction from a crystal lattice. Consider a cubic crystal lattice with a period d. We have a family of parallel atomic planes labeled as 1, 2, 3, etc., making an angle θ with a beam of primary X-ray rays with wavelength λ. When X-rays interact with atoms, which are located in these planes, they can be reflected and interfere with each other under certain conditions. The fulfillment of the necessary condition for interference required that the path difference Δ between two rays reflected from adjacent planes must be equal to an integer multiple k of the wavelength λ, i.e., $\Delta = k\lambda$.

From the geometry of the crystal lattice (Fig. 6.5), the path difference Δ between two adjacent planes can be calculated as $\Delta = 2AB = 2d\sin\theta$, where $AB = d\sin\theta$. Setting $\Delta = k\lambda$, we get:

$$2d\sin\theta = k\lambda \tag{6.49}$$

This is the basic Bragg-Wulf formula, which relates the wavelength of the incident X-rays λ, the lattice spacing d of the crystal, the angle of incidence θ, and the order of the diffraction peak k. The formula (6.47) is fundamental in X-ray crystallography and is used to determine the spacing between crystal lattice planes from the observed diffraction pattern.

Photoeffect. The phenomenon of the photoelectric effect consists in the fact that under the influence of a sufficient amount of electromagnetic energy flow of different wavelengths (we are interested in *X*-ray radiation at the moment), the process of ionization of neutral atoms can occur with the appearance of free carriers of electric current (for example, electrons), which have a certain speed and, accordingly, kinetic energy.

It is clear that the phenomenon of the photoeffect will be observed when the energy of the *X*-ray photon $E = h\nu$ is sufficient to perform the work A_i of ionizing the atom or the work for the release of an electron from the depth of a material (for example, a metal) to its surface. When the inequality $h\nu > A_i$ is fulfilled, the photon is absorbed, that is, its energy goes to perform the ionization work, and the excess energy of the photon $h\nu - A_i$ is transformed into the kinetic energy of the electron. The described phenomenon is called the external photoeffect. Figure 6.6 gives a schematic illustration of the external photoeffect process.

energy of an incident *X*-ray photon is absorbed and used to do the work of ionizing an atom and imparting kinetic energy to an electron.

The theory of the photoeffect phenomenon was created by Albert Einstein in 1905, and Einstein's corresponding formula for the external photoeffect is as follows:

$$h\nu = A_i + mv^2/2. \tag{6.50}$$

Formula (6.48) is confirmed qualitatively and quantitatively by reliable experimental data. It should be noted that Einstein's formula, which directly used Max Planck's quantum hypothesis regarding the minimum portion of the energy of an absolutely black body, confirmed the correctness of one of the most important formulas of quantum mechanics $E = h\nu$ also for quanta-photons of the electromagnetic field.

Incoherent (inelastic) scattering (or the Compton effect) occurs when photons of sufficiently high energies ($h\nu >> A_i$) interact with electrons that are in orbits far from the atomic neucleus and, accordingly, are weakly bound to it. At the same time, the primary (incident) photon transfers part of its energy $h\nu$ to these electron, which are detached from an atom. Such electrons are called secondary or Compton electrons. Together with the secondary electrons, which have a kinetic energy of $mv^2/2$, a secondary photon with energy $h\nu'$ less than that of the primary photon, which had energy $h\nu$, is also emitted. The direction of motion of the secondary photon is also different from the direction of motion of the primary photon.

In the case of incoherent inelastic scattering (or the Compton effect), the law of conservation of energy can be written as follows:

$$h\nu = A_i + mv^2/2 + h\nu'. \quad (6.51)$$

Thus, such processes, which are described by Eq. (6.49) and illustrated by Fig. 6.7, are reduced to the following events:

1. **Ionization**: The X-ray photon can ionize the atom, meaning it can knock an electron out of its orbit around the nucleus, giving the electron kinetic energy.
2. **Photoelectric Effect**: This is described by Einstein's equation, where the photon's energy is used to detach an electron from an atom and give it kinetic energy.
3. **Compton Effect**: This is another process where the X-ray photon scatters off an electron, transferring some of its energy to the electron and causing it to recoil.
4. **Incoherent Scattering**: Part of the energy of the primary X-ray photon can be used to create a secondary photon through incoherent scattering. In these processes, the energy of the secondary photon is less than that of the primary photon, leading to a longer wavelength (reddening) in the scattered photon compared to the incident photon.

As the energy of the X-ray photon increases, the relative importance of the Compton effect compared to the photoelectric effect also increases. This is because the probability of the Compton effect occurring depends on the energy of the photon. At lower energies, such as 10^4 eV, the photoelectric effect dominates, and the Compton effect is negligible. However, as the energy increases, such as to 8×10^4 eV, the probabilities of the Compton effect and the photoelectric effect become equal. Finally, at even higher energies, such as 2×10^5 eV, the Compton effect becomes much more likely, with a 99% probability, while the photoelectric effect becomes less probable, with only a 1% probability.

As the energy of the X-ray photon increases, the relative importance of the Compton effect compared to the photoelectric effect also increases. This is because the probability of the Compton effect occurring depends on the energy of the photon. At lower energies, such as 10^4 eV, the photoelectric effect dominates, and the Compton effect is negligible. However, as the energy increases, such as to 8×10^4 eV, the probabilities of the Compton effect and the photoelectric effect become equal. Finally, at even higher energies, such as 2×10^5 eV, the Compton effect becomes much more likely, with a 99% probability, while the photoelectric effect becomes less probable, with only a 1% probability.

The Bouguer-Lambert law formula for the attenuation of X-rays in matter. When a beam of X-rays propagates through matter, their intensity is weakened due to absorption and scattering. As is quite often believed, the intensity I of a parallel beam of electromagnetic radiation (light, X-rays, and gamma rays) when passing through a layer of substance of thickness d satisfies the Bouguer-Lambert law:

$$I = I_0 \exp(-\mu d), \tag{6.52}$$

where I_0 is the initial intensity of X-rays falling on the front surface of the substance layer, and μ is the linear attenuation coefficient. This law was first discovered in 1729 by the French physicist Pierre Bouguer (1698–1758), and then derived in 1760 by the German mathematician and physicist Jogann Heinrich Lambert (1728–1777).

It should be noted that formula (6.52) of the Bouguer-Lambert law requires the fulfillment of a number of approximations, the content of which becomes clear when considering the integro-differential transport equation and the consistent electrodynamic approach to the problem of multiple scattering of electromagnetic waves (see, for example, the publications [19, 20] and links there).

In the Bouguer-Lambert law, the attenuation of X-ray intensity is described by the linear attenuation coefficient μ, which accounts for the combined effects of absorption and scattering of X-rays as they pass through a material. Mathematically, it can be expressed as the sum of the linear absorption coefficient τ, and the linear scattering coefficient σ:

$$\mu = \tau + \sigma \tag{6.53}$$

This equation reflects the fact that the total attenuation of X-ray intensity is the sum of the attenuation due to absorption processes (photoelectric effect and Compton scattering) and the attenuation due to scattering (changes in the direction of propagation of X-rays).

Taking (6.53) into account, formula (6.52) can be rewritten as an approximation of the Bouguer-Lambert law in the form:

$$I = I_0 \exp[-(\tau + \sigma)]. \tag{6.54}$$

The following formula indicates that the linear absorption coefficient τ is proportional to the product of the density ρ, the atomic number Z^4 of the absorbing medium, and the wavelength λ^3 of X-rays:

$$\tau = k\rho Z^4\lambda^3, \tag{6.55}$$

where k is the proportionality factor.

(a) ***Primary processes of interaction of radioactive rays and particle flows with matter.*** As noted in Sect. 6.2.4 of Section A, the interaction of α- and β-radioactive radiation with matter is associated with such processes as the excitation and ionization of atoms, as well as at sufficiently high energies of α- and β-particles with the appearance of characteristic X-ray radiation. The primary mechanisms of action of relatively low-energy γ-radiation on matter (the specific values of such energies for various environments will be given below) are reduced to the photoeffect and the Compton effect.

For high-energy gamma radiation, another significant interaction mechanism becomes important: the formation of electron–positron pairs. This process can occur when a gamma-ray photon with energy greater than or equal to twice the rest energy of an electron $2E_0 = 2m_0c^2 = 1.02$ MeV interacts with matter.

Here, we note that the fundamental question of the nature of mass in classical and quantum mechanics, as well as in the theory of relativity, including the important problem of the limits of application of Einstein's famous formula $E = mc^2$, will be specially considered in Sect. 7.3.4 of the next chapter.

Exactly, the attenuation of gamma radiation can be described by the Bouguer-Lambert law, where the total linear attenuation coefficient μ is the sum of three components:

$$\mu = \mu_{ph} + \mu_C + \mu_{e-p}, \tag{6.56}$$

where μ_{ph} is the linear attenuation coefficient due to the photoelectric effect, μ_{C} is the linear attenuation coefficient due to the Compton effect, $\mu_{\text{e-p}}$ is the linear attenuation coefficient due to the formation of electron–positron pairs. Each of these mechanisms contributes to the overall attenuation of the gamma radiation beam as it passes through a material. The relative importance of each component depends on the energy of the gamma radiation and the properties of the absorbing material.

In substances with high atomic numbers like lead, the dominant interaction mechanism for gamma photons is as follows:

At energies lower than 1 MeV, the photoelectric effect mainly occurs.

At medium energies (2–5 MeV), the Compton effect is dominant.

At high energies, the formation of electron–positron pairs becomes important.

As a result, the overall attenuation coefficient μ initially decreases with increasing photon energy due to the reduction of the photoelectric effect, but then increases due to the increasing importance of pair production.

For substances with low atomic numbers like water:

At gamma photon energies below 100 keV, the photoelectric effect predominates.

At around 1 MeV, the Compton effect becomes dominant.

At energies above 10 MeV, the process of electron–positron pair formation prevails.

In air, the situation is as follows:

The photoelectric effect is significant only at photon energies up to 0.01 MeV.

The Compton effect mainly occurs at higher energies.

The formation of electron–positron pairs is possible only at energies around 30–40 MeV.

To further understand how radioactive radiation interacts with a substance, several parameters are used:

1. **Specific ionization.** This parameter quantifies the number of ion pairs formed per unit length (typically per 1 cm) along a particle's trajectory within a substance. It provides information about the ability of a radiation particle to ionize the medium along its path.
2. **Ionization losses.** These losses indicate the amount of energy lost by a particle per unit length (typically per 1 cm) along its path within the substance. It reflects the ability of the particle to transfer energy to the medium through ionization processes.
3. **Total traversal distance (L).** The total traversal distance L provides an indication of how far a particle can travel in a medium before losing sufficient energy to thermalize, i.e., its energy E matches the thermal energy kT of the substance. At room temperature T (approximately 293 K), the thermal energy is $kT \approx 4 \times 10^{-21}$ J or about 2.5×10^{-2} eV. The total traversal distance L provides an indication of how far a particle can travel in a medium before losing sufficient energy to thermalize.

These parameters are essential for understanding the interaction of radioactive radiation with matter and are used in various fields such as radiation protection, dosimetry, and nuclear physics.

Of course, these values depend on the type of radiation. For alpha particles with an energy of 8 MeV, specific ionization, ionization losses, and total traversal distance values can be significant. The maximum value of specific ionization in air for 8 MeV alpha particles is approximately 8×10^4 ion pairs per 1 cm of path. This value is not reached immediately upon entry of the alpha particle into the air but after it has traveled approximately 6 cm. At this point, the alpha particle's speed has decreased enough for its ionizing capacity to become maximal. At the point where the specific ionization is maximal (around 6 cm of path in air for 8 MeV alpha particles), the ionization losses reach the value of 2.7 MeV/cm. This means that the alpha particle loses approximately 2.7 MeV of energy for every centimeter it travels through air at this point. These values highlight the significant ionizing capacity of alpha particles, which is why they are commonly used in applications such as certain types of radiation therapy.

In liquids such as water and biological fluids, the total path length of alpha particles is typically much smaller than in air, approximately $L \approx 0.1 - 0.01$ mm. This is because these liquids contain many hydrogen atoms, whose mass is close to that of alpha particles. When the masses of colliding particles are similar, kinetic energy is transferred more efficiently from one particle to another. This efficiency is described by the accommodation coefficient, which approaches 1 in such cases. The same effect explains the rapid deceleration of alpha particles in the human body, which consists of 70–80% liquids similar in consistency to water. Similarly, heavy water (D_2O) is used to slow down neutrons in nuclear reactors due to its ability to efficiently transfer kinetic energy from the neutrons to the deuterium nuclei.

For beta particles, their attenuation in matter also follows approximately the Bouguer-Lambert law. The linear attenuation coefficient μ_β for beta radiation is influenced by processes such as ionization and excitation of atoms and molecules, deceleration leading to X-ray radiation, and scattering by electrons of atoms and molecules.

In the case of positron beta decay, where a proton transforms into a neutron with the emission of a positron (Eq. 6.5), the subsequent collision of the positron with an electron leads to an annihilation reaction.

$$^{0}_{-1}\beta + ^{0}_{+1}\beta \rightarrow \gamma + \gamma \tag{6.57}$$

This nuclear reaction results in the production of two γ photons, which are emitted in opposite directions due to the law of conservation of momentum. This process, known as electron–positron annihilation, is the reverse of the process where high-energy gamma photons transform into electron–positron pairs.

Beta particles (electrons) in the human body are characterized by significantly larger values of the total traversal distance L_β compared to alpha particles, approximately $L_\beta \approx 10 - 20$ mm. This greater penetration depth of beta particles is due to their lower mass compared to alpha particles ($m_\alpha/m_\beta \approx 7345$), as well as their significantly lower ionization losses and specific ionization.

For relativistic beta particles, which are electrons moving at speeds close to the light velocity, the specific ionization is approximately 50 ion pairs per 1 cm of path. This lower specific ionization compared to alpha particles contributes to their greater penetration depth in matter.

The interaction of gamma radiation, protons, and neutrons with matter involves various processes, as discussed earlier. In inelastic interactions of nuclei with neutrons, the nucleus can become excited and emit one or more gamma photons. When a neutron is absorbed by a nucleus, a nuclear reaction occurs, often leading to the transformation of the nucleus into a radioactive isotope followed by beta decay and emission of gamma photons. In biological tissues, neutrons can lead to the formation of deuterium (heavy hydrogen) through the reaction $^{1}_{0}n +^{1}_{1} H \rightarrow^{2}_{1} H + \gamma$, as well as radioactive isotopes of sodium through the reaction $^{23}_{11}Na +^{1}_{0} n \rightarrow^{24}_{11} Na + \gamma$. Radiocarbon can be formed through various reactions, such as $^{14}_{7}N +^{1}_{0}n \rightarrow^{14}_{6} C +^{1}_{1}p$. Ionizing radiation can also disrupt the structure of molecules in a substance. Water radiolysis, for example, involves the ionization and subsequent decay of an ionized water molecule, leading to the formation of unsaturated H and OH radicals. These radicals are highly chemically active due to their unsaturated valences.

6.4 Conclusions

In this chapter, an attempt was made to outline the basic physical principles of radiation medicine in a fairly concise form, taking into account the following circumstances, which appeared to be quite important to the author. First, as already noted in the introductions to Chaps. 1 and this chapter, the author did not pursue the task of writing another book on medical physics, devoted to a detailed presentation of the physical and medical aspects of radiation medicine, at least because such a book would hardly be able to withstand competition with excellent monographs and textbooks created by well-known authors on this topic (see links to some of them in the literature list for this chapter). Secondly, the author could not ignore one of the most fundamental topics of medical physics, devoted to the exposition of the basic physical principles of the impact of ionizing radiation on medical and biological systems. That is why this chapter was written, in which the presentation of the basic physical principles of radiation medicine was carried out on two levels, using a fairly popular approach in Sect. 6.2 and a more rigorous approach in Sect. 6.3 using mathematical apparatus and methods of theoretical physics.

In Sect. 6.2.1 of Section A, a popular explanation of the physical foundations of the formation of continuous deceleration and linear characteristic spectra of *X*-ray radiation was given. The discovery of *X*-rays, as one of the types of ionizing radiation widely used in medicine, was marked by the awarding of the first Nobel Prize in Physics, which was awarded to Wilhelm Konrad Roentgen in 1901. In the same Sect. 6.2.1, the contribution to the discovery of X-rays was noted, which was made by Ivan Puluj, a famous electrical scientist and physicist, an Ukrainian by origin, who constructed a cathode ray tube, which was essentially the first source of *X*-rays, but who did not pay due attention to this invention.

The use of the features of the spectra of X-ray radiation played an extremely important role in the development of the science of the microworld—quantum mechanics. As is known, the immediate and direct sign of any quantum mechanical regularity is the presence in the corresponding formula of Planck's constant $h = 6.626.10^{-34}$ J.s, which was introduced by Max Planck at the end of 1900 as a characteristic of the minimum portion of radiation energy (quantum) of an absolutely black body. It turned out that one of the most accurate methods for determining the numerical value of Planck's constant h is the use of formula (6.1) of Sect. 6.3.1 for the short-wavelength boundary of the continuous deceleration spectrum of *X*-ray radiation. Formula (6.3) for the flow of *X*-rays, as well as formulas of Moseley's law for the dependence of the frequency $\nu(Z)$ of characteristic *X*-ray radiation on the serial number Z of chemical elements, were also considered here, which was a direct confirmation of the correctness of Mendeleev's periodic law.

Further, in Sect. 6.2.2 and especially in Sect. 6.3.2, the necessary information about (a) the properties of alpha, beta, and gamma radioactive radiation was presented; (b) differential and integral forms of the law of radioactive decay (see formulas (6.8) and (6.14), respectively): and (c) formulas for the half-life period $T_{1/2}$ and its value for a number of radionuclides. In particular, for such radioactive isotopes that are capable of lingering in the body of humans and animals for a long time (for example, isotopes of strontium and cesium with a half-life period of 28.6 and 30.2 years). The concept of activity A of a radioactive isotope is introduced, which characterizes the rate of radioactive decay. Formula (6.20) makes it possible to draw an important conclusion that the activity of A increases with an increase in the number N of radioactive nuclei at a given time and with a decrease in the half-life period $T_{1/2}$. In Sect. 9.3.2, the theoretical method of estimating the age of organic substances based on the calculation of the concentration of a radioactive isotope of carbon with a half-life period $T_{1/2}$ of 5600 years, which is constantly formed as a result of the nuclear reaction (6.27) during the interaction of cosmic radiation neutrons with atmospheric nitrogen nuclei, is also analyzed.

In the subsequent Sects. 6.2.3 and 6.2.4 of Section A and in Sects. 6.3.3 and 6.3.4 of Section B, the general provisions and theoretical foundations of ionizing radiation dosimetry, as well as the primary physical and biophysical mechanisms of the interaction of ionizing radiation with matter, were presented. Speaking more specifically, in Sect. 6.2.3, exact definitions of exposure, absorbed and equivalent (biological) doses of ionizing radiation were formulated, as well as the power of these three doses and its units. In addition, Sect. 6.2.4 presented a popular description of all the primary processes of the interaction of ionizing radiation with matter, namely: (a) for X-ray radiation, such three main processes as coherent (elastic) scattering of X-rays, which played a decisive role in one of the most outstanding experimental discoveries of the twentieth century—the spiral structure of DNA by Rosalind Franklin in 1952; the photoelectric effect, the theory of which was created by Albert Einstein and marked by his awarding the Nobel Prize in Physics in 1921; incoherent (inelastic) scattering (or the Compton effect), discovered by Arthur Compton in 1923, who was awarded the Nobel Prize in Physics in 1927 for this discovery.

Important aspects of the interaction of ionizing radiation with matter were discussed in Sects. 6.2.4, 6.3.4. In particular, the main primary processes of interaction with matter of X-ray beams, gamma radiation, and particle streams are described at the popular level and studied theoretically. The main primary physical mechanisms of the interaction of all types of ionizing radiation with an increase in quantum energy are discussed, which manifest themselves in the following sequence: coherent (elastic) scattering, the photoeffect, incoherent (inelastic) scattering and the generation of electron–positron pairs; a comparative analysis of the ionizing and absorbing capacity of alpha-, beta-, and gamma-radioactive radiation was also conducted; considered features and approximate nature of application of the Bouguer-Lambert law to the problem of attenuation of ionizing radiation in various environments; some physico-chemical mechanisms and biological consequences of radiation effects are discussed, and special attention was paid to examples of calculation of various doses of ionizing radiation, statistical data at various stages of radiation damages, as well as other physical problems of radiation medicine.

References

1. J. Van Dyk, *True Tales of Medical Physics. Insights into a Life-Saving Specialty*, 1st edition, Springer Cham (2022)
2. S. Keevil, R. Padevani, S. Tabakov, C. Lewis, *T*, 1st edn. (Greener, Introduction to Medical Physics, CRC Press, 2022)
3. J.T. Bushberg, J.A. Seibert, *The Essential Physics of Medical Imaging. Study Guide*, LWW (2022)
4. I.R. Cameron, I.G. Skofronik, *Medical Physics* (Wiley, New York, 1878)
5. A. Martin, S.A. Harbison, *An Introduction to Radiation Prorection* (Chapman and Hall, London, 1979)
6. *Graduate Studies in Medical Physics*. University of Wisconsin-Madison, Madison (1996)
7. K. Russel, *Hobbie* (New York, Berlin, Heidelberg, Springer, Intermediate Physics for Medicine and Biology, 1997)
8. M. Nelkom, P. Parker, *Advansed Level Physics*, 5th edn. (Heinemann Educational Books Ltd., London, 1982)
9. *Medical and Biological Physics*, vol. 1, Ed. A.V. Chalyi. Bogomolets National Medical University, Kyiv (1999)
10. *Medical and Biological Physics*, vol. 2, Ed. A.V. Chalyi. Bogomolets National Medical University, Kyiv (2001)
11. *Medical and Biological Physics*, Ed. A.V. Chalyi. Knyga Plus, Kyiv (2005)
12. J. Puluj, *Nachtrag zur Abhandlung "Uber die Entstehung der Rontgen'schen Strahlen und ihre photographische Wirkund* (1896)
13. V.A. Shenderovsky, He belonged to those whonshaped the world. To the 150th anniversary of the birth of I. Puluy. Bulletin of the National Academy of Sciences of Ukraine. № 1–2, 56–60, (1995)
14. R. Goida, R. Plyatsko, *Ivan Puluy/Life and Creation* (Research and Publishing Chenter of Shevchenko Scientific Society (NTSh), Lviv, 2019)
15. P. Spiegel, The first clinical X-rays made in America—100 years. American J. Röntgenol. **164**, 241–243 (1995)
16. W.L. Bragg, The diffraction of short electrrjomagnetic waves by a crystal. Proc. Cambridge Philis. Sco. **17**, 43 (1914)
17. G.V. Wulf, Die Krystallrontgenagramme, *Fhysikalische Zeitschrift*, Bd 14 (1913)
18. W. Georgy (Yuri) Victorovych/ Y.M. Randyuk// *Encycclopedia of Modern Ukraine (electronic resource)/Editorial board: I.M. Dzyuba, A.I. Zhukovskiy, M.G. Zheleznyuk [et al.], Nat. Acad. Sci. of Ukraine,* NTSh, Kyiv, Institute of Encyclopedic Studies of Nat. Acad. Sci. of Ukraine (2006). *https:// esu. com.ua/article—30112*
19. A.V. Chalyi, Analysis of the microstructure of scattering and absotbing media taking into account multiple scattering of radiation. Ukr. J. Phys. **11**, 67–74 (1966)
20. A.V. Chalyi, E.L. Lakoza, Multiple light scattering near the critical point. Adv. Phys. Sci. **26**, 573 (1983)

Chapter 7
Application of Quantum Mechanics Methods in Medical Diagnostics

Abstract This chapter is devoted to the presentation of the basics of quantum mechanics—one of the main branches of physical science, which studies the laws of the micro and nano world. These laws find their effective application in medical physics for the creation of modern precision methods of medical diagnostics. First of all, this concerns such widely used tomographic methods as X-ray computed tomography (CT) using a layer-by-layer scanning to obtain medical imaging due to the gradual passage of an X-ray beam through a thin layer of tissues of the human body; magnetic resonance imaging (MRI) based on the phenomenon of nuclear magnetic resonance (NMR); positron emission tomography (PET), being a method of medical radioisotope diagnostics based on the phenomenon of electron and positron annihilation. As in all previous chapters, not only a brief general description of the basic ideas underlying the tomographic methods of medical diagnostics (see Sect. 7.2), but also a strict mathematical and physical justification of basic theoretical provisions of quantum mechanics is offered to the attention of the reader. As well as the mentioned tomographic methods of medical diagnostics (see Sect. 7.3). In addition, in connection with the growing interest in the use of PET not only as a diagnostic method, but also as a therapeutic radioisotope method for the treatment of oncological diseases, in the last Sect. 7.3.4 the intriguing problem of the law of conservation of mass in the reaction of annihilation of matter and antimatter is discussed (in particular, electron and positron in the method of positron emission tomography), as well as the general most mysterious problem of natural science about the existence of dark energy and dark matter.

Keywords Quantum mechanics · X-ray computed tomography · Magnetic resonance imaging · Positron emission tomography · Law of mass conservation · Dark matter and dark energy

A. V. Chalyi, *What is Medicine?*, https://doi.org/10.1007/978-3-031-64979-0_7

'...Just look along the road, and tell me if you can see either of them."

"I see nobody on the road," said Alice.

"I only wish I had such eyes," the King remarked in a fretful tone.

"To be able to see Nobody! And at that distance, too!!"

"Through the Looking-Glass" by Lewis Carroll, writer and mathematician [1]

7.1 Introduction

Quantum mechanics, which originated at the turn of the nineteenth and twentieth centuries, was from the very beginning and now, more than a hundred years later, remains the main science of atomic phenomena at the nanoscopic level [2–7]. The essence of the processes taking place in the nanoworld is beyond direct human perception. Therefore, attempts to intuitively describe quantum mechanical objects with the help of some classical images based on our everyday experience lead to rather brutal representations, which, most often, have no relation to reality.

A well-known example illustrating this statement is related to the history of the discovery of electron spin [7]. In an attempt to explain the experimental observation of a doublet in the spectrum of sodium, which appeared as two closely spaced yellow spectral lines, the Danish physicists Samuel Goudsmit (1902–1978) and George Uhlenbeek (1900–1988) assumed in 1925 that this doublet was connected with their own rotational motion of the electron cloud. However, such a visual representation of the electron as a rotating body, i.e., as a kind of spin, turned out to be wrong. This happened as a result of their conversation with the Danish theoretical physicist Niels Bohr (1885–1962), the 1922 Nobel Prize winner in Physics, to whom they told about their discovery and who calculated the linear speed of rotation of such an electron cloud at its periphery. This speed turned out to be greater than the speed of light, which is of course physical nonsense. However, Uhlenbeck and Goudsmith's introduction of the concept of spin as some internal characteristic of an electron, similar to its charge or mass, became very useful for the further development of quantum mechanics.

It is natural that the task of creating in-depth understanding of the quantum mechanical laws of the world of microparticles among medical students and doctors is fundamentally unrealistic due to the lack of time allocated to the course of medical physics and, as a rule, to the study of this important section as well. We can only hope that over time this most profound science of the physical world of animate and inanimate nature surrounding us (as well as all other fundamental sciences) will be given more attention in higher medical schools. This is necessary at least so that a qualified doctor has at least an initial idea of the principles underlying such diagnostic methods as Computed Tomography (CT), Magnetic Resonance Imaging (MRI), Positron Emission Tomography (PET), with the help of which now the majority of precise medical diagnoses are established [8–20].

This chapter contains brief theoretical information about the elementary foundations of quantum mechanics and their application in medical diagnostics. Per aspera ad astra!

7.2 Section A

The place of quantum mechanics in the system of sciences about the movement of bodies. In general, there are four mechanics:

1. ***The classical mechanics of Newton-Galileo*** (XVI–XVII centuries) is valid for cases when the speed of the body υ is significantly less than the speed of light propagation c ($\upsilon << c$), and the mass of the body m is significantly greater than the mass of the electron m_{el} ($m > > m_{el}$);
2. ***Mechanics of the theory of relativity*** (beginning of the twentieth century), describes the movement of a body with a high speed υ, which can approach the speed of light $c(\upsilon \approx c)$ and the mass of the body m is much greater than the mass of the electron m_{el} ($m >> m_{el}$);
3. ***Quantum mechanics (non-relativistic quantum mechanics, i.e., without taking into account the effects of the theory of relativity)*** describes the movement of bodies with a small mass m of the order of the mass of an electron m_{el} ($m \approx m_{el}$) and the speed of the body υ significantly less than the speed of light propagation $c(\upsilon << c)$;
4. ***Relativistic quantum mechanics***, that is, taking into account the effects of the theory of relativity, is valid for cases when the mass m is of the order of the mass of an electron m_{el} ($m \approx m_{el}$), and the speed υ can approach the speed of light $c(\upsilon \leq c)$.

7.2.1 Basic Concepts of Quantum Mechanics

Quantum mechanics is a science that describes the movement of microparticles, i.e., elementary particles, atomic nuclei, molecules, and systems composed of them. The laws of quantum mechanics form the basis for studying the structure of matter, allow us to find out the structure of atoms, the nature of chemical bonds, explain the periodic system of chemical elements, and create a reliable theoretical basis for the most modern precise methods of medical diagnosis, which are widely used in practical medicine [9–14, 17, 18].

At the beginning of the XX century it became obvious that classical physics is unable to describe the properties of systems consisting of microparticles. It is believed that the beginning of the creation of quantum mechanics occurred at the end of 1900, when the German theoretical physicist Max Planck (1858–1947), 1918 Nobel Prize winner in Physics, wrote down his famous formula for the minimum portion (quantum) of energy, which he used to explain the spectrum of the emission of an absolute black body [20] (see formula (7.3) in section B of this Chapter, as well as paragraph 5.3.2, which is devoted to the laws of radiation of an absolutely black body).

In addition, it turned out that the planetary model of the atom by the British physicist Ernst Rutherford (1871–1937), 1908 Nobel Prize laureate in Chemistry, and Niels Bohr, which includes a nucleus and electrons rotating in orbits around it, was unable to explain the linear nature of atomic spectra and the very fact of the stability of atoms. From the point of view of classical electrodynamics, the exceptional stability of atoms contradicts this model of their structure, since charged particles (electrons) moving with centripetal acceleration must radiate their energy and eventually fall on the atom's nucleus.

A solution was proposed by Niels Bohr in 1913, as a result of which contradictions were eliminated by introducing assumptions that did not satisfy both the laws of classical mechanics and the laws of classical electrodynamics. We are talking about Bohr's famous postulates [21], the essence of which consists in the following provisions.

In accordance with ***the first postulate of Niels Bohr***, from the set of possible electron orbits, only such discrete "stationary" orbits are realized, during the rotation of the electron along which the angular momentum of the electron in its orbit is determined by the formula (7.4) and is a multiple of the value $\hbar = h/2\pi$, where h is called Planck's constant (see Max Planck's quantum formula (7.3) for the minimum portion (quantum) of energy). According to this postulate of Niels Bohr, electrons do not emit or absorb energy when moving along stationary orbits.

In accordance with ***the second postulate of Niels Bohr***, radiation (absorption) of energy by an atom occurs only when an electron moves from one energy level to another. Formula (7.5) determines the dependence of the quantum energy (and its frequency) on the difference in atomic energies at those levels between which electronic transitions occur.

Thus, while preserving Rutherford's planetary model, Bohr introduced the ideas of Planck's quantum theory into it. This is how a new theory was created—the theory of Niels Bohr, which became the next step in the development of the theory of the structure of the atom. Bohr's theory not only gave a visual picture of the movement of an electron in a hydrogen atom, but also made it possible to calculate the possible values of its energy. At the same time, Bohr obtained the formula (7.6) for the frequencies emitted by hydrogen atoms. This Bohr formula generalized Balmer's formula, which was obtained by him on the basis of experimental data for the spectral lines of hydrogen atoms.

However, the success of Bohr's theory could not be extended to more complex atomic and molecular spectra. The weak side of this theory, which led to its subsequent failures, was its internal logical contradiction—it was neither a consistent classical nor a consistent quantum theory. A new quantum theory was needed that would describe the behavior and properties of microparticles. The basis for the creation of such a new quantum mechanical theory was the works of de Broglie, Heisenberg and Schrodinger.

Hypothesis of Louis de Broglie

Quantum mechanics, which was developed in the works of Max Planck, Albert Einstein, Niels Bohr and other scientists during the first two decades of the twentieth century, is often called old quantum mechanics. Its certain shortcomings were mentioned above. The first step on the way to the creation of a new quantum theory was the hypothesis proposed in 1924 by the French theoretical physicist Louis de Broglie (1892–1987), 1929 Nobel Prize laureate in Physics, that particle-wave dualism is a feature not only of optical phenomena, but also inherent in all material moving objects [22]. In other words, any particle (for example, an electron) is not only a particle (corpuscle) itself, but also a certain wave, that is, it also has wave properties, and the nature of this wave was not yet known.

As for the dualism of optical phenomena, it means that ***the light*** has at the same time a dual nature, being (a) ***electromagnetic waves*** of a certain frequency and (b) ***a flow of particles (photons)*** that have certain values of energy and momentum. The photon mass issue will be specifically discussed later in Sect. 7.3.4 of this chapter.

In 1927, de Broglie's hypothesis received brilliant experimental confirmation. American physicists Clinton Davison and Lester Germer observed the diffraction of electrons when they were reflected from a single crystal of nickel, in which the atoms and the spaces between them played the role of a diffraction grating. The coincidence of the experimental and theoretical value of the electron wavelength, which was obtained, respectively, from the analysis of the experimental diffraction pattern according to the Bragg-Wulf formula (6.49) and from the theoretical formula of Louis de Broglie (7.7), was characterized by an error equal to only 1.2%. In 1929, Louis de Broglie received the Nobel Prize in Physics "for the discovery of the wave nature of the electron".

The discovery of the wave properties of electrons contributed to the creation of the electron microscope, which turned out to be extremely important for theoretical and practical medicine, in particular for conducting medical-biological and physicochemical research. In contrast to optical microscopes, image formation in an electron microscope is carried out using electrostatic or magnetic lenses.

Werner Heisenberg's Uncertainty Principle. A characteristic feature of the classical mechanics of Newton-Galileo was the determinism of the concept of the trajectory of a body's movement, according to which, by solving the equation of Newton's second law, which satisfies certain initial conditions, it was possible to accurately predict the coordinate and speed of the body at any moment in time in the future and in the past. This principle of the determinism of the movement of celestial bodies is especially evident in astronomical observations, where we are talking about the movement of bodies with large masses compared to the mass of an electron (such as, for example, the planets of the Solar System), which move at a relatively low speed compared to the speed of light.

A completely different situation occurs in quantum mechanics, where this determinism disappears due to the realization in the microcosm of the relationship of uncertainties between the coordinate and the momentum (here it should be recalled that the momentum in the classical mechanics of Newton-Galileo is considered to be the product of the body's mass and its speed of movement). This pronciple of uncertainties in quantum mechanics was proved in 1927 by the German theoretical physicist Werner Heisenberg (1901–1975), laureate of the Nobel Prize in Physics in 1932 [19, 23].

The verbal meaning of Heisenberg's uncertainty principle is that it is impossible to accurately determine the coordinate and momentum (velocity) of a microparticle (for example, an electron in its orbit) at the same time. Thus, quantum mechanics necessarily abandons the deterministic concept of the trajectory of motion inherent in classical Newton-Galilean mechanics. Therefore, the forced transition from a deterministic to a probabilistic description of the position of microparticles in space is one of the essential and fundamental features of quantum mechanics as a science of processes in the microcosm and nanocosm that actually occur inside the human body at the atomic and molecular levels of the organization of living systems.

Since the real exact language of natural sciences is not so much the language of words, but rather, in my opinion, the language of mathematical formulas, the author asks his dear readers to pay attention to the formula (7.10) in Sect. 7.3.2, which is a mathematical form recording the Heisengberg uncertainty principle for the coordinate and momentum of a microparticle, as well as formula (7.11), which characterizes the uncertainty principle (in other words, the impossibility of accurate simultaneous measurement) for the energy of a microparticle and its life time in a fixed energy state.

7.2.2 *Tomographic Methods of Medical Diagnostics*

A great achievement of modern medicine is the introduction of tomographic diagnostic methods into medical practice. Tomography (from the Greek words "$\tau o \mu \eta$"—layer, cross-section and "$\gamma \rho \alpha \phi \omega$"—write) refers to research methods that allow you to make layer-by-layer images of a certain part of the object under study (in particular, a certain part of the human body).

This paragraph will provide general information regarding the following main tomographic methods that are widely used in modern medical diagnostics:

1. X-ray computed tomography (CT).
2. Magnetic resonance imaging (MRI), based on the phenomenon of nuclear magnetic resonance (NMR),
3. Positron emission tomography (PET).

X-ray Computed Tomogaphy (CT)

The essence of *X*-ray computed tomography (CT) is reduced to obtaining a layer-by-layer picture in the method of medical imaging based on *X*-ray layer-by-layer scanning (tomography) [8–14, 17, 18, 24, 25]. The specified purpose of this method is based on the gradual passage of a beam of *X*-rays through a thin layer of tissues of the human body. At the same time, two of the following three parts of an *X*-ray computer scanner move relative to each other: (a) X-ray tube; (b) detector on which the X-ray image is captured; (c) the patient as an object of research. In modern X-ray computed tomography, such a situation is implemented, in which the object of research remains stationary, and the X-ray tube and detector rotate around the patient. At the same time, the X-ray tube and the detector are moved in opposite directions so that the local position of a certain section of a thin layer of biological tissue has its own clear image in a fixed location of the detector. In order to obtain an X-ray image of the adjacent layer, the distance between the tube, the layer and the detector must be changed.

Computer processing of the results of the weakening of X-ray radiation as a result of its absorption and scattering (see the theoretical basis of physical processes that occur in the human body under the action of X-rays in Sect. 6.3), allows to detect the difference in the densities of the object under study (for thin layers of human organs and tissues with a thickness of about 0.5 mm), which is unreachable during an ordinary X-ray examination.

The first successful development of this research method belongs to the British electrical engineer and physicist Godfrey Hounsfield (1919–2004), who created in 1968 the "prototype" of the X-ray computer tomograph, which received the name EMI-scanner. It has been clinically approved since 1968 to 1972 by J. Ambrose at Atkinson Morley Hospital (USA). The first official report on the use of a new method of X-ray research was made by Hounsfield and Ambrose on April 19, 1972 at the annual congress of the British Institute of Radiology. They obtained the world's first clear image of a tumor inside the human brain.

Independently of Godfrey Hounsfield, the South African and American physicist Allan Cormack became interested in a similar problem starting in 1956. He realized that in order to determine the total absorbed dose of X-ray radiation in the treatment of malignant tumors, it is necessary to have information about the absorption of X-rays in different layers of the patient's tissues, which can be obtained by irradiating a certain internal area at different angles. As a result, Allan Cormack became interested in the mathematical problem of how to obtain the necessary information about the details of the internal structure of a certain two-dimensional layer of tissue and then find the three-dimensional structure of the entire object. He successfully coped with this difficult mathematical problem by using the mathematical transformations in the integral geometry of the Austrian mathematician Johann Radon (1887–1956) and the three-dimensional transformations of the French mathematician and physicist Jean-Baptiste Fourier (1768–1830), which now form the basis of the mathematical support of tomographic methods medical diagnostics. In fact, it was this problem that boiled down, as Allan Cormack said, "to a tomogram or a sequence of tomograms, although I did not know the word itself for many years."

In 1979, Godfrey Hounsfield (1919–2004) and Allan Cormack (1924–1998) received the Nobel Prize for the introduction of X-ray computer tomography in medicine and "for the development of computer assisted tomography". In almost half a century that has passed since the appearance of the first X-ray CTs, X-ray computed tomography has revolutionized the diagnosis of various diseases. With the help of CT, studies of the brain and neck, angiography of the arteries of the brain, pulmonary arteries, main arteries of the body, all parts of the spine, bones of the skeleton, joints, chest cavity, abdominal cavity, pelvic organs are carried out. In addition, such studies make it possible to detect inflammation in the lungs, kidney stones, gall bladder, bile ducts, brain tumors, adrenal glands and organs of the genitourinary system, etc.

Magnetic Resonance Imaging (MRI)

The basis of the ***Magnetic Resonance Imaging (MRI)*** diagnostic method, or as it is commonly called in Ukraine and some other countries—the ***Magnetic Resonance Tomography (MRT)*** method, is the phenomenon of ***nuclear magnetic resonance (NMR)***. The physical basis of the NMR phenomenon will be theoretically described below in Sect. 7.3.2.

As it turns out, protons that are in the nuclei of atoms and molecules of substances, which are part of biological fluids and tissues of the human body (in particular, water molecules) in the presence of an external magnetic field, can resonantly absorb the energy of a high-frequency electromagnetic field. Such resonant energy absorption processes occur in those cases when the frequency of electromagnetic radiation lies in the range of ultra-high frequencies ν electromagnetic waves, which are approximately equal to $\nu \approx 3-30$ GHz (1 GHz $= 10^9$ Hz) and correspond to centimeter wavelengths of the order of $\lambda \approx 1-10$ cm. A resonant absorption of energy by protons of nuclei is called ***proton magnetic resonance (PMR)***, which is a partial case of NMR (see the resonance condition defined by formula (7.28) in Section B of this Chapter). Figure 7.1 shows a block diagram of a magnetic resonance tomograph.

The principle of operation of a magnetic resonance tomograph, in which PMR takes place, can be explained simply as follows. An external magnetic field (3) acts on a biological object (1), being a certain part of the human body, which is placed in an inductance coil of a sufficiently large diameter (2). The influence of an external magnetic field manifests itself in the splitting of the energy level of a microparticle (in this case—a proton of the nucleus) into two (sometimes more than

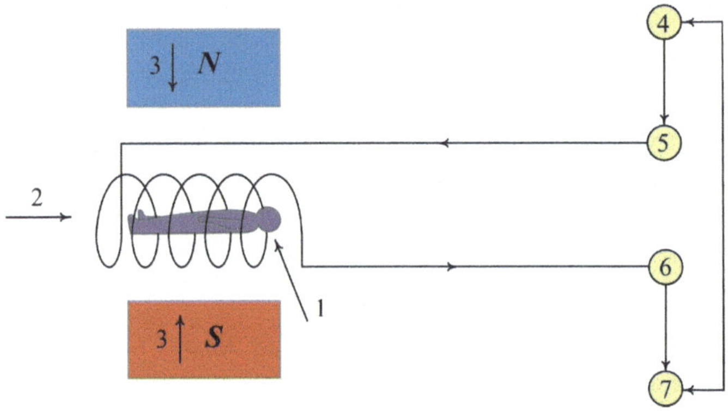

Fig. 7.1 Block diagram of a magnetic resonance tomograph: 1—object under investigation, 2—coil (solenoid), 3—magnet, 4—pulse generator, 5—radiofrequency transmitter, 6—receiver, 7—computer

two) energy sublevels. Such a splitting is a result of a physical effect discovered by the Danish physicist Peter Zeeman in 1897 [25] (see the corresponding explanation of the essence of this effect in Sect. 7.3.3). The interesting history of this discovery is that the splitting of spectral lines in a magnetic field was predicted even earlier (as Peter Zeeman wrote in his publication [25]) by the English physicist and chemist, well-known experimenter Michael Faraday (1791–1867). The second in chronological order—1902 Nobel Prize in Physics (as already noted, the winner of the first Nobel Prize in Physics in 1901 was Wilhelm Konrad Röntgen) was awarded to two Danish physicists Peter Zeeman (1865–1946) and Hendrik Lorentz (1853–1928) for the discovery and explanation of the effect of the splitting of spectral lines in a magnetic field (to be more precise, "for outstanding merits in the research of the influence of magnetism on radiation phenomena"). A modern correct explanation of this effect requires the application of the methods of quantum mechanics using ideas about the presence of spin in the proton and other elementary particles. It was the Zeeman effect that created the physical basis for the development of the magnetic resonance imaging (MRI) as a precision method of medical diagnostics by the American chemist and medical biophysicist Paul Lauterbur (1929–2007) [26, 27] and the British theoretical and medical physicist Sir Peter Mansfield (1933–2017) [28, 29], for which they were awarded the Nobel Prize in Medicine or Physiology in 2003.

In the MRI method, the biological object (1) is also affected by high-frequency radio pulses from the generator (4), which influence the biological object through the radio frequency transmitter (5). As a result of the phenomenon of proton magnetic resonance, which is characterized by the fulfillment of condition (7.28), the transition of protons to a higher energy level occurs with the absorption of the energy of the radio frequency electromagnetic field. This signal is measured by the receiver (6) and fed to the computer (7), which processes the signal and simultaneously determines the mode of operation of the radio pulse generator (4).

The practical implementation of the MRI-tomography method requires the creation of a magnetic field of the necessary spatial configuration, as well as visualization of the image of the biological object under study. To solve the first problem, the investigated object is placed in a magnetic field, the induction B of which varies linearly in one, two, and most often in all three mutually perpendicular directions (Fig. 7.2). The vertical lines conventionally represent the turns of the coil, through which an electric current is passed, which creates a magnetic field.

As is known (see Chap. 4), the product of the current I by the number of turns n per unit length of the coil determines the induction B on the axis of such a solenoid ($B \sim I$ x n). It is clear that it is possible to wind a metal wire in such a way that the number of turns n and, accordingly, the induction B change linearly along the x-axis, thus ensuring the fulfillment of the dependence (7.43). Since the magnetic field changes in the x direction, based on the resonance condition (6.26), the resonance frequency ν_{res} will also change in this direction. Further, taking into account the fact that the area under the absorption curve in the PMR spectrum determines the number of protons included in the nuclei of certain chemical elements, it is possible to obtain the spatial distribution of the density of protons (that is, atomic nuclei) along the x axis. In this way, a one-dimensional projection of the biological object is obtained.

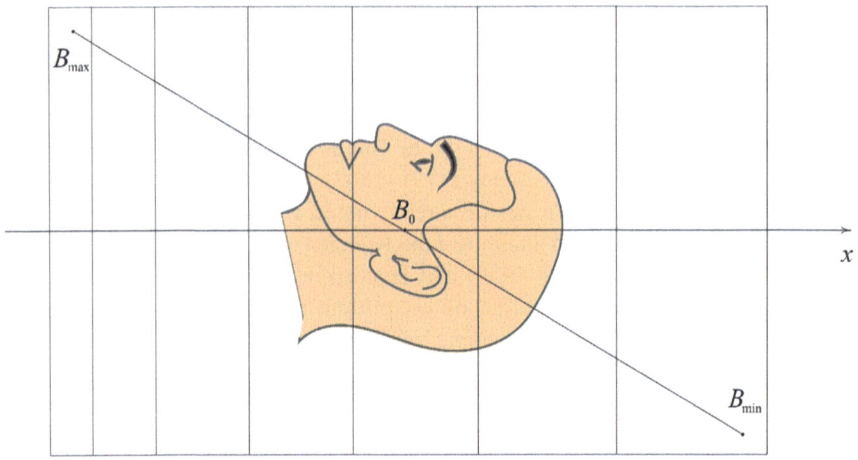

Fig. 7.2 Configuration of the magnetic field in the NMR-tomography method

The solution to the second problem—restoration of a three-dimensional image—became possible thanks to the work of Paul Lauterbur and other scientists, who used the so-called projection restoration method for this [26, 30]. The essence of this method is to obtain many one-dimensional projections of the object under study. This is achieved by changing the direction of the magnetic field induction gradient due to a change in the current in three mutually perpendicular coils (solenoids). In contrast to *X*-ray tomography, this method allows to exclude mechanical movements of the investigated biological object or hardware parts of the tomograph.

Another feature of the computer tomography method is the possibility of obtaining information (images) from thin layers of a three-dimensional biological object. In the MRI-tomography method, this is achieved in two ways [27, 28]:

(1) due to the use of special mathematical image processing (in particular, the so-called Rodon transformation and three-dimensional Fourier transformations);
(2) due to selective excitation of a thin layer of the biological object while fulfilling the quantum mechanical condition (7.26) for the implementation of PMR.

It should be noted that the signal (more precisely, the signal-to-noise ratio) in NMR tomograms is affected not only by the density of protons, but also by a number of internal and external factors (temperature, tissue composition, diamagnetic, paramagnetic, and ferromagnetic impurities, equipment parameters, specific features of computer programs, etc.). By changing these factors (for example, the parameters of the radio frequency pulse sequence), it is possible to achieve a significant improvement in image contrast in the NMR tomography method. This method is increasingly used in medicine due to its advantages, such as the absence of dose loads (compared to X-ray tomographs) and the possibility of obtaining contrast images for the purpose of effective diagnosis of various pathologies.

Fig. 7.3 Electron and positron annihilation reaction: two direct lines correspond the collision of an electron and positron pair, while two wavy lines represent the gamma-quants propagating in opposite directions at an angle of 180° relative each other

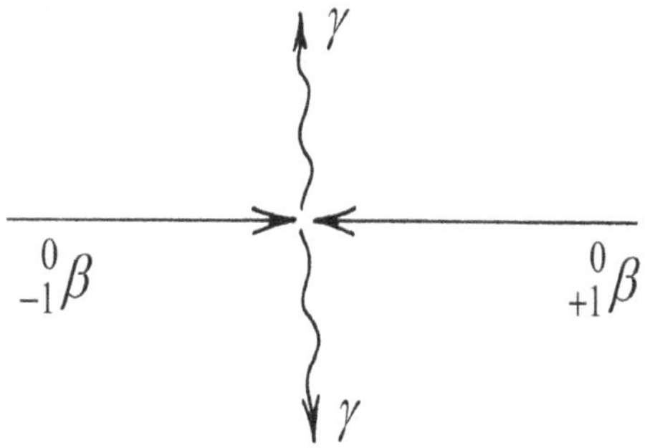

Positron Emission Tomography *(PET)*

Positron emission tomography is a method of medical radioisotope diagnostics (see Sect. 7.3.3). The physical principle of using positron emission tomography (PET) in medicine is based on the phenomenon of electron and positron annihilation, i.e. particles and antiparticles. The reaction characterizing this phenomenon looks as follows:

$$^{0}_{-1}\beta + ^{0}_{+1}\beta \to \gamma + \gamma \tag{7.1}$$

Schematically, this reaction between elementary particles can be represented by Fig. 7.3.

The annihilation reaction of the electron–positron pair was discovered in 1933, a year after the experimental discovery of the positron in cosmic rays. The very term “annihilation” in translation from Latin means “disappearance, transformation into nothing.” Of course, this term in its literal sense is incorrect, since the interaction of a particle and an antiparticle (in particular an electron $^{0}_{-1}\beta$ and a positron $^{0}_{+1}\beta$,) obeys all the fundamental laws of nature—the laws of conservation of energy, momentum, electric charge, spin, etc. At the same time, matter does not disappear and only transforms into other types of it, namely, into photons of electromagnetic radiation of the γ-range, or gamma-quanta. It should be noted that due to the law of conservation of the so-called charge parity, upon the disappearance (annihilation) of an electron and a positron, which have zero total spin, only an even number of γ-quanta can appear (most often there will be 2 of them).

At relatively low energies of the particle-antiparticle pair, the annihilation process is accompanied by the birth of lighter particles. An example of such an annihilation reaction is precisely the reaction with the formation of two γ-quanta, since the rest mass of the γ-quantum is zero. In the opposite case, that is, at significant energies, a light particle-antiparticle can annihilate with the formation of a pair of heavier particles and antiparticles. An example of such a reaction is the following reaction of the formation of two mesons from an electron and a positron:

$$^{0}_{-1}\beta + ^{0}_{+1}\beta \to \pi_{-} + \pi_{+}. \tag{7.2}$$

Mesons (this term means "intermediate, average") have rest masses that are between the rest mass of an electron ($m_0 = 9.1 \cdot 10^{-31}$ kg, or in energy units $E_0 = m_0c^2 \approx 0.51$ MeV) and the rest masses of a proton and a neutron ($m_{0p} \approx m_{0n} \approx 1.66 \times 10^{-27} kg, E_{0p} = 938.2\,\text{MeV}, E_{0n} = 939.5\,\text{MeV}$). Thus, for π—mesons, the rest mass (in energy units) is equal to $E_0 = m_0c^2 \approx 140MeV$, i.e. approximately 275 times greater than the rest mass of an electron and 6.7 times less than the rest mass of a proton. Complex processes of particle transformation, similar to reaction (7.2), are studied by the modern theory of strong nuclear interactions—quantum chromodynamics [29].

The essence of the PET method can be formulated as follows: first of all, radioactive isotopes are produced on special devices, which have a fairly short half-life $T_{1/2}$ (typically $T_{1/2} \approx$ several hours). Most often, these are isotopes of the so-called "authentic elements" (oxygen, nitrogen, carbon)—those elements contained in the human body. So, for example, in the Department of Medical Physics of the University of Wisconsin-Madison (USA), isotopes ${}^{15}_{8}O$ and others produced on a cyclotron are used to implement the PET method [(1)]. As mentioned earlier, for radioactive isotopes of similar light nuclei, where the number of protons and neutrons in the nucleus is approximately the same, positronic β-decay is inherent, as a result of which a positron is emitted from the atomic nucleus.

7.3 Section B

7.3.1 Latest Stages of the Development of Quantum Mechanics

Let us consider the main stages of the development of quantum mechanics throughout the twentieth century and the first quarter of the twenty-first century.

1900—Max Planck's concept of energy quantization.

1905 quantum nature of light to explain the photoelectric effect by Albert Einstein.

1913—Niels Bohr's quantum theory explained the emission and absorption of electromagnetic radiation for the spectral lines of the hydrogen atom.

1924—corpuscular-wave dualism of Louis de Broglie.

1925—creation of wave quantum mechanics by Erwin Schrödinger, who obtained a differential equation for the wave function of a microparticle. This wave equation of Schrödinger did not take into account the effects of the theory of relativity, that is, it was non-relativistic, because in its derivation the relationship between energy E, momentum p and mass m from classical mechanics was used, namely $E = p^2 / 2\,m$.

1925—development of matrix quantum mechanics by Werner Heisenberg.

1937—Werner Heisenberg's formulation of the quantum–mechanical uncertainty principle.

1928—obtaining the relativistic wave equation by Paul Dirac, in which quantum mechanics and the special theory of relativity of Albert Einstein were combined.

This Dirac equation takes into account the effects of the theory of relativity in the following relation $E^2 = p^2c^2 + m^2c^4$, where c is the speed of light in a vacuum. The Dirac equation is used for the electron and other particles with spin ½, which are described by Fermi–Dirac statistics. It should be noted that for particles with zero spin, the relativistic Klein-Gordon-Fock wave equation should be used, which was obtained for the first time by Oskar Klein and Walter Gordon in 1926.

1932—proving the equivalence of Erwin Schrödinger's wave quantum mechanics and Werner Heisenberg's matrix quantum mechanics was made by John von Neumann, an American mathematician and physicist of Hungarian origin [18].

1944—the discovery of the phenomenon of electron paramagnetic resonance (EPR) by the experimental physicist Evgeny Zavoiskyi (1907–1976), who was born in the family of a military doctor in the Ukrainian city of Mogilev-Podilskyi.

1946—the discovery of nuclear magnetic resonance (NMR) by the Swiss and American theoretical physicist Felix Bloch (1905–1983) and the American nuclear physicist Edward Purcell (1912–1997), who received the Nobel Prize in Physics in 1952 for the discovery of NMR in liquids and solid bodies.

At the beginning of the second half of the twentieth century, two scientists independently began research on the creation of X-ray computed tomography, namely: the physicist and electrical engineer Godfrey Hounsfield (1919–2004), as well as the South African and American physicist Allan Cormack (1924–1998), who in 1979 received the Nobel Prize in Medicine or Physiology for the introduction of X-ray computer tomography into medicine and "for the development of computer assisted tomography".

The subsequent history of modern methods of medical diagnosis was directly related to the Nobel Prize in Physics, which was awarded in 1902 to two Dutch physicists Peter Zeeman (1865–1946) and Hendrik Lorentz (1853–1928) for the experimental discovery (Zeeman) and a theoretical explanation (Lorentz) of the effect of the splitting of spectral lines in a magnetic field. It was this Zeeman effect that created the basis for the development of the magnetic resonance imaging (MRI) method by the American medical physicist Paul Lauterbur (1929–2007) and the English mathematician Peter Mansfield (1933–2017), for which they were awarded the Nobel Prize in Medicine or Physiology in 2003.

7.3.2 Basic Theoretical Provisions of Quantum Mechanics

In this section, we will focus on the main ideas and most important theoretical aspects of non-relativistic quantum mechanics discussed earlier in Sect. 7.2.1.

Max Planck's formula for the energy of a quantum. Max Planck's famous formula, which marked the beginning of quantum mechanics, defines the minimum portion (quantum) of energy E, which is equal to

$$E = h\nu, \tag{7.3}$$

where $h = 6.63 \cdot 10^{-34}$ J·s is Planck's constant, and ν is the frequency of radiated energy [20].

Albert Einstein's formula for the photoelectric effect. The German and American theoretical physicist Albert Einstein (1979–1955), 1921 Nobel prize winner in Physics "for his services to theoretical physics, and especially for his discovery of the law of the photoelectric effect" [31]. The Einstein's formula of the photoelectric effect has already been written down (see previous Chap. 6, where the primary physical mechanisms of the interaction of X-rays and gamma radiation with matter were discussed).

Niels Bohr's quantum theory, emission and absorption of electromagnetic radiation in the hydrogen atom. As already noted above in Sect. 7.2.1, Niels Bohr proposed the following quantum postulates to explain the linear spectrum of emission and absorption of electromagnetic radiation in a hydrogen atom [22].

Bohr's 1st postulate. Electrons rotate in stationary electronic orbits, for which the following quantization condition must be fulfilled: only discrete "stationary" orbits exist, for which the angular momentum is a multiple of a reduced Planck's constant value $\hbar = h/2\pi = 1.05 \cdot 10^{-34}$ J·s, i.e.

$$m\upsilon r = n\hbar, \tag{7.4}$$

where m is the mass of the electron, υ is the speed of rotation of the electron in an orbit of radius r, and the integer $n = 1, 2, 3, \ldots$ specifies the number of the orbit. The physical meaning of the quantization condition (7.4) is that not any electron orbits are realized, but only those discrete (or quantized) orbits for which the momentum of the electron motion, equal to the product $m\upsilon r$, becomes a multiple of reduced Planck's constant. The quantum nature of the movement of electrons in discrete orbits, according to Bohr's first postulate, is also manifested in the fact that it is not accompanied by the emission (or absorption) of electromagnetic energy.

Bohr's 2nd postulate. Electromagnetic waves are emitted or absorbed by atoms when electrons move from one stationary orbit to another. At the same time, radiation and absorption occur in portions (quanta) of energy. The frequency of waves emitted by atoms during such transitions is determined by the difference in energies of stationary states before and after radiation according to the equation:

$$h\nu = E_n - E_k, \tag{7.5}$$

where n and k are the numbers of stationary orbits between which electron transitions occur.

According to Niels Bohr's theory, the frequencies emitted by hydrogen atoms can be determined by the formula

$$\nu = (E_n - E_k)/h = R(\frac{1}{n^2} - \frac{1}{k^2}), \tag{7.6}$$

where R is the Rydberg constant and the condition $k > n$ is fulfilled.

This formula is a generalization of the formula of the Swiss physicist and mathematician Johann Balmer (1825–1898), obtained by him in 1885 based on the generalization of experimental data on spectral lines for the hydrogen atom:

Louis de Broglie's hypothesis of particle-wave dualism. In 1924, the French theoretical physicist Louis de Broglie (1892–1987), 1929 Nobel Prize laureate in Physics, hypothesized that particle-wave dualism is a feature not only of optical phenomena, but is inherent in all moving material bodies [23]. According to the de Broglie hypothesis, the wave properties of bodies can be characterized by the wavelength λ associated with the speed of movement υ and the mass of the body m by the following formula:

$$\lambda = h/m\upsilon. \tag{7.7}$$

De Broglie's hypothesis was experimentally confirmed in the experiments of Davison and Germer on electron diffraction on a Ni single crystal. The experimental value of the electron wavelength $\lambda_{\text{exp}} = 1.65 \times 10^{-10}$m was found based on the analysis of the diffraction pattern using the Bragg-Wulff formula (6.47). The theoretical value of the electron wavelength, which was calculated according to the de Broglie formula (7.7) for the experimental value of the voltage $U = 54$ V in Davison and Germer's experiments, was $\lambda_{\text{theor}} = h/mu == 1.67 \times 10^{-10}$m, that is, the discrepancy between the experimental and theoretical values of the de Broglie wavelength of the electron was only $2 \cdot 10^{-12}$ m or about 1%.

The wave properties of electrons have found wide application in electron microscopy due to the significant improvement of the resolution limit by hundreds and thousands of times compared to the optical microscope.

Substitute the wavelength λ for an electron according to the Louis de Broglie formula (7.7) into the following formula for the resolution limit of an optical microscope:

$$Z_{\text{min}} = \lambda/2n \sin\theta. \tag{7.8}$$

Formula (7.8), where n is the relative refractive index, and θ is the angular aperture, was obtained by the German optical physicist and astronomer Ernst Abbe (1840–1905), who calculated the limits of the microscope's resolution.

As a result, we get the following formula for the resolution limit of an electron microscope, that is, for the minimum distance between two points that can be seen separately in an electron microscope:

$$Z_{\text{min}} = h/2 \sin\theta \tag{7.9}$$

To find the numerical value of the resolution limit Z_{min} of an electron microscope, the value of Planck's constant $h = 6.63 \cdot 10^{-34}$ J s, the mass and charge of the electron, which are equal to $m = 9.11 \cdot 10^{-31}$ kg and $e = 1.60 \cdot 10^{-19}$ C, should be substituted into (7.9), and also, as an example, the values of the accelerating voltage $U = 100$ kV and the angular aperture $\theta = 10^{-2}$ radian are typical for an electron microscope. As a result, we get the following value of the resolution limit of the electron microscope: $Z_{min} \approx 10^{-10} m$. This value is typical for the size of objects in the nanoworld (e.g., the diameter of water molecule is $d \approx 2.8 \times 10^{-10}$ m or $d \approx 0.28$ nm).

Theoretical foundations of the Heisenberg uncertainty principle. The German theoretical physicist Werner Heisenberg (1901–1976), 1932 Nobel Prize winner in Physics, formulated quantum mechanics in terms of matrics in 1925, as well as his uncertainty principle in 1927 [24]. The wave properties of microparticles indicate the limitation of applying to them some concepts that characterize bodies in classical mechanics. So, in classical mechanics, we can simultaneously specify the position (x-coordinate) of a body in space and its momentum $p = mv$, which allows us to specify the spatial position and speed of movement of the body at the next moment in time, thus determining the trajectory of its movement. This becomes impossible for a microparticle. There are always uncertainties in the values of its coordinate Δx and momentum Δp, related by a certain ratio, which is called the Heisenberg uncertainty principle and can be written in the form of the following formula:

$$\Delta x \times \Delta p \geq \hbar/2, \tag{7.10}$$

It follows from this relationship that the more precisely we try to determine the particle's coordinate, the less accurately we can characterize its momentum. Indeed, on the basis of formula (7.10), it is clear that when the uncertainty in the coordinate measurement goes to zero ($\Delta x \to 0$), the uncertainty in the momentum measurement goes to infinity ($\Delta p \to \infty$).

Example: according to the classical idea, an electron in an atom moves in a circular orbit with a speed of $v = 10^6 m/s$. This result for the speed of an electron in an atom is easy to determine, provided that the force according to Coulomb's law plays the role of centripetal force ($F_{ky\pi} = F_{\partial 0y}$), $(1/4\pi\varepsilon_0)\, Ze^2/R^2 = mv^2/R$.

The electron's belonging to an atom requires that the uncertainty in the value of its coordinate Δx corresponds to the atomic dimensions, i.e. $\Delta x \sim 10^{-10}$ m. Then from the ratio $m\Delta v \geq \hbar/\Delta x$ we have: $\Delta v \geq \hbar/m\Delta x = 1.05 \cdot 10^{-34} / 9.1 \cdot 10^{-31} \cdot 10^{-10} \approx 10^6$ m/s. It follows from this that the uncertainty in the value of the velocity Δv of the electron is equal to the velocity v. itself. Thus, it is impossible to preserve the idea of an orbit along which an electron moves with a certain speed, that is, we cannot apply classical ideas in this case.

Similarly, the uncertainty of measuring the energy of a microparticle ΔE and the uncertainty of measuring the time Δt of this microparticle staying in a given energy state turn out to be related to each other using the following formula:

$$\Delta E \times \Delta t \geq \hbar/2. \tag{7.11}$$

The given formulas (7.10) and (7.11) represent ***the Heisenberg's uncertainties principle***. They constitute one of the basic provisions of quantum mechanics.

Werner Heisenberg once characterized the situation with the uncertainty's principle using the following comparison (the words below do not claim literal accuracy, but approximately convey the meaning of what was said): Coordinate and momentum are similar to the figures of a man and a woman in an antique barometer. When one figure appears, the other disappears.

The Schrödinger wave equation of quantum mechanics. The Austrian theoretical physicist Erwin Schrödinger (1887–1961), 1933 Nobel Prize laureate in Physics together with the British theoretical physicist Paul Dirac (1902–1984) "for the discovery of new productive forms of atomic energy" [32–35]. The equation describing the movement of a microparticle should reproduce its wave properties, that is, it should be similar to the differential wave equation describing the propagation of electromagnetic or acoustic waves

$$\frac{\partial^2 S}{\partial x^2} = \frac{1}{\upsilon^2} \cdot \frac{\partial^2 S}{\partial t^2}, \tag{7.12}$$

where S is the electric field strength or magnetic field induction in an electromagnetic wave, and for an acoustic wave it is the displacement of particles of an elastic medium from the equilibrium position.

In the case of quantum mechanics, one should imagine ae follows: if a moving micro-particle has wave properties and can be characterized by a wavelength, then its state can be described by some function Ψ, which must satisfy the wave equation, which is similar to (7.12):

$$\frac{\partial^2 \Psi}{\partial x^2} = \frac{1}{\upsilon^2} \cdot \frac{\partial^2 \Psi}{\partial t^2}. \tag{7.13}$$

Next, let us present the function Ψ (x,t), which satisfies the wave Eq. (7.13), in the following form: Ψ (x,t) $= \psi(x)$ sinω t, that is, we will assume that the time dependence of the function Ψ (x,t) is harmonic.

Further we will find the corresponding partial derivatives with respect to the spatial coordinate x and with respect to the time variable t, namely:

$$\frac{\partial^2 \Psi}{\partial t^2} = -\omega^2 \psi(x) \cdot \sin \omega\, t, \quad \frac{\partial^2 \Psi}{\partial x^2} = \sin \omega\, t \cdot \frac{\partial^2 \psi(x)}{\partial x^2}$$

and substitute them in the Eq. (10.11), which gives as a result

$$\frac{\partial^2 \psi(x)}{\partial x^2} = -\frac{\omega^2}{\upsilon^2} \psi(x). \tag{7.14}$$

Taking into account the relationship $\omega = 2\pi\nu = 2\pi/T$ between the cyclic frequency ω and the period T, as well as the relationship $\lambda = \upsilon\, T$ between the wavelength, speed and period T, and also, what is very important, the formula (7.7) for the de Broglie wavelength $\lambda = h / m\upsilon$, the ratio ω^2/υ^2 can be written in the following form

$$\frac{\omega^2}{\upsilon^2} = \frac{(2\pi\nu)^2}{\upsilon^2} = \frac{4\pi^2 m(m\upsilon^2)\cdot 2}{h^2\cdot 2} = \frac{2m}{\hbar^2}\cdot E_k. \quad (7.15)$$

Then Eq. (7.15) takes the following form:

$$\frac{\partial^2\psi(x)}{\partial x^2} + \frac{2m}{\hbar^2}E_k\psi(x) = 0, \quad (7.16)$$

where in Eqs. (7.15) and (7.16) the well-known formula of from classical mechanics for kinetic energy $E_{k=} = m\upsilon^2/2$ was used.

Equation (7.16) is ***the Schrödinger wave equation of quantum mechanics*** describing the one-dimensional motion of a particle along the x-axis. In the case when a particle moves in three-dimensional space, Eq. (7.16) will have the following for

$$\Delta\psi + \frac{2m}{\hbar^2}E_k\psi = 0, \quad (7.17)$$

where $\Delta\psi = \frac{\partial^2\psi(x)}{\partial x^2} + \frac{\partial^2\psi}{\partial y^2} + \frac{\partial^2\psi}{\partial z^2}$ is the so-called Laplace operator, which acts on the wave function ψ and is equal to the sum of the second spatial derivatives of ψ with respect to coordinates x, y, z. ***The Schrödinger wave*** Eq. (7.17) describes the free movement of a microparticle (for example, an electron) in the absence of an external field.

If a particle moves in a force field, its total energy E is equal to the sum of kinetic Ek and potential E_p energies: $E = E_k + E_p$. In this case the kinetic energy $E_k = E–E_p$, and Eq. (7.17) is written as follows:

$$\Delta\psi + \frac{8\pi^2 m}{h^2}(E - E_p)\psi = 0, \quad (7.18)$$

The differential Eq. (7.18) is called ***the stationary Schrödinger wave equation of quantum mechanics in an external field*** with the E_p. The Austrian theoretical physicist Erwin Schrödinger proposed the Eq. (7.18) in 1926. This equation describes the spatial behavior of an electron in an atom (that is, an electron moving in the field of the nucleus). The wave function ψ (x, y, z), which is a solution of the Schrödinger Eq. (7.18), does not depend on time and characterizes the stationary states of the system.

The relationship between the ψ-function and the particle, which it describes, is analogous to the relationship between a light wave and a photon. The square of the amplitude of the light wave A^2 determines the probability of the photon hitting the corresponding point in space. As Max Born showed, the square of the amplitude of the wave function $|\psi|^2$ characterizes the probability of finding a particle at a given point in space, and $|\psi|^2 dV$ corresponds to the probability of finding a particle in an element of volume dV. Thus, the physical content is not the wave function itself, but its square. It should be noted that, unlike optical waves, the ψ-function does not characterize an electromagnetic wave, but a wave of the probability of finding a microparticle at a certain point.

Quantum numbers. The wave function ψ(r), which is a solution of the Schrödinger equation for a hydrogen atom, has the following form:

$$\Delta\psi + \frac{8\pi^2 m}{h^2}(E + e^2/4\pi\varepsilon_0 r)\psi = 0, \tag{7.19}$$

contains three integer parameters called quantum numbers.

The principal quantum number n, which characterizes the quantization (change in discrete portions) of the energy E of a microparticle (for example, an electron in a hydrogen atom) in accordance with the following formula:

$$E = -\frac{me^4}{8\varepsilon_0^2 h^2 n^2}, \tag{7.20}$$

as well as the shapes of its orbitals (the dimensions of the electron cloud), can take the values n = 1, 2, 3, 4, ..., which correspond to the energy levels *K, L, M, N*, ...

The orbital quantum number l characterizes the quantization of the angular momentum $L = m\,\upsilon\,r$, that is, it characterizes various possible forms of orbitals for a microparticle (in particular, for an electron in a hydrogen atom) in accordance with the following formula:

$$L = \hbar\sqrt{l(l+1)}, \tag{7.21}$$

where the orbital quantum number l can take the value l = 0, 1, 2, 3, ..., (n−1).

The magnetic quantum number m_l characterizes spatial quantization (discrete change of spatial orientation) of orbitals along which microparticles (for example, an electron in a hydrogen atom) can move in the presence of an external magnetic field. It turns out that in a magnetic field, only such orientations of the momentum vector L to the direction of the external field B are possible (for the sake of certainty, we will assume that the magnetic field is directed along the z axis), for which the projection L_z on the direction of the magnetic field is determined by the following formula:

$$L_Z = m_l\hbar, \tag{7.22}$$

where the magnetic quantum number $m_l = -l, -l+1, \ldots, -1, 0, +1, \ldots, l\text{–}1, +l$, i.e. can acquire $2l - 1$ integer value

The magnetic spin number m_S. There is another quantum number whose introduction is not related to Schrödinger's wave equation. According to the concepts of quantum mechanics, a microparticle (in particular, an electron) has another property inherent in it, which is called ***spin***. Spin is the internal moment L_s of the amount of movement of a microparticle, which has a quantum nature and is not caused by the rotational movement of this entire microparticle. It should be noted that the internal moment L_s of the amount of movement of a microparticle is also quantized, that is, it takes discrete values in accordance with the following formula:

$$L_S = \hbar\sqrt{S(S+1)}, \tag{7.23}$$

where S is called ***the spin quantum number***, or simply ***spin***. So, for an electron, its numerical value is $S = 1/2$.

In the presence of an external magnetic field B, which acts along the z axis, the L_{SZ} projection in this direction can take on the value:

$$L_{SZ} = m_S\hbar, \tag{7.24}$$

where ***the magnetic spin number*** $m_S = +\,1/2, -1/2$. Thus, the magnetic spin number m_S determines the projection of the internal moment L_{SZ} of the amount of movement of the microparticle in the direction of the magnetic field.

The Pauli exclusion principle. The distribution of electrons in an atom by orbitals is subject to the fundamental ***Pauli exclusion principle***, which was formulated by an Austrian theoretical physicist Wolfgang Pauli (1900–1958) in 1925 for electrons and further in 1940 for all fermions, i.e., elementary particles, which obey Ferms-Dirac quantum statistics. He received the Nobel Prize in Physics in 1945 for "decisive contribution through his discovery of a new law of Nature, the exclusive principle or Pauli principle" [36, 37]. This principle consists in the fact that an atom cannot have two or more electrons with four identical quantum numbers n, l, m_l and m_S.

Wolfgang Pauli (1900-1958)

7.3.3 Physical Basis of Computer Tomography and Their Medical and Biological Applications

Nowadays, along with the traditional methods of optical spectroscopy, magnetic spectroscopy methods have become widely used in medicine, which allow obtaining valuable information not only about the structure of organic molecules, the nature of intermolecular interactions, the nature of molecular movements, but also allow using these methods for the purpose of precise diagnostics many dangerous human diseases. These methods are based on the same principle—energy absorption by the system of paramagnetic particles during transitions between energy levels induced by high-frequency electromagnetic radiation, into which the energy levels of the system are split in the external magnetic field [25–30].

Nuclear magnetic resonance method. Let us consider the basic principles underlying the nuclear magnetic resonance (NMR) method. As is known, nuclei containing an odd number of nucleons (protons and neutrons) have non-zero spin and, accordingly, magnetic moments. Such nuclei have paramagnetic properties.

Let us consider a special case of NMR, namely proton magnetic resonance (PMR). In this case, the magnetic moment of the nucleus P_n equals

$$P_n = g_n \mu_n J. \tag{7.25}$$

The following notations are used in formula (7.25): J is the spin quantum number of the nucleus (for a proton, $J = \pm 1/2$, and the sign depends on the mutual orientation of the magnetic field induction vector and the proton magnetic moment vector), and the other two factors are constant values, which have the names: nuclear magneton $\mu_n = e\ \hbar/2$ M, where e is the charge of the proton, $\hbar$ is the reduced Planck constant, M is the mass of the proton, and g_n is the so-called "g-factor" for the nucleus.

It turns out that due to the Zeeman effect in the external magnetic field, the energy levels in the system of magnetic nuclei break up into subsystems that correspond to two possible orientations of the nuclear magnetic moments relative to the z-direction of the external magnetic field and, accordingly, to two values of the magnetic spin number $m_\mathrm{z} = + 1/2, -1/2$.

Figure 7.4 illustrates the Zeeman effect, which is due to the splitting of one energy level with energy E_0 in the absence of a magnetic field, that is, at zero magnetic induction ($B = 0$), into two energy sublevels at nonzero magnetic induction ($B \neq 0$).

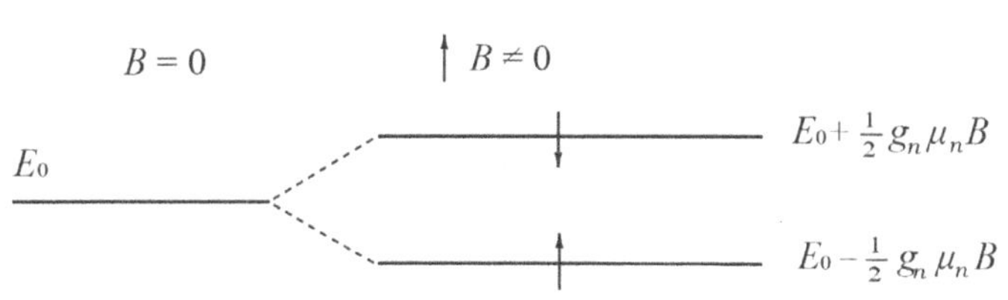

Fig. 7.4 The splitting of the energy level of protons in an external magnetic field

The proton energy at both energy sublevels at a non-zero value of the magnetic induction B is given by the following formula:

$$E = E_0 \pm g_n \mu_n J B. \tag{7.26}$$

The condition for resonant absorption of a quantum of high-frequency electromagnetic eenergy $h\nu$ by a system of protons in the PMR method is as follows:

$$h\nu = E_2 - E_1. \tag{7.27}$$

Taking into account the formula (7.28) and the value of the spin quantum number for the proton J = ± 1/2.where the sign + corresponds to a larger energy value E_2, and the sign − to a smaller value E_1, allows rewriting the condition of resonance absorption using the following formula:

$$h\nu = g_n \mu_n B. \tag{7.28}$$

As noted earlier in Sect. 7.2.2, the practical implementation of the MRI-tomography method requires the creation of a magnetic field of the necessary spatial configuration. Under the condition of proton magnetic resonance (7.28), the resonance frequency ν is directly proportional to the induction B of the magnetic field and can vary within fairly wide limits.

From the above, it could be concluded that at a fixed frequency of the high-frequency field, all protons that are part of the molecules will give an absorption signal at one and the same value of the induction of the magnetic field B. In a real situation, the condition of nuclear resonance absorption has the form:

$$h\nu = g_n \mu_n (B + B_{loc}), \tag{7.29}$$

where B_{loc} is an additional local magnetic field, which is created at the location of the resonating nucleus by the surrounding nuclei and electrons.

One of the main reasons for the appearance of such fields is the effect of diamagnetic shielding: the external magnetic field induces electron currents in the molecule, which cause the appearance of magnetic fields directed in the opposite direction to the external magnetic field according to the Lenz rule, i.e. $B_{\text{loc}} = -\sigma B$, where σ is the shielding constant.

Thus, each proton is in some effective field, which is characterized by induction:

$$B_{eff} = B - \sigma B = B(1 - \sigma). \tag{7.30}$$

Differences in the electronic shielding of protons that make up the molecule can be caused by different electron densities. So, for example, protons attached to electro negative groups and atoms (OH, O, halogens) or located near them are shielded more weakly and give a PMR signal at lower values of induction B of the external mfgnetic field.

The NMR method is successfully used for the study of biological membranes, since the analysis of NMR spectra of membranes allows not only to determine how many of these or other functional groups are contained in the object under study, but also to establish the degree of mobility of the corresponding groups in the membranes by the width of the absorption bands.

In recent years, along with X-ray tomography, the NMR tomography method has become widely used, which provides the possibility of studying different parts of a macroscopic object based on the differences in the NMR signal caused by magnetic field gradients in different directions.

In such a case, we speak of a linear gradient of the magnetic field, for which the change in induction B with the coordinate x occurs according to the law

$$B(x) = B_0 + ax, \tag{7.31}$$

where $a = dB/dx = const$ is a constant value that characterizes the gradient of the magnetic field, B_0 is a certain value of induction that is achieved inside the studied volume (Fig. 7.2). Vertical lines of variable density conventionally depict turns of a coil through which a current is passed, creating a spatially inhomogeneous magnetic field in accordance with the formula (7.31).

An important advantage of NMR tomography is the absence of radiation damage and the ability to obtain clear images of internal organs (for example, the skull, chest, eetc.) due to the greater transparency of these organs for radio waves.

7.3.4 The Concept of Mass in Physical and Astrophysical Theories

The concept of mass is, of course, fundamental, since mass is one of the most important characteristics of matter. This paragraph will use the results of studies [38, 39], which considered the fundamental issues of the development of mass since the formation of classical mechanics in the 17th century in the works of Galileo [40], Newton [41] and the creation of relativistic mechanics at the beginning of the twentieth century in works of Einstein, Lorentz, Poincaré, Minkowski to modern ideas and controversies about whether mass depends on speed, whether the law of conservation of mass is always fulfilled, whether a system of photons has mass, and, finally, what questions regarding the nature of mass remain unsolved on today in this fundamental problem [42–47].

In this connection, it should be noted once again (see also Chap. 1) that for a long period of time the natural sciences were not yet significantly differentiated. Therefore, the terms "natural philosophy" and "physics" characterized the totality of knowledge about both living and non-living nature. Subsequently, during the XVIII—XIX centuries, processes of differentiation began in the natural sciences, which led to the emergence of new sciences with their own research approaches and methods.

Later, especially in the twentieth century, transitional, or complex, sciences arose, in which the methods of one science began to be effectively used to study the objects of another science. This is how physical chemistry, biological physics, biological chemistry and other sciences arose. This shows that there are no clear boundaries between different natural sciences (and thanks to the synergistic approach—and not only between natural sciences).

A striking example of the fact that, at the same time as the differentiation of sciences, their unification (integration) is taking place, is the modern development of medical physics—an applied branch of physics, in which the achievements of physics are successfully used in medicine for the purpose of developing new methods of diagnosis and therapy (see, for example, [17, 18, 25–30, 38]). Such powerful diagnostic methods as magnetic resonance imaging (MRI) and positron emission tomography (PET) are increasingly used in medical practice. This is evidenced by the awarding of the Nobel Prize in Medicine in 2003 for achievements in the field of magnetic resonance imaging to two scientists—physicist Paul Lauterbur (USA) and mathematician Peter Mansfield (Great Britain). It should be noted that both MRI and PET methods are based on physical processes involving photons—quanta of the electromagnetic field—in the range of radio waves (in the MRI method) or gamma radiation (in the PET method). In particular, the basic physical principle of positron emission tomography is based on the phenomenon of annihilation of an electron and a positron with the formation of an even number (usually two) of photons of gamma-range electromagnetic radiation. Therefore, the question of whether a photon has mass is of fundamental importance not only for physical science, but also for all natural sciences, including medicine.

The concept of mass in classical mechanics. In classical mechanics, the mass of a body acts as a measure of its gravitational and inertial properties. Therefore, two masses are usually distinguished—1) gravitational mass and 2) inertial mass.

To define the concept of "gravitational mass", the law of universal gravitation is used, which was written down by Newton in 1687 in his main work "Mathematical principles of natural philosophy" [41] in the following form:

$$F = G\frac{m_1 m_2}{r^2} \tag{7.32}$$

Here F is the force with which two bodies with masses m_1 and m_2 are mutually attracted, and the centers of these masses are at a distance r from each other, and the coefficient G is called the gravitational constant, which is equal to $G = 6,67 \cdot 10^{-11}$ $m^3\ kg^{-1}\ s^{-2}$.

Another way of introducing the gravitational mass is connected with the use for the gravitational force P of the body of the mass m in the gravitational field of another body (for example, the Earth, the mass of which will be denoted by M) by the following formula, which directly follows from the law of universal gravitation (7.32):

$$P = mg,\ g = GM/r^2, \tag{7.33}$$

where g is the acceleration of free fall in the Earth's gravitational field. The masses m_1 and m_2 in formula (7.32) and the mass m in formula (7.33) are called gravitational masses.

The inertial mass was introduced in classical Newtonian mechanics as a coefficient of proportionality between the momentum $\vec{p}$ of a body and the speed $\vec{\upsilon}$ of this body in accordance with the formula

$$\vec{p} = m\vec{\upsilon}. \tag{7.34}$$

Another way of introducing inertial mass is based on the use of the basic law of dynamics—Newton's second law

$$\vec{F} = m\vec{a}. \tag{7.35}$$

In this formula, the inertial mass acts as a coefficient of proportionality between the force $\vec{F}$ acting on the body and the acceleration $\vec{a}$ caused by this force.

At the beginning of the twentieth century, Etvesh, who in 1888 designed a torsion balance, proved with high accuracy (up to 10^{-8}) the equality of gravitational and inertial masses. In the following years, in the experiments of Dicke, Braginsky and other scientists, this equality was proved with a higher accuracy, which reaches 10^{-12}. Currently, the upper limit for the mass of a photon is considered not to exceed 10^{-63} kg. The issue of the non-zero mass of a quantum of electromagnetic radiation was discussed in review articles [42, 48].

In Newtonian mechanics, the law of conservation of mass is fulfilled, namely: the mass of a system of bodies is equal to the sum of the masses of each of the bodies that make up the system, regardless of the changes that occur in this system.

The concept of mass in the theory of relativity. The theory of relativity, or relativistic mechanics, which was created in works [50, 51], made revolutionary changes to the concept of body mass. Let us consider the most important results obtained in the theory of relativity for the mass of bodies moving with velocities, the magnitude of which is of the order of the speed of light.

Longitudinal and transverse masses. Even before Einstein's creation of the special theory of relativity in 1905, Abraham [47] and Lorenz [49] obtained the following formulas for the dependence of body mass on speed depending on the direction of the force vector $\vec{F}$ that causes the body's movement in relation to the body's velocity vector $\vec{\upsilon}$:

$$m_{II} = m/\sqrt{1 - \upsilon^2/c^2}, \tag{7.36}$$

if these vectors are parallel, and

$$m_{\perp} = m/\sqrt{(1 - \upsilon^2/c^2)^3}, \tag{7.37}$$

if these vectors are perpendicular to each other. In formulas (7.36) and (7.37), the mass of the body is denoted by m, provided that the ratio $\upsilon^2/c^2 \to 0$, which gives reason to call the mass m the rest mass (often it is also denoted by m_0).

Kinetic energy in the theory of relativity. In his main works [50, 51], devoted to the construction of the special theory of relativity, Einstein wrote down the formula for the kinetic energy W of an electron with a mass μ moving at a high speed, which approaches the speed of light (note that for the speed of light in these works, Einstein used the notation V):

$$W = \mu V^2\{\frac{1}{\sqrt{1-(\upsilon/V)^2}} - 1\}. \tag{7.38}$$

In addition, he obtained a formula for the so-called longitudinal mass in the case of parallelism of the vectors $\vec{F}$ and $\vec{\upsilon}$, which completely coincides with formula (7.36). It is clear that from formula (7.37) after expanding the first term in curly brackets into a Taylor series in the approximation of small velocities ($\upsilon << V$), the formula for kinetic energy $W = \mu\upsilon^2/2$ in Newtonian mechanics can be easily obtained.

Lorentz transformation. The 4-dimensional formulation of the theory of relativity, proposed by Minkowski in 1908, was a fundamental step in understanding the question of the dependence of mass on velocity (see also his work [52] and Einstein's review article [53]). This formulation is based on the Lorentz transformation for the spatial coordinates x, y, z and time t when passing from one inertial system K to another inertial system K', which moves with a speed V relative to the first one parallel to the X axis. We write the Lorentz transformation in the following form, using the notation from the Landau and Lifshitz monograph [54]:

$$x \to \frac{x' + Vt'}{\sqrt{1 - V^2/c^2}}. \tag{7.39}$$

$$y \leftarrow y\prime \tag{7.40}$$

$$z \to z\prime, \tag{7.41}$$

$$t \to \frac{t\prime + (V/c^2)x\prime}{\sqrt{1 - V^2/c^2}}. \tag{7.42}$$

Note that the Lorentz transformation (7.39)–(7.42) leaves the interval $ds^2 = c^2dt^2 - dx^2 - dy^2 - dz^2$ between two events in the inertial systems K and K' unchanged (invariant). This important requirement of the theory of relativity is not fulfilled in classical mechanics, because the intervals between events are not invariant with respect to the Galilean transformation

$$x \to x\prime + Vt', z \to z', t \to t'. \tag{7.43}$$

for this non-invariance lies, as is known, in the absoluteness of time ($t = t'$), which is postulated in classical Newton-Galilean mechanics.

Later, in the article [55], Einstein received important formulas for the transformation of the 4-dimensional energy–momentum vector, the components of which are momentum $\vec{p} = p_x\vec{i} + p_y\vec{j} + p_z\vec{k}$ and energy E. When moving from one inertial system to another, the components p_x, p_y, p_z, E are transformed similarly to spatial coordinates x, y, z and time t according to Lorentz transformations (7.39)–(7.42), namely (see [54]):

$$p_x \to \frac{p_{x\prime} + (V/c^2)E\prime}{\sqrt{1 - V^2/c^2}}, \tag{7.44}$$

$$p_y \to p_{y\prime}, \tag{7.45}$$

$$p_z \to p_{z\prime}. \tag{7.46}$$

$$E \to \frac{E\prime + Vp_{x\prime}}{\sqrt{1 - V^2/c^2}}. \tag{7.47}$$

From formulas (7.39)–(7.42) and (7.44)–(7.47), it is easy to see their equivalence when replacing $x, y, z \leftrightarrow p_x, p_y, p_z$ and $t \leftrightarrow E$.

Connection of rest mass, energy and momentum. Another fundamental result of the special theory of relativity (relativistic mechanics) consists in obtaining a relationship between the rest mass m of a body moving at a speed $\vec{\upsilon}$ with its energy E and momentum $\vec{p}$, which is given by the following formula:

$$\frac{E^2}{c^2} = p^2 + m^2c^2. \tag{7.48}$$

Taking into account the importance of this formula for understanding the general problem of mass under discussion, we will provide a brief derivation of formula (7.48) in accordance with [54, 56]. The energy E of the body, which is related to the vectors of velocity $\vec{\upsilon}$, momentum $\vec{p}$ and the Lagrange function L by the following relation:

$$E = \vec{p}\vec{\upsilon} - L. \tag{7.49}$$

Substituting the expressions for the Lagrange function and momentum into (7.49).

$$L = -mc^2\sqrt{1 - \upsilon^2/c^2}, \tag{7.50}$$

$$\vec{p} = m\vec{\upsilon}/\sqrt{1 - \upsilon^2/c^2}, \tag{7.51}$$

we get the well-known relativistic formula for the total energy of the body

$$E = \frac{mc^2}{\sqrt{1 - \upsilon^2/c^2}}. \tag{7.52}$$

Squaring expressions (7.51) and (7.52) and comparing them gives

$$\frac{E^2}{c^2} - \frac{E^2 m^2 \upsilon^2}{c^2 p^2} = p^2. \tag{7.53}$$

Taking into account this formula (7.53) and the expression $E^2/p^2 = c^4/\upsilon^2$, the sought relation (7.48) directly follows.

Formula (7.48) can be rewritten in an equivalent form

$$m^2 = \frac{E^2}{c^4} - \frac{p^2}{c^2}, \tag{7.54}$$

according to which the right-hand side is invariant with respect to Lorentz transformations (7.44)–(7.47). Therefore, the rest mass m must also remain unchanged under Lorentz transformations, that is, mass m is a scalar invariant. In other words, rest mass m acts as the main characteristic of the body.

As noted by L. Okun, "... what remains unclear: why for half a century, during which Einstein discussed the connection between mass and energy, he never once gave in his writings and letters the formula $E^2 - p^2c^2 = m^2c^4$ that defines the Lorentz-invariant mass?".

An important formula (7.54) gives additional grounds for asserting that it contains rest mass m, and not some other mass. Indeed, at zero momentum $\vec{p} = 0$, that is, when the body is at rest, the following result follows from formula (7.48): $E_0^2/c^2 = (mc)^2$, where $E_0 = E(\vec{p} = 0)$ is the rest energy and $m = m(\vec{p} = 0) = m_0$ is the rest mass. Then, by taking the square root of this expression, one can obtain an expression for the rest energy

$$E_0 = mc^2, \tag{7.55}$$

which was probably first given by Einstein in 1914 [56]. Thus, formula (7.54) agrees exactly with formula (7.55) at zero momentum $\vec{p} = 0$, i.e., for rest energy and rest mass. ***Thus, Einshtein's famous formula E*** $= mc^2$ ***turns out to be true only in a state of rest***. Moreover, ***the formula*** $E_0 = mc^2$ ***is a partial case of the more general formula*** $E^2/c^2 = p^2 + m^2c^2$, which connects the energy E, mass m, and momentum p of a moving body.

As already noted above, the same formula (7.48) agrees with formula (7.52) for the total energy of a body moving at high speed. On the other hand, formula (7.52) can be obtained using formula (7.51) for momentum. Indeed, substituting (7.51) into (7.48) gives

$$E^2 - \frac{m^2 \upsilon^2 c^2}{1 - \upsilon^2/c^2} = m^2 c^4, \tag{7.56}$$

from where, as a result of simple algebraic transformations, we get the formula

$$E^2 = \frac{m^2 c^4}{1 - \upsilon^2/c^2}, \tag{7.57}$$

which confirms the consistency of the formulas $\frac{E^2}{c^2} = p^2 + m^2 c^2$ and $E = \frac{mc^2}{\sqrt{1-\upsilon^2/c^2}}$, in which the rest mass of the body is denoted by m.

Contradiction between the concept of "relativistic mass" and the requirements of Lorentz invariance of mass. The false concept of the relativistic mass, which depends on the speed, arose due to at least two reasons: (1) the appearance of longitudinal and lateral masses in the first works of Abraham [47] and Lorentz [49], (2) the presence of a multiplier $m/\sqrt{1 - \upsilon^2/c^2}$ in many formulas of the theory of relativity (see, for example, formulas (7.51), (7.52)). Now it becomes clear that there are various unrelated reasons for the occurrence of such a fraction of the rest mass m and the value $\sqrt{1 - \upsilon^2/c^2}$. The rest mass arises naturally in formulas (7.51) for momentum and (7.48) for the relation between energy and momentum, while the factor $\sqrt{1 - \upsilon^2/c^2}$ is a direct consequence of the Lorentz transformations (7.39) and (7.42). Therefore, combining these factors into a single formula for the relativistic mass (which we denote by m_{rel}, so as not to be confused with the rest mass m)

$$m_{rel} = m/\sqrt{1 - \upsilon^2/c^2} \tag{7.58}$$

is an artificial device that does not correspond to any reasonable physical reason. Moreover, the analysis of formula (7.54) shows that expression (7.58) for the relativistic mass completely contradicts the requirements of the Lorentz invariance of the mass, that is, its invariance when moving from one inertial system to another using the Lorentz transformations (7.39) and (7.42).

Thus, it should be recognized that as a result of an increase in the speed of the body in accordance with formulas (7.51), (7.52), the momentum $\vec{p}$ of the body and its energy E increase, while the mass m of the body, which is included in formulas (7.48) and (7.54) of the relationship between $\vec{p}$ and E, remains a constant value, being equal to the rest mass of the body.

In support of this conclusion, it should be quated the words that Einstein wrote in 1948 in a letter to Barnet, the author of the book "The Universe and Dr. Einstein", and which were reproduced in the article [48]: "It is not good to introduce such a concept of the mass $M = m/\sqrt{1 - \upsilon^2/c^2}$ of a moving body, for which there is no clear definition. It is better not to introduce any other concept except "rest mass" m.

Instead of entering M, it is better to give an expression for the momentum and energy of a moving body."

Considering the importance of the above-mentioned provisions, the author would like to suggest to dear readers, if possible, once again focus their attention on the following conclusions regarding the origins of the false concept of the "relativistic mass" of a body, which depends on the speed of its movement:

(1) the mass of the body remains unchanged ($m = \text{const}$), being a scalar invariant, which does not change its value when moving from one inertial system to another;
(2) the fallacy of the concept of "relativistic mass", which depends on the speed, is easy to understand, because the multiplier

$$\gamma(\upsilon) = 1/\sqrt{1 - \upsilon^2/c^2} \tag{7.59}$$

was mistakenly attributed to the rest mass, while it characterizes the transformation of the 4-dimensional energy–momentum vector in accordance with formulas (7.44)–(7.47).

In other words, this error is rather ideological in its nature. But the fallacy of ideology, as is known, is fundamental. In this case, the incorrect interpretation of the multiplier $\gamma(\upsilon)$ led to the appearance of the incorrect concept of "relativistic mass"

$$m\gamma(\upsilon) = m/\sqrt{1 - \upsilon^2/c^2}, \tag{7.60}$$

which grows and becomes infinite as the speed υ of the body moves closer to the speed c of light.

However, the correct interpretation is that as the speed υ of the body increases, the total energy E of the body increases, which is determined by Einstein's famous formula

$$E = mc^2\gamma(\upsilon). \tag{7.61}$$

At the same time, the total energy E of the body consists of rest energy, which does not depend on the speed υ of the body's movement and is given by the following formula:

$$E_0 = mc^2 \tag{7.62}$$

and kinetic energy

$$E_\kappa = mc^2[\gamma(\upsilon) - 1], \tag{7.63}$$

which depends on the speed υ of movement of the body.

The law of conservation of mass and its possible violations, non-additivity of mass in the theory of relativity. Consider the issue of the law of conservation of mass in classical mechanics and rest mass in relativistic mechanics. In classical (Newtonian) mechanics, the law of conservation of mass is fulfilled, which was already discussed above. Unlike classical mechanics, in relativistic mechanics (that is, in the mechanics of Einstein's theory of relativity), the law of conservation of the sum of inert rest masses for a system of bodies ceases to be fulfilled. To confirm this statement, consider the following example.

It is well known from nuclear physics that the total rest energy of a system of interacting bodies (for example, nucleons in the nucleus of an atom $E_{0\,nucleus}$) is less than the sum $\sum_i E_{0i}$ of the rest energies of individual nucleons by the value of the binding energy ΔE_{bind}, i.e.,

$$\sum_i E_{0i} - E_{0\,nucleus} = \Delta E_{bind}, \tag{7.64}$$

Hence, taking into account Einstein's formula $E_0 = mc^2$ for the rest energies of individual nucleons of the nucleus $\sum_i E_{0i} = \sum_i m_i c^2$, the bound nucleus $E_{0\,nucleus} = m_{nucleus}c^2$ and their difference $\Delta E_{bind} = \Delta mc^2$ one receives the following result:

$$\sum_i m_i - m_{nucleus} = \Delta m. \tag{7.65}$$

Here Δm is a nonzero mass defect, which is equal to the difference in the rest masses of individual nucleons and bound nucleons in the nucleus of an atom. Formula (7.64) confirms the fact, which is realized in the mechanics of relativity theory, that the inertial rest mass of a system of bodies is not equal to the sum of the inertial rest masses of each of the bodies that make up this system.

Let us consider the issue of the additivity of mass in classical mechanics and in the mechanics of the theory of relativity. In classical (Newtonian) mechanics, mass is an additive quantity, that is, the mass of a system of bodies is equal to the sum of the masses of each of the bodies that make up this system, regardless of the changes that occur in it. Unlike classical mechanics, mass is not an additive quantity in relativistic mechanics (see, for example, [48]).

It is easy to verify this by using formula (7.54) and the additivity of energy E and momentum $\vec{p}$. These quantities are stored in a closed system, which, for example, consists of two non-interacting particles with total energy and momentum

$$E_{12} = E_1 + E_2, \vec{p}_{12} = \vec{p}_1 + \vec{p}_2. \tag{7.66}$$

Indeed, on the basis of (7.54) and (7.66), it turns out that the mass m_{12} of such a complex of two particles is determined by the following formula:

$$m_{12} = \sqrt{(E_1 + E_2)^2/c^4 - (\vec{p} + \vec{p})^2/c^2}. \tag{7.67}$$

An important conclusion regarding the non-additivity of mass in the theory of relativity directly follows from (7.67), namely:

$$m_{12} \neq m_1 + m_2. \tag{7.68}$$

However, the case that is an exception to this formula in the theory of relativity will be considered below.

The problem of photon mass and its connection with the medical diagnostic method of positron emission tomography (PET). Interest in the problem of photon mass is also caused by the important fact that this problem is directly related to the reaction of the annihilation of an electron and a positron with the formation of two photons of electromagnetic radiation of the gamma range, which is the basis of one of the modern and precise methods of medical diagnostics, namely: positron emission tomography (PET). Indeed, the electron and positron that enter into the annihilation reaction have a total rest mass of $1,82 \cdot 10^{-30}$ kg and a total rest energy of 1.02 MeV. As for the products of the annihilation reaction of an electron and a positron, as a result of which two gamma quanta are formed, each of them has zero mass and zero rest energy. The answer to the question of whether the law of conservation of mass is fulfilled in this annihilation reaction, as well as a consequence in the PET diagnostic method, will become clear below.

Photon mass. In modern physical science, it is generally accepted that photons are quanta of the electromagnetic field, which are massless elementary particles. This means they have zero rest mass and their speed of movement coincides with the the speed of light. A photon has energy E and momentum (impulse) $\vec{p}$, which are determined by the following formulas:

$$E = h\nu = hc/\lambda, \tag{7.69}$$

$$\vec{p} = h\nu\vec{n}/c = h\vec{n}/\lambda. \tag{7.70}$$

Let us remind that h is Planck's constant, ν and λ are the frequency and wavelength of electromagnetic radiation, c is the speed of light in a vacuum, and $\vec{n}$ is a unit vector that characterizes the direction of photon movement. It is easy to check that using the relativistic formula (7.54) for a photon after substituting expressions (7.69) and (7.70) for photon energy and momentum gives a zero result for the photon mass: $m = 0$.

It is interesting to note that the concept of gravitational mass does not make sense for a photon, because if we use this concept for a relativistic particle, which is a photon, it turns out that for a photon moving horizontally, the gravitational mass turns out to be twice as large as for a photon falling vertically. In addition, processes involving particles with zero rest mass (photons, and possibly gravitons) demonstrate additional facts of failure of the law of conservation of rest mass in relativistic mechanics.

The mass of the system, which consists of two photons. Let us continue the consideration of the issue of non-additivity of mass using the example of a system consisting of two photons. As was shown in [39, 42], the application of formula (7.67) to a two-photon complex gives the following consequences, taking into account the zero mass of the photon and the known relationship $E = c|\vec{p}|$ between the energy E and the modulus of the photon momentum $|\vec{p}|$:

1. If the momentum vectors of both photons $\vec{p}_1 = p_1\vec{n}_1$ and $\vec{p}_2 = p_2\vec{n}_2$ are ***parallel*** $(\vec{n}_1 = \vec{n}_2 = \vec{n})$, then

$$(\vec{p}_1 + \vec{p}_2)^2 = \vec{n}^2(p_1 + p_2)^2 = (E_1 + E_2)^2/c^2. \tag{7.71}$$

Substituting the results obtained in (7.71) into formula (7.67), we get for this case

$$m_{12}^{paral} = 0. \tag{7.72}$$

Thus, ***if the momenta (impulses) of two photons are parallel, then not only is the mass of each photon equal to zero, but there is also no mass of the two-photon complex***.

2. If the Photon Momentum Vectors Are ***Antiparalle***l $(\vec{n}_1 = -\vec{n}_2)$, then

$$(\vec{p}_1 + \vec{p}_2)^2 = (p_1\vec{n}_1 + p_2\vec{n}_2)^2 = (p_1\vec{n}_1 - p_2\vec{n}_1)^2 = \vec{n}_1^2(p_1 - p_2)^2 = (E_1 - E_2)^2/c^2. \tag{7.73}$$

In this case, substututing the expression (7.73) into formula (7.67), we directly obtain the following result:

$$m_{12}^{antiparal} = \sqrt{(E_1 + E_2)^2/c^4 - (E_1 - E_2)^2/c^4} = 2\sqrt{E_1 E_2}/c^2 = 2h\sqrt{\nu_1\nu_2}/c^2. \tag{7.74}$$

It is taken into account here that in the general case of different photon frequencies, photon energies are equal to $E_1 = h\nu_1$, $E_2 = h\nu_2$. Thus, ***if the momenta (impulses) of two photons are antiparallel, the mass of the two-photon complex turns out to be nonzero***. As far as the author knows, this result was consistently obtained for the first time in [39].

The law of conservation of mass in the process of annihilation of the electron–positron pair and in the PET diagnostic method. The case of antiparallel moments (impulses) of two photons discussed above *is* directly related to the well-known process of annihilation of an electron and a positron, which, as already noted in Chap. 6 and [39], is the basis of the modern diagnostic method of positron emission tomography (PET) (see e.g. [17]).

The initial stage of the annihilation reaction consists in the collision of a relativistic electron with energy $E_e = m_e c^2\gamma(\upsilon)$ and momentum $\vec{p}_e = m_e\upsilon\gamma(\upsilon)\vec{n}_1$ and a relativistic positron with energy $E_p = m_p c^2\gamma(\upsilon)$ and momentum $\vec{p}_p = m_p\upsilon\gamma(\upsilon)\vec{n}_2$ under conditions $E_e = E_p$, $\vec{p}_e = -\vec{p}_p$ $(\vec{n}_1 = -\vec{n}_2)$.

Using formula (7.67) for a complex of two photons with energy $E_{ph} = h\nu$ for each of them and with antiparallel momentum vectors $\vec{p}_{ph1} = -\vec{p}_{ph2}$ gives the following result, respectively, for the mass m_{12}^{ep} of the electron–positron complex and the mass m_{12}^{ph} of the two-photon complex:

$$m_{12}^{ep} = (E_e + E_p)/c^2 = 2m_e\gamma(\upsilon), \tag{7.75}$$

$$m_{12}^{ph} = (E_\phi + E_\phi)/c^2 = 2h\nu/c^2 \tag{7.76}$$

On the basis of formula (7.76), it should be emphasized that the mass m_{12}^{ph} of the two-photon complex is nonzero, although the mass of each of the photons is individually equal to zero. In addition, it follows from Eqs. (7.74) and (7.76) that the law of conservation of total mass is fulfilled in the process of electron and positron annihilation, that is, the mass of the electron–positron complex is exactly equal to the mass of the two-photon complex: $m_{12}^{ep} = m_{12}^{ph}$ Thus, ***in the case of the electron–positron pair annihilation reaction in the medical diagnostic method of PET, not only the laws of conservation of energy and momentum, but also the law of conservation of mass is fulfilled***.

In connection with the problem of fulfilling (or not fulfilling) the law of conservation of mass, it should be noted that this problem is solved by Einstein in the theory of relativity as follows: the law of conservation of mass lost its independence and merged with the law of conservation of energy.

The same idea is essentially used by Albert Einstein and Leopold Infeld. According to the theory of relativity, there is no significant difference between mass and energy. Energy has mass, and mass is energy. Instead of two laws of conservation, we have only one: the law of conservation of mass-energy.

Certainly, there is no doubt that the law of conservation of energy in an isolated system is fulfilled in all the above (and in all other not given) cases.

Unsolved problems of the nature of mass. The problem of understanding the very nature of the mass, which has a rather long history, still remains complex and unsolved.

One of the first attempts to find out the nature of mass probably belongs to Lorentz, who, even before the creation of the theory of relativity, expressed the opinion that the mass of an electron has an electromagnetic nature. His reasoning was based on the completely understandable idea that the movement of an electron creates a current around which a magnetic field arises.

Later, Feynman created a consistent theory of the electromagnetic mass of the electron (see [57]). In classical mechanics, as already noted earlier, inertial mass, as a property inherent in a body, represents the proportionality coefficient between the body's momentum and its speed. Feynman calculated the momentum of a non-relativistic electron moving with speed $\upsilon << c$, obtaining the following formula:

$$\vec{p} = \frac{2e^2\vec{\upsilon}}{3ac^2}, \tag{7.77}$$

where e is the charge of the electron, a is its radius. Thus, the mass m_e of the electron according to Feynman, which is determined by the energy of the electromagnetic field, is equal to

$$m_e = \frac{2e^2}{3ac^2}. \tag{7.78}$$

The field nature of mass was reflected in Einstein's general theory of relativity [58], an essential element of which was the principle of equivalence between the movement of bodies in a gravitational field and the free movement of bodies in a non-inertial frame of reference. The physical content of the mass and gravity of bodies is reduced to the change in the properties of the curved 4-dimensional space–time around these bodies. Using more precise words, the equations describing the nature of gravity in general relativity relate the characteristic of curved space–time, which is the curvature tensor $\hat{R}$, and the mass, which is included in the energy–momentum tensor $\hat{T}$ (a certain analogue of the 4-dimensional energy vector -momentum in the special theory of relativity).

Since 1915, when the basic principles of the general theory of relativity were created, and to this day, the efforts of scientists to build a unified field theory with the aim of unifying all known fields (gravitational, electromagnetic, strong and weak nuclear) and thus a deeper understanding the nature of the mass. The principle stages of work on this extremely difficult problem are as follows:

In 1917, Einstein introduced the cosmological constant λ into the equations of the general theory of relativity, which characterizes the non-zero energy of the physical vacuum, which is uniformly distributed in all points of the universe [59];

In 1922, Friedman, using the ideas of Einstein, developed the theory of the expanding universe [60];

7 years later, in 1929, Hubble [61] discovered the effect of a "redshift" of spectral lines in the spectra of distant galaxies, which experimentally confirmed the process of expansion of the universe;

In 1964, the British physicist Peter Higgs and the Belgian physicist Francois Englert independently and almost simultaneously theoretically introduced the concept of hypothetical particles that have zero spin and were later called Higgs bosons. These bosons characterize an additional field at all points in space, the interaction with which causes the appearance of mass in other particles, with a strong interaction causing the appearance of heavy particles, and a weak interaction—light particles;

In 1967, Glashaw, Salam, and Weinberg created the theory of the electroweak interaction, which combined the electromagnetic interaction and the weak nuclear interaction;

Modern versions of the theory of the unification of all interactions use the concept of additional measurements in relation to the three spatial and one temporal measurements of Einstein's theory of relativity. In particular, a fifth measurement is added to the Kalutza-Klein theory, which allows us to introduce Maxwell's equations into the general theory of relativity, with the help of which in the second half of the nineteenth century it was possible to consistently describe electrical, magnetic, and optical phenomena;

At the very end of the twentieth century, in 1998–1999, it was established that contrary to Hubble's law, which describes the uniform expansion of the universe over time, this process turns out to be accelerated. Moreover, new astrophysical data (see, for example, [7, 32]) indicate that the matter that has been investigated so far contains only approximately 5% of the total energy of the universe. The last 95% of the matter in the universe is divided into the following components: about 27% of all the energy of the universe is contained in dark matter of unknown nature, and the rest of the energy of the universe (about 68%) is hidden in dark energy;

The Nobel Prize in Physics for 2011 was awarded to American astrophysicists Saul Perlmutter and Adam Reiss, as well as Australian astrophysicist Brian Schmidt for the discovery of the accelerated expansion of the universe through experimental observations of supernovae [45, 46]. This discovery led to the assertion that the accelerated expansion of the universe is related to the existence of a hypothetical dark energy, and also provides possibly additional information regarding the physical nature of the cosmological constant λ of Einstein's general theory relativity;

Great hopes for further clarification of the problem of the nature of mass are associated with experimental studies of Higgs bosons, which were discovered in 2012 at the Large Hadron Collider at the European Center for Nuclear Research (CERN) in Geneva, Switzerland. This experimental confirmation of the existence of the Higgs boson was marked the following year in 2013 by the awarding of the Nobel Prize in Physics to the two authors of the theoretical discovery of the Higgs boson, a British theoretical physicists Peter Higgs and a Belgian theoretical physicist Francois Englert, who 60 years ago predicted the existence of this fundamental elementary particle to explain the nature of the masses of other elementary particles.

Peter Higgs (1929-2024)

Francois Englert (born in 1932)

The true nature of dark matter and dark energy is currently unknown. The first theoretical hypothesis about the possibility of the existence of dark matter belongs to the Swiss and American astrophysicist Fritz Zwicky (1898–1974), who expressed an assumption about 90 years ago about the influence of some invisible matter on certain anomalies in the movement of galaxies [43]. An unusual property of dark matter is the experimental fact that it cannot emit electromagnetic radiation. Dark matter is associated, perhaps, with unknown elementary particles that interact with already studied matter most likely by means of gravitational and weak nuclear interactions, that is, two of the four known interactions (the other two interactions are electromagnetic and strong nuclear interactions). Of course, there may be a fifth and even greater number of fundamental force interactions that are responsible for the still unknown properties of 95% of the dark matter and dark energy of the universe.

According to the scientific community, the first experimental confirmation of the existence of dark matter was received in 1970 by the American astrophysicist Vera Rubin (1928–2016), who made this discovery while studying the rotation curve of the Andromeda galaxy together with her scientific colleague, the American astrophysicist Kent Ford (1931–2023) [44]. The so-called Rubin-Ford effect discovered by them consisted in the fact that the rotational speeds of stars on the periphery and in the center of the galaxy turned out to be close in measured value in an unexpected contradiction with Kepler's laws. The resulting anomalies of the curve of the Andromeda galaxy's rotational movement and some other results of astronomical observations by Vera Rubin and Kent Ford testified to the presence of invisible dark matter in accordance with the theoretical hypothesis of Fritz Zwicky.

7.4 Conclusions

When writing this chapter, the author set himself these two main goals. The first of them was to present the foundations of quantum mechanics—one of the most important branches of physical science, which studies the fundamental principles and laws of the micro and nanoworld. These laws and principles underlie the processes that occur in the macrocosm around us. The second goal is directly related to the use of quantum mechanics in medical physics for the creation and practical application of tomographic methods for precision medical diagnostics. The main attention was paid to the three most common modern diagnostic methods: X-ray computed tomography (CT), magnetic resonance imaging (MRI), and positron emission tomography (PET).

As in previous chapters, Sect. 7.2 provided a general (if possible, popular) description of quantum mechanical ideas and principles of these medical diagnostic methods, while Sect. 7.3 offered a more precise physical and mathematical description of not only the fundamental principles of quantum mechanics, but also the theoretical foundations of CT, MRI and PET tomographic methods.

Particular attention was paid to the following problems underlying the use of X-ray computed tomography, magnetic resonance imaging and positron emission tomography methods:

Application in the CT method of a layer-by-layer scanning to obtain medical imaging due to the gradual passage of a X-ray beam through a thin layer of human body tissues (1979 Nobel Prize in Physiology and Medicine was awarded to Godfrey Hounsfield and Allan Cormack "for the development of computer assisted tomography"), as well as (the first 1901 Nobel Prize in Physics in Physics, which was awarded to Wilhelm Konrad Röntgen for the discovery of X-rays).

The tomographic method MRI in medicine (2003 Nobel Prize in Physiology and Medicine was awarded to Paul Lauterbur and Peter Mansfield "for their discoveries concerning magnetic resonance imaging") uses (a) the effect of proton energy level splitting in an external magnetic field (1902 Nobel Prize in Physics in Physics was awarded to Peter Zeeman and Hendrik Lorentz for the discovery and explanation of the effect of the splitting of spectral lines in a magnetic field) and (b) the phenomenon of nuclear magnetic resonance (NMR) (1952 Nobel Prize in Physics was awarded to Felix Bloch and Edward Purcell for the discovery of NMR in liquids and solid bodies).

The positron emission tomography PET, being a method of medical radioisotope diagnostics, uses the phenomenon of electron positron annihilation as its physical basis. The existence of the positron was theoretically predicted by Paul Dirac as a result of solving the relativistic (taking into account the effects of the theory of relativity) quantum equation of electron motion, which Paul Dirac derived in 1928. Soon after the theoretical discovery of the positron by Dirac, its experimental confirmation took place. The positron was discovered in experimental studies by Karl Anderson in 1932 by observing the tracks of microparticles of cosmic radiation in Wilson chambers, which were placed in an external magnetic field. Note that the Dirac equation is a generalization of the nonrelativistic Schrödinger wave equation. Both equations are the basic equations of quantum mechanics. The Nobel Prize in Physics for 1933 was awarded to Erwin Schrödinger and Paul Dirac "for the discovery of new productive forms of atomic energy."

At the end of Sect. 7.3, the problem of mass in the classical mechanics of Newton-Galileo, as well as in the theory of relativity, was discussed. In particular, an explanation was presented for the fallacy of the concept of relativistic mass, which depends on the speed of movement. In addition, the relevance of the latter problem is dictated by the fact that the diagnostic method of positron emission therapy (PET) uses the process of annihilation of matter and antimatter during the interaction of an electron and its antiparticle—a positron. As a result of this annihilation reaction, two gamma quanta appear, which have opposite direction vectors of photons of short-wave electromagnetic radiation. Despite the lack of mass of an individual photon, it has been proven that the mass of a two-photon complex with antiparallel momentum vectors is nonzero in the PET method. Moreover, in this diagnostic method, the law of conservation of mass is implemented at all its stages. In connection with the above, it is interesting to note that in classical Newtonian mechanics, the law of conservation of mass is fulfilled, while in relativistic mechanics (i.e., in the mechanics of Einstein's theory of relativity), the law of conservation of the sum of inert masses for a system of bodies ceases to be fulfilled. Additionally, the hitherto unresolved problem of the existence of dark energy and dark matter was discussed.

References

1. L. Carroll. *Through the Looking-Glass*. eBook, The Millennium Fulcrum Edition (2008)
2. R.P. Feynman, R.B. Leighton, M. Sands, *The Feynman Lectures on Physics*, vol. 3 (London. Addison-Wesley Publishing Company, Reading, Massachusetts, Palo Alto, 1965)
3. L.D. Landau, E.M. Lifshits, *Quantum Mechanics* (Frankfurt. Pergamon Press, Non-Relativistic Theory, Oxford, New York, Toronto, Sydney, Paris, 1977)
4. A.S. Davydov, *Biology & Quantun Mechanics* (New York, Pergamon Press, Oxford, 1982)
5. I.R. Yukhnovskiy. *Fundamentals of Quantum Mechanics*, Lybid, Kyiv (2002)
6. IO Vakarchyk, *Quantum Mechanics* (Ivan Franko Lviv National University, Lviv, 2012)
7. S. Goudsmit, G. Uhlenbeck, Ersetzung der Hypothese von unmechanischen Zwang durch eine Forderung bezuglich des innen verhaltens jedes einzelnen elektrons. Naturwissehschaften - Springer Science **13**(47), 953–954 (1925)
8. J. Van Dyk, *True Tales of Medical Physics. Insights into a Life-Saving Specialty*, 1st edition, Springer Cham (2022)
9. S. Keevil, R. Padevani, S. Tabakov, C. Lewis, *T*, 1st edn. (Greener, Introduction to Medical Physics, CRC Press, 2022)
10. J.T. Bushberg, J.A. Seibert, *The Essential Physics of Medical Imaging. Study Guid*e, LWW (2022)
11. I.R. Cameron, I.G. Skofronik, *Medical Physics* (Wiley, New York, 1878)
12. A. Martin, S.A. Harbison, *An Introduction to Radiation Protection* (Chapman and Hall, London, 1979)
13. *Graduate Studies in Medical Physics*. University of Wisconsin-Medison, Medison (1996)
14. K. Russel, *Hobbie* (New York, Berlin, Heidelberg, Springer-Verlag, Intermediate Physics for Medicine and Biology, 1997)

15. M. Nelkom, P. Parker, *Advansed Level Physics*, 5th edn. (Heinemann Educational Books Ltd., London, 1982)
16. *Medical and Biological Physics*, vol. 2, Ed. A.V. Chalyi. Bogomolets National Medical University, Kyiv (2001)
17. *Medical and Biological Physics*, Ed. A.V. Chalyi. Knyga Plus, Kyiv (2005)
18. J. von Neumann, *Mathematische Grundlagen der Quantenmechanik* (Springer, Berlin, 1932)
19. W. Heisenberg. *Collected Papers*. Berlin, Heidelberg, New York, London, Paris, Tokyo, Hong Kong; Springer-Verlag, 1985–1989
20. M. Planck, On the theory of energy distribution law of the normal spectrum radiation. Verhandl. Dtsch. Phys. Ges. **2**, 237–245 (1900)
21. N. Bohr, On the constitution of atoms and molecules. The London, Edinburgh, and Dublin Philos Mag J Sci **26**(151), 1–25 (1913)
22. L. de Broglie, Researcheson the quantum theory. Thesis, Paris, 1924, *Ann de Physique* **3**(10), 22 (1925)
23. W. Heisenberg, *The physical principles of the quantum theory* (Courier Dover, Stuttgart, 1930)
24. E. Schrödinger, *Abhandlungen zur Wellenmechanik*. Leipzig (1927)
25. P. Zeeman, The effect of magnetization on the nature of light emitted by a substance. Nature **55**, 347 (1897)
26. P. Lauterbur, Reflections on MRI-leadership reflections. *IEEE: Signal Processing Magazine* 22(1) 1407710 (2005)
27. P. Mansfield. *Proton Magnetic Resonance Relaxation in Solids by Transient Methods*. University of London (Queen Mary College) (1963)
28. . P. Mansfield, Snap-shot MRI. *The Nobel Prizes 2003. Nobel Foundation*, 266–283 (2004)
29. F. Indurine, *Quantum Hromodynamics. Introduction to theory of quarks and gluons* (1986)
30. P.C. Lauterbur, All science is interdisciplinary. *The Nobel Prizes 2003. Nobel Foundation*, 245–251 (2004)
31. A. Einstein. On a heuristic viewpoint concerning the production and transformation of light. *Annalen der Physik* 17(322) (1905)
32. E. Schrödinger, *Vier Vorlesungen uber Wellenmechanik*. Berlin (1928)
33. P.A.M. Dirac, *The Principles of Quantum Mechanics*. Oxford University Press, Oxford (1930)
34. P.A.M, Dirac. *The Quantum Theory of the Emission and Absorption of Radiation*. 114(767): 247–65 (1927)
35. *Wolfgang Pauli—Biographical*. NobelPrize.org. Nobel Prize Outreach AB 2924. https://www.nobelprize.org/prizes/physics/1945/pauli/biographical/
36. W. Pauli, The connection between spin and statistics. Phys. Rev. **58**(8), 716–722 (1940)
37. E. Seeram. *Computed Tomography: Physical Principles, Patient Care, Clinical Applications, and Quality Control*. 5th edition, Maryland Heights, Missouri (2023)
38. A.V. Chalyi, N.L. Grytsenko, Historical development of the concept of mass in physical theories of the XVII-XXI centuries. 1. Classical mechanics and the theory of relativity. *Physics and astronomy at school*, №3, 43–47 (2012)
39. A.V. Chalyi, N.L. Grytsenko, Historical development of the concept of mass in physical theories of the XVII-XXI centuries. 2. The law of mass conservation in the process of electron-positron annihilation and in the PET diagnostic method. *Physics and astronomy at school*, №6, 43–47 (2012)
40. G. Galilei, *Dialogo sopra i due massimi sistemi del mondo Tolemaico e Copernicano*. Fiorenza, Per Gio: Batista Landini (1632)
41. I. Newton, *Philosophiae Naturalis Principia Mathematica*. Landini: Jussu Societatis Regiae (1687)
42. A.M. Gabovych, N. Gabovych, How to explain the non-zero mass of electromagnetic radiation consisting of zero-mass photons? Europ. J. Phys. **28**(4), 649 (2007)
43. F. Zwicky, On the frequency of Supernovae. Astrophys. J. **96**, 28 (1942)
44. V.S. Rubin, W.K. Ford, Rotation of the Andromeda Nebula from a spectroscopic survey of emission regions. Astrophys. J. **159**, 379 (1970)

45. A.G. Reiss et al., Observational evidence from supernovae for an accelerating universe and a cosmological constant. Astrophys. J. **116**, 1009 (1998)
46. S. Perlmutter et al., Measurements of Omega and Lyambda from 42 high-redshift supernovae. Astrophys. J. **517**, 565 (1999)
47. M. Abraham. Prinzipien der dynamik des elektrons. *Ann. Phys.* (Leipzig) **10**:105 (1902)
48. L.B. Okun, The concept of mass. Phys. Today **42**, 31 (1989)
49. H. Lorentz, Electromagnetic phenomena in a system moving with any velocity smaller than that of light. Proc. Acad. of Sci. Amsterdam. **6**, 809 (1904)
50. A. Einstein, Zur electrodynamik bewegter korper. *Ann. Phys.* (Leipzig) **17**, 891 (1905)
51. A. Einstein. Ist die tragheit eines korpers von seinem energiegieihalt abhangig? *Ann. Phys.* (Leipzig) **18**, 639 (1905)
52. H. Minkowski, Raum und Zeit. Phys. Z. **10**, 104 (1909)
53. A. Einstein, Le principle de relativite et ses consequences dans la physique modern. Arch. Sci. Phys. Naturelles **29**, 5 (1910)
54. L.D. Landau, E.M. Lifshits, *The classical theory of fields* (Pergamon Press, Oxford, 1971)
55. A. Einstein Lecture notes for course on relativity at the University of Berlin, winter semester 1914/1915. *The collected papers of Albert Einstein*. Princton, NJ: Princton Univ.Press, 1987–2006
56. L.D. Landau, E.M. Lifshits, *Mechanics* (Pergamon Press, Oxford, 1976)
57. R.P. Feynman, R.B. Leighton, M. Sands, *The Feynman Lectures on Physics*, vol. 2 (London. Addison-Wesley Publishing Company, Reading, Massachusetts, Palo Alto, 1965)
58. A. Einstein. Die Feldgleichungen der Gravitation. *Sitzungber. Konigl. Preub. Akad. Wissenschaft.* (Berlin) **48**, 844 (1915)
59. A. Einstein. Kosmologische Betrachtungen zur allgemeinen Relativitatstheorie., *Sitzungber. Konigl. Preub. Akad. Wissenschaft.* (Berlin) **1**, 142 (1917)
60. A. Friedmann, Uber die Krummung des Raumes. Z. Phys. **10**, 377 (1922)
61. E. Habble, A relation between distance and radial velocity among extra-galactic Nebulae. Proc. Natl. Acad. Sci. **15**, 68 (1929)

Chapter 8
Physics and Possibilities of Human "Extrasensory" Perception

Abstract In this chapter, the physical and mathematical foundations of the unusual, so-called "extrasensory" human perception will be considered. I confess that the author enters "the minefield of extrasensory abilities" with a risk. The reason for this caution is absolutely understandable, since most scientists have a sharply negative attitude to this topic, considering this direction a false science. In this regard, I would like to warn the dear reader from the very beginning that I in no way want to act as an "apologist for extrasensory human abilities." At the same time, I also do not want to look like a supporter of dogmas like "This can never be!". Further, the main attention will be paid to physical and mathematical ideas, laws and principles that have a reliable scientific basis and make it possible to try explaining the unusual human perception that is currently considered "extrasensory".

Keywords The rule of three sigma · Human physical fields · Radiotaxis · Foresight and intuition · Quantum teleportation · Anticipatory solutions of wave equations · Quantum world and "cause-consequence" uncertainty principle

> *Imagination is more important than knowledge. Knowledge is limited. Imagination encircles the world.*
>
> Albert Einstein

8.1 Introduction ("The Rule of Three Sigma")

First of all, it should be noted that the phenomenal abilities of a person in any field are not something strange and incomprehensible. From the point of view of mathematical statistics, it can be assumed that human abilities in any field obey the normal law (Gaussian distribution). For random variables characterized by the Gaussian distribution or other distributions (Bernoulli, binomial, exponential, etc.), there are the most important characteristics of these distributions for random variables, namely: the mean value (mathematical expectation) $M(x)$, variance $D(x)$ and the mean square

A. V. Chalyi, *What is Medicine?*, https://doi.org/10.1007/978-3-031-64979-0_8

deviation (standard deviation) $\sigma(x)$. The corresponding formulas, by which these characteristics of the distributions of random variables are determined, are given in Sect. 8.3B.

Let us emphasize one important result, which is known in mathematical statistics as "The rule of three sigma". This rule can be formulated as follows: according to the distribution of random cases (which include, in particular, any human abilities), obeying the normal law or, what is the same, the Gaussian distribution, the overwhelming majority, or to be mathematically precise, 99.7% of all random cases with a certain human ability fall into the interval $M(x) \pm 3\sigma(x)$. In other words, a very small percentage of random cases (or more precisely only 0.3% of them) falls outside this interval, which has as its center the average value (mathematical expectation) $M(x)$, and the right and left ends, respectively, $M(x) + 3\sigma(x)$, i.e., the average value $M(x)$ plus three mean square deviations $3\sigma(x)$, and $M(x) - 3\sigma(x)$, i.e., the average value $M(x)$ minus three mean square deviations $3\sigma(x)$. In addition, from these 0.3% cases, 0.15% of random cases fall towards larger (relatively speaking, "better") cases, while other 0.15% of random cases fall towards smaller or "worse" cases. Applying "The rule of three sigma" to human abilities, it can be tentatively stated that only 0.15% of representatives of the entire human population are "geniuses", while the other 0.15% are "morons". Of course, such a gradation is completely conditional, since it works for a certain human ability (for example, the ability to run quickly, or to count instantly, or to draw beautifully, or to foresee something unusual, etc.). For another ability, some of the "geniuses" and "morons" can be swapped!

8.2 Section A

8.2.1 Scientific Research of Human Physical Fields

The results of theoretical and experimental studies [1–14] of the physical fields of the human body are of considerable interest from the point of view of the manifestations of possible "extrasensory" capabilities of a person. For this purpose, special equipment is used, which ensures the registration of the following types of radiation from the human body: (a) electromagnetic radiation in the infrared range, which physicists also call equilibrium thermal radiation; (b) electromagnetic radiation in the radio range; (c) electromagnetic radiation in the ultraviolet range; (d) radiation caused by non-stationary processes of electrical activity of human body organs; (e) radiation caused by non-stationary magnetic fields; (f) acoustic radiation of organs and tissues, (g) chemiluminescent radiation, which is caused by various chemical and biochemical reactions accompanying metabolic processes, etc.

The strongest energy channel among all the above-mentioned physical fields of the human body is the channel caused by infrared thermal radiation (see the explanation of the basic laws of thermal radiation and the specifics of thermal radiation of the human body in Chap. 5, Sections A and B). It turns out (see, for example, [1]) that 1 cm^2 of the human body radiates a power of electromagnetic waves equal to about 10 mW, while the entire surface of the human body with an average surface area of 1.0–1.5 m^2 radiates a significant power of electromagnetic thermal radiation of up to 100 W. The wavelength of this radiation is approximately 10 microns in the infrared range.

It is important to emphasize that human skin is quite sensitive to thermal radiation. More precisely, its sensitivity is tenths of a milliwatt per 1 square centimeter of the skin surface, which is ten times less than the surface power of thermal electromagnetic radiation in the infrared range. Thanks to the thermal receptors located on it, the skin feels thermal radiation from the finger (the area of its pad is approximately equal to 1 cm^2), which is located at a distance of several centimeters from the surface of the skin. At the same time, the skin is not able to sense signals from other physical fields (for example, electromagnetic radiation of the radio range), which by their surface power are below the sensitivity limit of human skin and are several orders of magnitude lower than the power of thermal infrared radiation.

As a result of the experimental studies, no significant fundamental differences between the thermal radiation capacities of different people were found. At the same time, some of those people who took part in the experiments demonstrated the ability to increase the temperature of the palms of their hands by several degrees, that is, they went beyond "The rule of three sigma" in terms of their thermal radiation. In terms of quality, ordinary people can develop a similar property of increasing the temperature of the palms of their hands through appropriate training.

In addition, it turned out that a greater physiological response to human thermal radiation in the infrared range is observed for patients in biologically active points or places on the body surface, which are also called Zakharyin-Head zones in honor of the Russian doctor-therapist Grigory Zakharyin (1829–1897), as well as the English neuropathologist and neuropsychologist Sir Henry Head (1861–1940). Such areas warm up the most compared to other areas of the human body surface. It should be noted that this fact has been known for a long time and primarily due to the medical practice of Chinese doctors.

A number of scientific studies (see for example [2–4] provide a definite explanation of the effectiveness of the low-energy impact of thermal radiation, say, from the human palms, compared to large thermal impacts, say, from a reflector, which are thousands of times higher than the sensitivity of thermal receptors of human skin. In other words, the human self-regulating system perceives the effect of large doses of thermal energy as a negative external influence, while the effect of weak thermal signals as a positive external influence.

Some of the previously conducted studies of the features of thermal radiation of the human body were reproduced in Chap. 5 [5, 6]. In particular, experimental measurements of the temperature relaxation of thermal radiation of the human skin were performed (see Fig. 5.4). This and other results provide reliable and unequivocal grounds for making an important conclusion that the so-called "extrasensory" capabilities of a person are explained solely by the action of the physical fields of the human body. At the same time, there is no reason to introduce a hypothetical "biological field" different from the physical fields of living and inanimate nature known to science.

8.2.2 Experimental Studies of Radiotaxis

Let us consider further the experimental studies of Yuriy N. Levchuk (1939–2005), Dr. Sci. (Phys & Math), Palladin Institute of Biochemistry, National Academy of Sciences (Kyiv, Ukraine). He used a precise experimental method of photon-correlation spectroscopy to study the possible "extrasensory" capabilities of a person. The idea of the experiments that were planned by Yuriy Levchuk was as follows: everyone who considered themselves so-called "extrasensor" and agreed to conduct scientific research on their abilities, was offered to raise their palms to a cuvette, in which there was a suspension of spherical polymer particles with an average size of several micrometers or microns (1 μm = 10^{-6} m).

Yuriy N. Levchuk (1939-2005)

The method of photon-correlation spectroscopy [7, 8] allows to study the spectrum of scattered light. More precisely, for this case—to calculate the spectral width of the central Rayleigh line. It is this that makes it possible to measure the speed and direction of movement of the polymer particles of the suspension, which is caused by the action of radiation emanating from the "extrasensor's" palms. In other words, Yuriy Levchuk proposed to use a called "radiotaxis" to determine the "extrasensory" human capabilities. Radiotaxis is the directed movement of particles, as well as cells, microorganisms, etc., under the influence of an external stimulus (in this case, external radiation).

Of course, quite a lot of similar directional movements (taxis) are known, first of all, of microorganisms and cells in biology, biophysics, biochemistry and other branches of scientific knowledge. Let us mention such examples: ***phototaxis***—movement under the influence of optical radiation (light), ***thermotaxis***—movement under the influence of a temperature gradient towards comfortable temperatures, ***chemotaxis***—movement under the influence of chemical (biochemical) reactions, etc.

Yuriy Levchuk performed a wide range of research on the radiotaxic capabilities of about a thousand people who considered themselves "extrasensors" using the method of correlation spectroscopy. The results of these studies turned out to be very interesting, namely: the characteristics of radiotaxis of polymer particles in suspension under the influence of radiation from the "extrasensor's" palms changed both in magnitude and in sign. The last fact means that radiotaxis, like other types of taxis (for example, thermotaxis), can be positive and negative. With positive radiotaxis, the polymer particles moved towards the palms, while with negative radiotaxis—away from the palms. Further, it was possible to rank all the participants of this study also by the magnitude of the effect, i.e., by the speed of the radiotaxis to or from the palms of each of them. Another result turned out to be non-trivial: a 9-year-old girl was able to demonstrate the greatest effect—the highest speed of movement of polymer particles *.

In his research, Yuriy Levchuk did not mention (and, unfortunately, he will no longer mention, because this talented scientist is no longer among us) the exact physical nature of the radiation from the "extrasensor's" palms, which caused the radiotaxis of polymer particles in the suspension. At the same time, the following hypothesis could be proposed: the physical cause of the radiotaxis of polymer particles is none other than thermal electromagnetic radiation of the infrared range, which was emitted by the palms of the participants in the experiment. Thus, everything that has been said about Levchuk's research gives reason to draw the same conclusion: the so-called "extrasensory" capabilities of a person in experiments on radiotaxis are explained by the action of ordinary physical fields of thermal radiation caused by the presence of temperature gradients.

* The following interesting fact should be noted here: Yuriy Levchuk told the author that the so-called "extrasensors" (for obvious reasons, we will not mention their names) who were well known to the Ukrainian general public, refused to participate in these experiments.

8.2.3 Foresight and Intuition, Cause and Consequence Relationships

A person's prediction of events in the future is not recognized by modern science, because it is believed that it uses hypothetical information channels, which are still unknown to science and whose existence is not confirmed by reliable experimental data. Another, more positive, attitude of the scientific community to intuition, as a person's ability to predict results, events based on his (her) acquired scientific knowledge, practical skills and previous experience.

An attempt will be made here to present some ideas and experimental results that, from the point of view of modern physics, can to some extent provide a basis for explaining intuitive abilities (the ability to predict).

Experimental studies of human intuitive abilities. One of the scientific experimental studies in this direction are the research of three scientists J. Mossbridge (Illinois, USA), P. Tressoldi (Padua, Italy) and J. Utts (California, USA), who published an article [9], as well as a review article [10] and an article [11] in the proceedings of the prestigious AIP conference. In these and many other publications, the authors proposed a number of experiments, among which were tasks to test the ability to predict with the subsequent display of correct answers. During the answers to the tests, various physiological parameters of the participants were monitored, namely: heart rate, electrical resistance in the electrode–skin contact, circulating blood volume, pupil dilation, electroencephalographic (EEG) brain activity, blood oxygen level, etc. Computer processing of the tests showed that the participants could predict events that were out of the ordinary and that should take place in the near future. The authors do not consider the results of their research to be extraordinary. They wrote approximately the following words. This is nothing more than an ordinary law of nature, which has not been investigated by other scientists before.

In our time, revolutionary changes are taking place in relation to phenomena that were previously considered absolutely fantastic and unreal. Let us consider some of them.

Quantum teleportation. Quantum teleportation refers to the phenomenon of rapid transfer of quantum objects to macroscopic distances. At the same time, it should be understood that what is transferred to the object itself is the superposition (that is, the sum with weighting coefficients) of a certain number of wave functions, which describes the quantum state of this object. To explain these words, it is necessary to take into account the real objects of quantum mechanics, which studies the processes that occur in the microcosm. These processes are described by so-called wave functions, which are solutions of Schrödinger's wave differential equation [15, 16] (or Pauli's wave equation for a relativistic case [17, 18]). At the same time, the physical meaning has the square of the wave function, which determines the probability of finding a microparticle (for example, electron, proton, neutron, and other microparticles) at a certain point in space and at a certain moment in time. The dear reader can find more detailed information about the basic ideas, laws and equations of quantum mechanics, as well as their application in theoretical and practical medicine, in Chaps. 6 and 7 and in [12–29].

Returning to the problem of quantum teleportation, it should be noted that one of the first experimental confirmations of quantum teleportation was obtained in 2002 by physicists V. Bowen, G. Bacher, T. Ralk et al. from the Australian National University in Canberra in experiments with laser beams. In Ukraine, the problems of quantum teleportation were successfully investigated by the well-known theoretical physicist, Prof. Ivan Vakarchuk and other scientists [14].

The great scientific importance of the problem of quantum deportation is emphasized by the awarding of the 2022 Nobel Prize in Physics to three physicists Alain Aspect (France), John Clauser (USA) and Anton Zeilinger (Austria) for the experimental study of the phenomenon of quantum entanglement, which is closely related to practical implementation quantum communication (teleportation). Strictly saying, they received this prize "for experiments with entangled photons, establishing the violation of Bell inequalities and pioneering quantum information science" [13].

Alain Aspect (born in 1947)

John Clauser (born in 1942)

Anton Zeiliger (born in 1945)

It should be noted that information on the first teleportation of the three-dimensional entangled state of the superposition of the wave functions of three photons, the dear reader may find in the article [14]. Until 2019, such teleportation transfers practically existed only for two-dimensional entangled states of the superposition of the wave functions of two photons.

In addition, it is interesting to cite the words of the famous theoretical physicist Michio Kaku, who wrote many interesting books, including "Visions" and "Physics of the Future" [19, 20], in which he expresses his predictions of the future. In particular, Michio Kaku believes that the practical implementation of quantum teleportation of real multidimensional objects, describing by superposition of the wave functions of large number of photons, will take place in about 50 years.

Michio Kaku (born in 1947)

Control of devices and robots with the help of human thought. A model of a rotorcraft was constructed at the University of Minnesota (USA) in 2013, which was controlled by signals of the electrical activity of the human brain. At the same time, an EEG signal converter was used without inserting electrodes into the pilot's cerebral cortex.

At the end of 2013, the President of Israel Shimon Peres presented a prize of 1 million dollars to the BrainGate scientific team, which created a robot that moved its limbs and walked, receiving electrical signals from the human brain.

Another example of such scientific and technical achievements is the creation of a mechanical hand—a robot, which was attached to the nerve endings of the amputated left forearm of the 26-year-old Italian P. Petruzziello. Over the course of one month, he learned, as Petruzziello himself says, "to alternately move the fingers of a mechanical hand, clench the palm into a fist, pick up various objects and perform other actions."

Currently, we are witnessing how, before our eyes, the fantastic and truly "extrasensory" ideas of Elon Musk and his teammates in the Neuralink project are gradually coming to life, when the blind will begin to see, and paralyzed people will begin to walk.

Let us now proceed directly to the examination of physical considerations in favor of the hypothesis regarding the intuitive abilities of a person, or the ability to predict.

Solutions of wave equations for the electromagnetic field. In 1865, the outstanding English physicist James Clerk Maxwell created the electromagnetic theory, which leaded to the well-known wave equations for the main characteristics of the electromagnetic field—the intensity of the electric field and the induction of the magnetic field. These equations describe the process of propagation of electromagnetic waves of various ranges (from radio waves to gamma radiation) in vacuum and material environments. The wave equations of Maxwell's electromagnetic field theory are differential equations of the 2nd order in partial derivatives with respect to coordinates x, y, z in 3-dimensional space and time t. The solutions of these differential equations represent electromagnetic waves of two types:

1. Waves proceeding from the source, or so-called ***solutions with a delay, which satisfy the principle of causality***. For these solutions, the effect (an electromagnetic wave propagating in space) is delayed in time with respect to its cause (the oscillation of the source that emits the electromagnetic waves).
2. Waves propagating from distant regions of space (in an unlimited environment—from infinitely distant regions) to the source, or the so-called ***anticipatory solutions***. These anticipatory solutions of the wave equations do not satisfy the principle of causality, because for them the effect (waves) precedes the reason that causes these electromagnetic waves.

It should be emphasized that from a mathematical point of view, both types of electromagnetic waves—both those that lag behind and those that precede—have the same right to exist. But from a physical point of view based on the principle of causality, anticipatory solutions are rejected, that is, they are actually not taken into account.

Anticipatory solutions and their possible connection with the intuitive abilities of a person. At the same time, it can be assumed that anticipatory solutions, which provide information about future events, must be somehow related to the intuitive abilities of a person. An extremely important role in understanding the anticipatory solutions of Maxwell's wave equations was played by results of the outstanding American physicist Richard Feynman, who created the quantum theory of the electron—one of the important sections of quantum electrodynamics. R. Feynman, together with his two colleagues D. Schwinger and S. Tomonago, received the 1965 Nobel Prize in Physics "for their fundamental work in quantum electrodynamics, with deep-ploughing consequences for the physics of elementary particles". [21] Almost a hundred years after the creation of Maxwell's electrodynamic theory, in the 50 s of the last century, R. Feynman noted the following non-trivial fact. If in the equations obtained in 1928 by the outstanding English physicist, Nobel laureate Paul Dirac, which describe not only the motion of the electron, but also predict the existence of the positron—the antiparticle of the electron, the sign of time and simultaneously the sign of the charge of the electron are changed, then Dirac's equations [28, 29] remain unchanged, that is, invariant with respect to such transformations.

This result gives grounds for asserting that an electron moving forward in time has the same right to exist as a positron moving backward in time. Thus, the time-reversed solutions of Maxwell's wave equations are nothing more than the movement of antimatter from infinitely distant points of the universe to us.

Undoubtedly, the hypothesis regarding the possible explanation of a person's intuitive abilities or his ability to predict requires a deeper understanding and consideration of a number of factors, namely:

(a) elucidation of hitherto unknown mechanisms of reception of the above-mentioned manifestations of antimatter, which ensures the presence in our material world, consisting of particles, not antiparticles, of the propagation of time-reversed (anticipatory) solutions of Maxwell's wave equations;
(b) the possibility of existence in the quantum world (microcosm) of the "cause-consequence" uncertainty principle, which is similar to Werner Heisenberg's uncertainty principle (see formulas (7.10) and (7.11) in Chap. 7) for coordinate-momentum, energy-time and other conjugate quantities;
(c) the effect of the existence of additional spatial dimensions. Indeed, modern versions of the theories of unification of physical interactions use the concept of additional dimensions in relation to three spatial and one temporal dimension. In particular, a fifth dimension is added in the theory of two German mathematicians Theodor Kaluza (1885–1954) and Felix Klein (1849–1925), making it possible to introduce Maxwell's equations, which successively describe electrical, magnetic, and optical phenomena, into Albert Einstein's general theory of relativity;
(d) the influence of wormholes in cosmology, which can provide a connection in time between the past and the future in the general science of nature as a whole (Science of Whole), which is currently developing.

It should be noted that there are many fundamental concepts in modern physics that we do not fully understand. For example, the concept of "mass", which was discussed in Sect. 7.3.4 [22–27]. There is a tentative estimate: it is believed that our knowledge about mass is limited to about 5%, and 1% of this knowledge is particles, 3%—physical fields, another 1%—for the recently discovered Higgs boson. Peter Ware Higgs and Francois Englert have shared the 2013 Nobel Prize in Physics "for the theoretical discovery of a mechanism that contributes to our understanding of the origin of mass of subatomic particles, and which recently was confirmed though the discovery of the predicted fundamental particle, by the ATLAS and CMS experiments at CERN's Large Hadron Collider". The last, still unknown 95% of our knowledge about the nature of mass is unevenly divided into the following parts: about 25–27% is "dark matter" (black holes, mole holes, and the like), and about 68–70% that remains, to "dark matter", which causes the accelerated expansion of the Universe (Nobel Prize in Physics in 2011 was awarded to Saul Perlmutter, Brian P. Schmidt and Adam G. Reiss "for the accelerating expansion of the Universe through observations of distant supernovae".) and the nature of which we still do not fully understand.

8.2.4 Physical and Biophysical Principles of the Influence of Weak External Fields on Biomedical Systems

Many vital processes in the human body are periodic. This kind of processes turned out to be the most adapted to long-term biological evolution. The presence of periodic processes in open systems on the non-thermodynamic branch of states is a necessary condition for the stability of such systems. In the field of thermodynamic instability, the existence of any system in the absence of coherent spatial or rhythmic (periodic in time) structures is impossible. The appearance of such structures is associated with the interaction of fluctuations of the order parameter at different points and at different moments of time. This is what makes open systems structurally stable.

Undoubtedly, thermodynamic parameters, as well as natural external fields, primarily the natural electromagnetic background and the human body's own electromagnetic and other physical fields, are of great importance in the implementation of the entire complex of conditions that ensure the stability of open biological systems.

Pathological processes occurring in the human body can lead it out of the area of structural stability in certain parameters (for example, in vital biochemical indicators). The most probable is the transfer of the medical-biological system to the region of such states, where the processes associated with biochemical reactions responsible for the release of mediators, the launch of adaptation mechanisms, etc., proceed as metastable. The main specificity of such metastable states is the realization of local extreme values for thermodynamic potentials and, as a result, exponential (rather than periodic) attenuation of the correlation function of fluctuations of the order parameter—the degree of completeness of biochemical reactions.

Theoretical studies show that external influences, primarily electromagnetic radiation (EMR) of the millimeter range, are able to return an open system to the region of structural stability, normalizing vital functions. At the same time, small external influences turn out to be not only effective, but also necessary. This important fact can be explained from the point of view of modern physics of phase transitions, assuming that biochemical processes are characterized by their proximity to the limit of stability, that is, to critical (bifurcation) points, where the susceptibility of these processes to changes in external factors is abnormally high. Too large powers of EMF and other irritating factors can be ineffective and even harmful, because they cause large disturbances that do not allow the medical-biological system to be in a sufficiently narrow area in terms of governing parameters, where states of structural stability are realized.

The study of the course of biochemical reactions near the points of phase transitions (bifurcation points) should be based on the following fundamental provisions: (a) the order parameter of the system with biochemical reactions is the degree of completeness, the change in which is associated with the change in the number of moles of reaction components and its stoichiometric coefficients; (b) the conjugate value for this order parameter is the affinity of the reaction, which is determined through the chemical potentials of the reacting components; (c) at bifurcation points, the sensitivity of the reaction to external influences, which is proportional to the derivative of the degree of completeness by affinity, increases abnormally; (d) external influences of the most diverse nature give additional contributions to the chemical potentials, thereby changing the affinity of the biochemical reaction (see Table 8.1). It should be emphasized that the course of vital biochemical reactions can change more significantly, the closer to the points of phase transitions (bifurcation points) is the system exposed to external influences.

A general consideration of the processes of formation of ordered structures in living systems gave grounds for assertion that the transition from a pathological to a healthy state is a kind of phase transition of the 2nd order, since this transition is associated with the crossing of the stability boundary line, i.e., the line of singular points, where a number of physical and chemical properties of the biological object undergo peculiarities [30–35].

Table 8.1 Change in chemical potential depending on the type of external influence

Type of external influence	Change in chemical potential $\Delta\mu$
Mechanical	$\Delta\mu = \Delta P/\rho$ (ΔP—the pressure change, ρ—the density)
Electromagnetic	$\Delta\mu = -\frac{1}{2}\left(\varepsilon_0 \frac{\partial\varepsilon}{\partial\rho} E^2 + \tilde{\mu}_0 \frac{\partial\tilde{\mu}}{\partial\rho} H^2\right)$ (ε_0 and $\tilde{\mu}_0$—the electric and magnetic constants, ε and $\tilde{\mu}$—the relative dielectric and magnetic permeabilities, E and H—the electric and magnetic field strengths)
Physicochemical effect of drugs	$\Delta\mu = RT \ln C$ (R—the universal gas constant, T—the absolute temperature, C—the drug concentration)

The pathological state corresponds to metastable regions in the space of thermodynamic and biochemical variables, while the normal state corresponds to the structurally stable behavior of the system with biochemical reactions.

One of the main consequences of the proposed synergistic approach is the establishment of the fundamental possibility of transition of a system with biochemical reactions from a metastable to a stable state under weak external influences. This allows us to use the attractive hypothesis of the universality of various phase transitions and critical phenomena occurring in the complex world around us to study the effect of weak physical fields on biological objects.

8.2.5 Real Science and False Science

Real Science

Real science is a sphere of human activity aimed at obtaining objective knowledge about nature and society, i.e., about the real world around and within us.

Criteria of Real Science

1. It operates with strictly defined concepts that have a precise meaning.
2. A new scientific theory is self-consistent in the sense that previously established theoretical results must follow from it, and it must also be confirmed by accurate and reproducible experimental data.

A Classic Example of Real Science

From the Einstein theory of relativity at a body velocity, which is much less than the speed of light, the Newton-Galileo mechanics follows, being consistent with numerous experimental data.

Main stages of scientific knowledge (see "Richard Feynman's Classification of the three stages of studying natural phenomena", which is proposed in [24, 25] and discussed in Chap. 2):

Stage 1: accumulation of experimental data.

Stage 2: generalization of experimental results in the form of strict laws.

Stage 3: establishing the underlying causes ("first principles") of knowledge obtained at the 1st and 2nd stages of scientific knowledge.

One example of such stages (see details in [24–27] and additional examples in Chap. 2):

Stage 1—obtaining experimental data on the refraction of light at the air–water boundary (Ptolemy, II century AD);

Stage 2—the establishment of the law of refraction of light (Snellius, 1621);

Stage 3—the formulation of Fermat's principle (Fermat,1660).

In connection with ***the processes of differentiation of sciences***, it should be noted that nature does not know anything about the processes of its fragmentation.

There are such first principles of nature:

"Nature knows better!" and "You have to pay for everything!"

You have to pay at a really high price—the loss of the integrity of the perception of the world around us.

Now there is another trend in the evolution of real scientific knowledge:

Simultaneously with the process of differentiation of sciences, there is also ***a process of integration of sciences***.

Example The emergence of a new science (scientific direction)—synergetics, created by Hermann Haken in 1969 (see details in Sect. 8.3) [30, 31].

False Science

False science is a misrepresenting of real and true science as a system of objective knowledge about reality.

Criteria of False Science

1. False science does not use exact concepts.
2. False science is not consistent with previous rigorous theories and well-established experimental facts.
3. False science is created by dilettantes (sometimes by absolute ignoramuses) or very ambitious persons.

Axiom of Mikhail Volkenstein

"A false scientist is the product of ignorance and ambition!"

Pseudoscience relies on the absence of ethical and moral prohibitions, bad faith and the desire for dishonest enrichment (sometimes outright cheating).

The Statement of Niels Bohr, 1922 Nobel Prize Winner in Physics

"The opposite of a correct statement is a false statement. But the opposite of a profound truth may be another profound truth" [32]

"An expert is a man who has made all the mistakes in a very narrow field" [33]

The postulate of Professor Sergei Kapitsa, son of Piotr Kapitsa, Nobel Prize winner in Physics: "A scientist may make mistakes to his students, but a false scientist insists on his mistakes."

Examples of False-Scientific Physical Theories Used in Medicine

1. **Torsion method of treating diseases**. The very concept of torsion fields arose in the general theory of relativity (GTR) of Albert Einstein. In GTR, the idea of combining gravitational and electromagnetic interactions was proposed. One of the results of GTR: a massive body can bend space, creating a very weak torsion field. False-scientific theories of torsion fields claim that these fields have

"miraculous" properties. In particular, the therapeutic effect in the treatment of the most dangerous diseases (cancer, AIDS, etc.). None of these effects has received rigorous experimental confirmation. In this case, the complex notion of a torsion field in GTR is used commercially as a basis for advertising the "torsion method" of treating diseases.

2. **The biological field**. The biological field (the field of living objects) is another example of a typical false-scientific concept from the point of view of modern physics and real science. This statement requires some clarification: the concept of "biological field" is false-scientific, if this concept is understood as a type of field (interaction), which cannot be reduced to the following known 4 types of real physical fields (interactions): (1) electromagnetic, (2) gravitational, (3) strong nuclear, and (4) weak nuclear.
3. **Long-term memory of water**. There are no reliably reproducible experiments. confirming the existence of long-term relaxation times (of the order of hours and longer) for water and aqueous solutions.

8.3 Section B

8.3.1 Elements of Mathematical Statistics

The distribution of a discrete random variable X is the set of its possible values X_1, X_2, X_3, ..., X_n and the probabilities P_1, P_2, P_3, ...,P_n, with which these values are reached. The distribution of a discrete random variable can be given in the form of the following Table 8.2 (see e.g. [34]).

Let us consider the main characteristics of such a distribution from the point of view of mathematical statistics.

1. ***The mathematical expectation M(X)*** gives an idea of the average value of a random variable X. ***The mathematical expectation M(X) of a discrete random variable X is the sum of products of possible values of the random variable X by corresponding probabilities P***:

$$M(X) = \sum_{i=1}^{n} P_i X_i = P_1 X_1 + P_2 X_2 + \cdots + P_n X_n. \tag{8.1}$$

If all random events occur with equal probability, then

Table 8.2 The distribution of a discrete random variable

X	X_1	X_2	X_3	...	X_{n-1}	X_n
P	P_1	P_{21}	P_3	...	P_{n-1}	P_n

$$P_1 = P_2 = \cdots = P_n = 1/n$$

where n is the total number of possible random events.

In this special case, the mathematical expectation is equal to the arithmetic mean value of the random variable, namely:

$$M(X) = \frac{X_1 + X_2 + \cdots + X_n}{n}.$$

For a random variable that takes continuous values, the mathematical expectation is calculated using the following formula:

$$M(X) = \int_{-\infty}^{+\infty} xf(x)dx,$$

where the function *f(x)*describes the probability that the random variable will receive a certain value x (this function *f(x)* is also called the distribution density) In this formula, for certainty, it is assumed that the random variable x changes continuously in an unlimited interval of its values.

Let us introduce two other important characteristics of the distribution of random variables—dispersion *D(X)* and standard deviation $\sigma(X)$, which give an idea of the deviation of possible values of a random variable relative to its mathematical expectation *M(X)*.

2. ***The variance D(X) of a random variable X is the mathematical expectation of the square of the deviation of the random variable X from its mean value M(X).*** According to this definition, the variance *D(X)* is given by the following formula:

$$D(X) = M(X - M(X))^2. \tag{8.2a}$$

For practical purposes of calculating the variance *D(X)*, you can also use another equivalent formula

$$D(X) = M\left(X^2\right) - M^2(X), \tag{8.2b}$$

from which it follows that ***the variance D(X) is the difference between the mathematical expectation of the square of the random variable M(X^2) and the square of the mathematical expectation [M(X)].***

For a continuous random variable with mathematical expectation $M(X) = a$, the variance value *D(X)* is calculated by the formula

$$D(X) = \int_{-\infty}^{+\infty} (x - a)^2 f(x)dx$$

3. ***The mean square deviation (standard deviation) σ(X).*** The dimension of the variance *D(X)* coincides with the dimension of the square of the studied quantity X. Therefore, to consider the deviation not of the square X^2, but of the random variable X itself, the concept of ***mean square deviation σ(X)*** is introduced, which is related to the variance *D(X)* by the following simple formula:

$$\sigma = \sqrt{D(X)}. \tag{8.3}$$

It is clear that the dimension of $\sigma(X)$ coincides with the dimension of the random variable X itself.

8.3.2 The Gaussian Distribution and Probability to Find Random Variables in a Fixed Interval

The Gaussian distribution (normal law) of random variables. The law of distribution of random variables is called ***normal***, or the ***Gaussian distribution***, if the density of this distribution *f(x)* is described by the following formula:

$$f(x) = \frac{1}{\sigma\sqrt{2\pi}} \exp[-(x-a)^2/2\sigma^2] \tag{8.4}$$

where $a = M(x)$ is the mean value (mathematical expectation) of a random variable, and σ is its mean square deviation.

The graph of the Gaussian distribution is characterized by a curve, which is symmetrical with respect to $a = M(x)$ (see Fig. 8.1). If $x = a$, then the ordinate of the density of the function of the Gaussian distribution *f(a)* is equal to $f(a) = 1/\sigma\sqrt{2\pi}$.

The normal law (Gaussian distribution) with parameters a = 0 and σ = 1 is called ***standard*** or ***normalized***. In this case, the distribution density function becomes equal $f(x) = \frac{1}{\sqrt{2\pi}}\exp\left(-\frac{x^2}{2}\right)$ and is depicted by the curve 1 in Fig. 8.2.

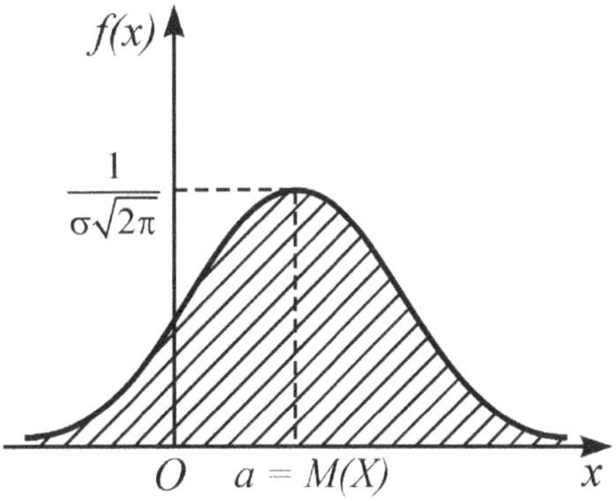

Fig. 8.1 The graph of the Gaussian distribution

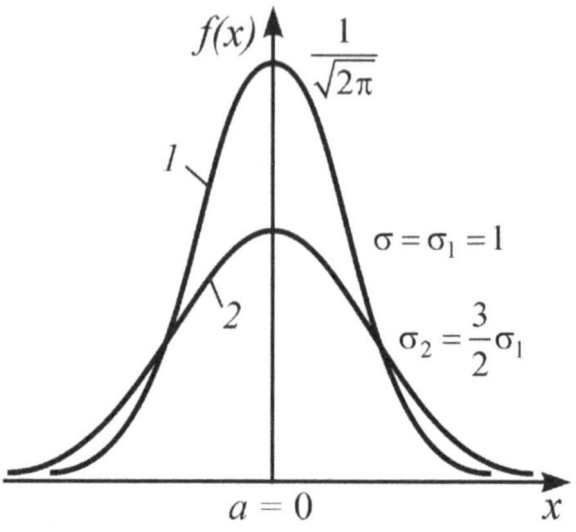

Fig. 8.2 The graph of the standard (normalized) Gaussian distribution

Let us consider an important question about the probability of finding the random variable in a given interval of its values. If the random variable X satisfies the normal law (Gaussian distribution), then the probability of this random variable falling into the interval (α, β) will be equal to

$$P(\alpha < X < \beta) = \frac{1}{\sigma\sqrt{2\pi}} \int\limits_{\alpha}^{\beta} e^{-\frac{(x-a)^2}{2\sigma^2}} dx$$

It is convenient to replace the old variable x by a new variable t у відповідності до такої формули: $\frac{x-a}{\sigma} = t$. Тоді для змінної x маємо $x = a + \sigma t$, а для її диференціалу, відповідно, $dx = \sigma\, dt$. As a result, the distribution density function acquires a standard (normalized) form, namely:

$$P(\alpha < X < \beta) = \frac{1}{\sqrt{2\pi}} \int\limits_{\frac{\alpha-a}{\sigma}}^{\frac{\beta-a}{\sigma}} e^{-\frac{t^2}{2}} dt. \tag{8.5}$$

Unfortunately, the resulting integral is not taken in elementary functions, so a special function is introduced to determine it

$$\Phi(x) = \frac{1}{\sqrt{2\pi}} \int\limits_{0}^{x} e^{-\frac{t^2}{2}} dt,$$

which is called ***the Laplace function*** (other names are ***the error function*** or ***the probability integral***). There are corresponding tables for calculating the Laplace function (see, for example, [1]).

Using the Laplace function, the probability of finding the random variable X in a given interval of its values (α, β) is given by the following formula:

$$P(\alpha < X < \beta) = \Phi\left(\frac{\beta - a}{\sigma}\right) - \Phi\left(\frac{\alpha - a}{\sigma}\right) = \Phi(t_2) - \Phi(t_1). \qquad (8.6)$$

Here, the values $t_1 = \frac{\alpha - a}{\sigma}$, $t_2 = \frac{\beta - a}{\sigma}$ correspond to the limit values of the specified interval (α, β) for the random variable X.

Mathematical justification of "The rule of three sigma". It is often necessary to find the probability that the deviation of a normally distributed random variable X from its mathematical expectation $a = M(x)$ does not exceed the modulus (absolute value) of some given positive number, which in our case can be related to the values σ, 2σ, 3σ, where σ is the mean square deviation. Then, taking into account the last formula given in a previous subsection, the numerical values of the Laplace function and its oddity, i.e. $\Phi(-x) = -\Phi(x)$, one has

$$P(|X - a| < \sigma) = P(a - \sigma < X < a + \sigma) = \Phi\left(\frac{\sigma}{\sigma}\right) - \Phi\left(-\frac{\sigma}{\sigma}\right) = 2\Phi(1) = 0,683;$$
$$P(|X - a| < 2\sigma) = P(a - 2\sigma < X < a + 2\sigma) = \Phi\left(\frac{2\sigma}{\sigma}\right) - \Phi\left(-\frac{2\sigma}{\sigma}\right) = 2\Phi(2) = 0,954;$$
$$P(|X - a| < 3\sigma) = P(a - 3\sigma < X < a + 3\sigma) = \Phi\left(\frac{3\sigma}{\sigma}\right) - \Phi\left(-\frac{3\sigma}{\sigma}\right) = 2\Phi(3) = 0,997. \qquad (8.7)$$

The last equality proves that almost reliably (with a probability of P = 0.997) the random variable will not deviate from the mathematical expectation by a modulus ("absolute value") of more than 3σ. This statement is called "***The rule of three sigma***". Figure 8.3 illustrates the above calculations. It is these considerations that were used earlier in Section A.

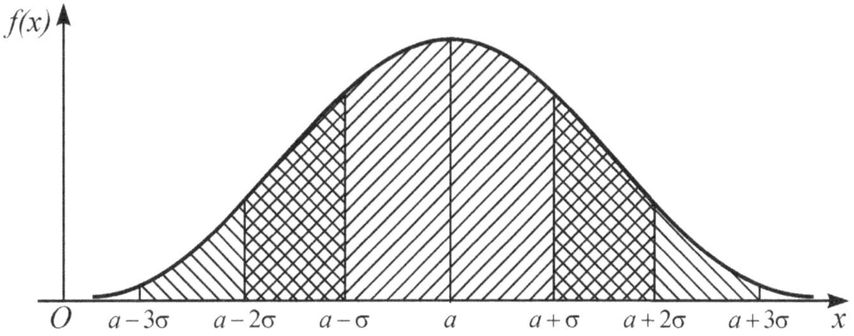

Fig. 8.3 This curve illustrates "The rule of three sigma"; the unshaded left and right regions, where this rule is not fulfilled, have an area of only 0.3% of the total area under the curve of the Gaussian distribution of random variables

8.3.3 Physical and Mathematical Principles of the Formation of Ordered Structures in Medical and Biological Systems in an External Electromagnetic Field

Statistal description of the formation of ordered structures in systems with biochemical reactions. As noted above in Sect. 7.2.5, a special role in synergetics belongs to the methods of the theory of phase transitions. This is dictated by the fundamental circumstance that systems undergoing phase transitions are characterized by an anomalously high susceptibility to external influences. The singularity of susceptibility at phase transition points (bifurcation points) is the direct reason why weak actions of external physical fields can cause a significant (sometimes decisive) effect on the properties of such systems. It is on this strict theoretical basis that we will consider the issues of the formation of coherent ordered structures at the level of biochemical reactions that are responsible for the synaptic mechanisms of information transmission [35, 36]. It will be shown that low powers of electromagnetic radiation (EMR) in the millimeter range can cause processes of ordering and self-organization induced by external fields.

Clarification and formulation of the basic physical and biophysical principles of ordering in biological objects requires the selection of the main control mode (in terms of synergetics) or, which is the same thing, the order parameter (in terms of the physics of phase transitions). The order parameter of a system with chemical (biochemical) reactions is the degree of completeness (coordinate) ξ, determined in accordance with the following relation:

$$d\xi(r) = dn_i(r)/\nu_i, \tag{8.8}$$

where dn_i is the change in the number of moles of the component i, ν_i are the stoichiometric coefficients. The conjugate value for the degree of completeness $\xi\ (r)$ (in the thermodynamic sense) is the affinity of the biochemical reaction $A(r)$, which is a linear combination of the chemical potentials $\mu_i\ (r)$ of the mixture components, namely:

$$A(r) = \sum_i \nu_i \mu_i. \tag{8.9}$$

The value of susceptibility χ of a biochemical reaction is of fundamental importance

$$\chi = -(\partial\xi/\partial A)_{P,T}. \tag{8.10}$$

At the boundary of stability (spinodal) of the reaction, the value of χ tends to infinity. The value of the reciprocal susceptibility χ^1 of a biochemical reaction is expressed according to the formula

$$\chi^{-1} = \sum_{i,j} \nu_i \nu_j (\partial \mu_i / \partial n_j)_{P,T} \tag{8.11}$$

through derivatives of chemical potentials by the number of moles of components participating in the reaction. To describe fluctuations of local values of the degree of completeness of a biochemical reaction, we expand the Gibbs thermodynamic potential (the choice of this thermodynamic potential is dictated by the real conditions of isobaricity and isothermally) into the Taylor functional series:

$$G[\xi] = \sum_{n \geq 0} \frac{1}{n!} \int \cdots \int \frac{\delta^n G[\xi]}{\delta\xi(\vec{r}) \cdots \delta\xi(\vec{r}_n)} \times \prod_{i=1}^{n} \xi(\vec{r}_i) d\vec{r}_i. \tag{8.12}$$

The functional derivative of the first order $\delta G/\delta\xi(r)$ determines the local value of the affinity $A(r)$ of the biochemical reaction. The second-order functional derivative sets the direct correlation function (DCF) $C(r)$ of the biochemical system

$$C(r) = \frac{1}{k_B T} \frac{\delta^2 G}{\delta\xi(\vec{r}_1)\delta\xi(\vec{r}_2)} \Big|_{\xi(\vec{r}_i)=0}, \qquad r = |\mathbf{r}_1 - \mathbf{r}_2|. \tag{8.13}$$

The role of the effective Hamiltonian of systems with biochemical reactions is played by the fluctuation part of the thermodynamic Gibbs potential, which is naturally chosen in the form of the Ginzburg–Landau Hamiltonian known from the theory of phase transitions:

$$\begin{aligned} \Delta G[\xi] &= \int d^d x [C_0 \xi^2 + a_4 \xi^4 + C_2 (\nabla \xi)^2 - A\xi] = \\ &= \sum_k (C_0 + C_2 k^2) \xi_k \xi_{-k} + \sum_{k,k',k''} a_4 \xi_k \xi_{k'} \xi_{k''} \xi_{-k-k'-k''} - L^{d/2} A \xi_0, \end{aligned} \tag{8.14}$$

where $\xi_k = V^{-1} \int \xi(r) e^{ikr} dr$ is the Fourier transform of the order parameter (degree of completeness) $\xi(r)$. The expansion coefficients C_0 and C_2 in (8.14) should be identified with the spatial moments of the direct correlation function $C(r)$ given by (8.13) as follows:

$$C_0 = (1/V) \int C(r) d\vec{r}, \; C_2 = (1/6V) \int C(r) r^2 d\vec{r} \tag{8.15}$$

Thus, the value of susceptibility χ, more precisely, the zero spatial moment of the DCF C_0 of a biochemical reaction, determines the coefficient at the square of the order parameter in the expansion of the fluctuation part of the thermodynamic Gibbs potential.

A consistent study of the process of structure formation in systems with biochemical reactions requires an analysis of the pair correlation functions $G_2(r)$ of fluctuations in the degree of completeness of the reaction. Accounting for the interaction of fluctuations ξ (composed of ξ^4 in the Ginzburg–Landau Hamiltonian) can be carried out by methods of quantum field theory using diagram techniques. In the approximation of a free field of fluctuations, i.e., taking into account the quadratic and gradient terms in the Ginzburg–Landau Hamiltonian (8.14), the pair correlation function $G_2(r)$ of fluctuations of the degree of completeness of the biochemical reaction has the following form in the space of wave vectors:

$$< \xi_{k_1} \xi_{k_2} >= \frac{L^d \delta_{k_1 k_2}}{C_o + C_2 k^2} \tag{8.16}$$

The question of the presence or absence of ordered structures at the level of biochemical reactions is associated with the elucidation of the analytical behavior of the pair correlation function $G_2(r)$ in the coordinate representation

$$G_2(r) =< \xi(\vec{r}_1)\xi(\vec{r}_2) >= \left(\frac{L}{2\pi}\right)^d \int\int d\vec{k}_1 d\vec{k}_2 \cdot \frac{exp[i(\vec{k}_1\vec{r}_1 - \vec{k}_2\vec{r}_2)]}{C_o + C_2 k^2} \delta_{k_1 k_2}, \tag{8.17}$$

where d is the space dimension. This study will be carried out below.

The impact of an external field on a system with biochemical reactions*: *changing the coordinates of bifurcation points and the correlation radius. In order to study the problem of the influence of millimeter-wave EMR on a system with biochemical reactions, we first consider a liquid dielectric, which is in an external electric field. As is known [37], the chemical potential of such a system is equal to

$$\mu = \mu_0 - \frac{\varepsilon_0}{2}\left(\frac{\partial \varepsilon}{\partial \rho}\right)_T E^2, \tag{8.18}$$

where μ_0 is the chemical potential of the medium in the absence of an electric field. The external electric field apparently gives an additional contribution to the value of the zero moment DCF C0, which is equal to

$$\Delta C_0^E = -\frac{\varepsilon_o E^2 N_A}{2k_B T}\sum_{s,j} \nu_{i\nu_j}\left(\frac{\partial^2 \varepsilon}{\partial \rho_i \partial \rho_j}\right)_T . \tag{8.19}$$

It can be seen that this contribution is proportional to the square of the electric field strength and the second derivative of the permittivity with respect to density. If you set the permittivity as save by the standard Lorenz-Lorentz approximation, then the value $\partial^2\varepsilon/\partial\rho_i\partial\rho_j$ will turn into zero. Therefore, it is necessary to take into account the correlation corrections to the permittivity from the effects of spatial dispersion. The calculation of these corrections in the vicinity of the critical point of a dielectric fluid leads to the following result:

$$\frac{\partial^2 \varepsilon}{\partial \rho_i \partial \rho_j} \sim \begin{cases} R_c^{1+2/\varsigma}, kR_c < 1 \\ R_c^{2/\varsigma-1}, kR_c > 1 \end{cases}, \tag{8.20}$$

where k is the wavenumber, R_c is the correlation radius of order parameter fluctuations (for liquids—density fluctuations). According to (8.20), with approaching the critical point in the region of strong spatial dispersion, where $kR_c > 1$, the second derivative $\partial^2\varepsilon/\partial\rho_i\partial\rho_j$ has a singularity, since the critical index $\varsigma \approx 1.8$ and $2/\varsigma - 1 \approx 0.11$.

Generalizing these results to the biological media of the human body, which, as a rule, have well-defined dielectric and diamagnetic properties, we write the contribution of the electromagnetic field to the zero spatial momentum of the DCF C_0 in the form

$$\Delta C_0^{EH} = -\frac{N_A}{2K_B T}\sum_{i,j} \nu_i \nu_j \times \left(\varepsilon_o \frac{\partial^2 \varepsilon}{\partial \rho_i \partial \rho_j} E^2 + \tilde{\mu}_0 \frac{\partial^2 \tilde{\mu}}{\partial \rho_i \partial \rho_j} H^2 \right), \tag{8.21}$$

where $\tilde{\mu}$ is the magnetic permeability of the biological medium, and H is the intensity of the magnetic component in the electromagnetic wave.

The zero moment of DCF C_0, as noted above, determines the inverse susceptibility χ^1 of the system with biochemical reactions. Together with the second moment C_2, it also sets the radius of correlation of fluctuations in the degree of completeness of the biochemical reaction. Indeed, as follows from the general theory of critical fluctuations of the order parameter, the correlation radius $R_c = \sqrt{C_2/C_o}$.

In the absence of an external electromagnetic field, the correlation radius

$$R_c^o = \left\{ \frac{N_A}{k_B T C_2} \sum_{i,j} \nu_i \nu_j \times \left[\frac{\partial \mu_{ic}}{\partial \rho_j} \left(\frac{\rho - \rho_{ic}}{\rho_i} - \frac{1}{\rho_i} \left(\frac{\partial \rho_i}{\partial T} \right)_\rho (T - T_{ic}) \right) \right] \right\}^{-1/2} \tag{8.22}$$

is determined only by the proximity of the temperature and pressure of the system with biochemical reactions to the critical (bifurcation) values T_{ic} and P_{ic} of the components involved in the reaction.

Under the influence of an external electromagnetic field, the correlation radius changes according to the formula

$$R_c^{E,H} = \left\{ C_2 / \left[C_0 - \frac{N_a}{2k_B T} \sum_{i,j} \nu_i \nu_j \left(\varepsilon_0 \frac{\partial^2 \varepsilon}{\partial \rho_i \partial \rho_j} E^2 + \tilde{\mu}_0 \frac{\partial^2 \tilde{\mu}}{\partial \rho_i \partial \rho_j} H^2 \right) \right] \right\}^{1/2}. \tag{8.23}$$

The obtained result can be interpreted in terms of changes in the critical (bifurcation) temperature and pressure values of the components involved in the biochemical reaction:

(a) in isobaric conditions, the critical temperature changes by an amount

$$\Delta T_{ic} = T_{ic}^* - T_{ic} = \frac{\rho_i}{2} \left(\frac{\partial T}{\partial P} \right)_\rho \left(\varepsilon_o \frac{\partial \varepsilon}{\partial \rho_i} E^2 + \tilde{\mu}_o \frac{\partial \tilde{\mu}}{\partial \rho_i} H^2 \right), \tag{8.24}$$

(b) in isothermal conditions, the critical pressure changes by an amount

$$\Delta P_{ic} = P_{ic}^{*} - P_{ic} = \frac{\rho_i}{2}\left(\varepsilon_o \frac{\partial \varepsilon}{\partial \rho_i} E^2 + \tilde{\mu}_o \frac{\partial \tilde{\mu}}{\partial \rho_i} H^2\right). \tag{8.25}$$

Critical (bifurcation) parameters P_{ic} and T_{ic} of the components of a biochemical reaction can differ greatly from each other. Therefore, the dependence of the susceptibility of the biochemical reaction to external influences on the temperature in isobaric conditions (or on the pressure in isothermal conditions), being equivalent to the variation of the parameters of the electromagnetic field, is characterized by a rather of complex function, which reaches local maxima when approaching the critical values P_{ic} and T_{ic} of each of the components. The main maxima in these dependences correspond, obviously, to the realization of the critical (bifurcation) value for the main order parameter—the controlling mode of biochemical reaction.

Transition between metastable and structurally stable states in biological systems under the action of the electromagnetic field. The study of the correlation functions of order-parameter fluctuations in systems with biochemical reactions near the boundary of stability makes it possible to solve such an important problem: under what conditions it becomes possible to form coherent spatial structures such as Turing structures or rhythmic phenomena such as autowaves of chemical activity in the Belousov-Zhabotinsky reaction in biological objects. As will be shown later, a transition from the state of a biochemical reaction, where there is no coherent behavior of the degree of completeness, to a state characterized by a synchronous change in the correlation function $G_2(r)$ of fluctuations in the degree of completeness of the biochemical reaction, slowly decaying at a distance and in time becomes real under the action of EMF.

Let us consider such an initial state of a system with biochemical reactions in which both spatial moments C_0 and C_2 of the direct correlation function $C(r)$ are positive. For this case, the pair CF $G_2(r)$ of fluctuations in the degree of completeness of the biochemical reaction has the following form in three-dimensional space (d = 3):

$$G_2(r) = \frac{V}{4\pi k_B T C_2} \exp\left(-\frac{r}{R_c^{E,H}}\right)/r, \tag{8.26}$$

where the correlation radius $R_c^{E,H}$ is given by formula (8.23). With approaching the stability limit (bifurcation point), at which the zero moment of the DCF is $C_0 \to 0$, the correlations between local fluctuations of the degree of completeness $\xi(r)$ slowly fade away with the distance $G_2(r) \sim 1/r$, that is, they become long-range. This result means that coherent structures cannot arise in such a biochemical system.

The situation changes radically if the sign of one of the spatial moments becomes negative, i.e., provided that either $C_0 < 0$, $C_2 > 0$ (thermodynamically unstable system) or $C_0 > 0$, $C_2 < 0$. Similar cases occur for a medium with the so-called ascending diffusion, that is, with a negative diffusion coefficient. The point in the space of thermodynamic variables where the sign of the diffusion coefficient or, which is the same, the second spatial moment C_2, changes sign is called the Lifshitz point.

We now turn to the consideration of the system with biochemical reactions on the non-thermodynamic branch of states, i.e. at $C_0 < 0$, $C_2 > 0$. For such a system at d = 3, the CF $G_2(r)$ of fluctuations of the degree of completeness of the biochemical reaction is described by a formula of the form

$$G_2(r) \sim N_{1/2}(r/R_c)/r^{1/2}. \tag{8.27}$$

Here $N_{1/2}(z)$ is the cylindrical Neumann function and the correlation radius $R_c = \sqrt{C_2/|C_0|}$. The result presented by formula (4.27) means that in a system with biochemical reactions there is a slowly decaying, spatially periodic correlation between local values of the degree of completeness $\xi(r)$, i.e. emerging structures $\lambda_0 = 2\pi\sqrt{C_2/|C_0|}$. With approaching the boundary of stability ($C_0 \to 0$), the characteristic size of periodic biochemical structures becomes quite large, reaching macroscopic sizes.

It should be noted that under the influence of an external electromagnetic field, a system with biochemical reactions can move from the region of thermodynamic stability $C_0 > 0$ to a thermodynamically unstable region, where $C_0 + \Delta\, {C_0}^{E,H} < 0$, since, in accordance with formula (4.21), the addition term $\Delta {C_0}^{E,H}$ is negative. At the same time, structural stability may be inherent in the latter state due to the effect of nonlocality (spatial dispersion) of fluctuations in the degree of completeness $\xi(\boldsymbol{r})$, which provides a positive value of the generalized zero moment of DCF $C(r)$ (in fact, the value of the generalized susceptibility to external influences): $C_0 + \Delta {C_0}^{E,H} + C_2 k^{\,2} \geq 0$. This directly implies the following relation for the characteristic spatial size of the corresponding coherent structures:

$$\lambda_0 = \frac{2\pi}{k} = \sqrt{C_2/\left|C_0 + \Delta C_0^{E,H}\right|}. \tag{8.28}$$

Such structure in a biochemical system can be identified, for example, with the size of spatial periodic zones due to the nature of singular points for the processes of intercellular interaction (synaptic transmission), being discussed in Chapter 4.

Consequently, the physical mechanism of the transition of a biochemical system to an ordered state under the action of EMF is based on an effective change in the coordinates of the bifurcation points (in particular, critical values of temperature or pressure) and, as a result, a change in the sign of the zero moment of the DCF C_0. Thus, a previously stable system without structures is brought under the boundary of stability (spinodal) of a biochemical system, where the thermodynamic instability is stabilized by the appearance of coherent ordered spatial structures.

Quantitative estimates of the influence of electromagnetic fields on the parameters of processes in biomedical systems. Let us perform quantitative estimates of the effects of millimeter-wave EMF exposure on a system with biochemical reactions and clarify the question of the dependence of the characteristics of the resulting ordered structures on the EMF parameters.

For the frequencies ω_{b-ch} of periodic chemical (biochemical) reactions, such as the emergence and decay of mediator-receptor complexes, and the frequencies $\omega_{\text{e-m}}$ of electromagnetic processes in the millimeter (EHF) range, there is a following strong inequality:

$$\omega_{elect-magn} >> \omega_{biochem}, \tag{8.29}$$

which allows evaluating the effects of exposure to external electromagnetic fields as averaged over the period of biochemical reactions.

Let us estimate the amount of energy absorbed per unit volume of the medium per unit time for a medium with dispersion using the following formula:

$$W = \omega(\varepsilon_0 \varepsilon'' E^2 + \mu_0 \mu'' H^2). \tag{8.30}$$

Here ε'', $\tilde{\mu}''$ are the imaginary parts of the dielectric and magnetic permeability, and the risk over the strengths of the electric and magnetic fields means averaging over time. Due to the obvious relationship $< E^2 > = {E_0}^2/2$, where E_0 is the amplitude of the electric field strength, we have

$$W = \omega \varepsilon_0 \varepsilon'' E^2. \tag{8.31}$$

To carry out a numerical assessment of the value of W, we will assume for definiteness that the wavelength of EMR in the millimeter range is equal to $\lambda = 3$ mm, i.e., the corresponding cyclic frequency $\omega_{elect-magn} = 2\pi/\lambda = 6,3 \times 10^{11} rad/s$. In what follows, we assume that the surface density power $|\boldsymbol{S}|$ supplied to the biological object in an electromagnetic wave (Pointing vector)

$$|S| = 2\varepsilon_0 \varepsilon'' c < E^2 >= \varepsilon_0 \varepsilon'' c E_0^2 \tag{8.32}$$

has order $|\boldsymbol{S}| \approx 1$ mW/cm^2 = 10 W/m^2, which seems realistic in terms of the EMF parameters used in the method of microwave resonance therapy method. Then the volumetric energy density of the electromagnetic wave is equal to

$$\varepsilon_0 \varepsilon'' E_0^2 = |S|/c \approx 3 \times 10^{-8} J/m^3 \tag{8.33}$$

Further, it should be taken into account that the imaginary part of the permittivity near the points of phase transitions (for example, the critical point of liquids), taking into account pairwise correlations of the order parameter, is approximately ten thousandth of the real part of the permittivity, i.e., $\varepsilon'' \approx 10^{-4}\varepsilon$. Finally, for the desired value of the EMP power absorbed per unit volume P^* one has the following result:

$$P^* = 2W/m^3. \tag{8.34}$$

Let us compare the result obtained in (8.34) with the numerical estimate of the power P^*, which must be provided to a unit volume of a system with biochemical reactions per unit time in order to reach the stability limit of this system in terms of its thermodynamic parameters (temperature and pressure). For this purpose, we use the real values of pressure and temperature deviations from their critical (bifurcation) values $(P–P_c) / P_c \approx (T–T_c) / T_c \approx 10^{-5}$, the dimensional estimate of the derivative $(\partial P/\partial T)_r \approx nP_c/T_c$ with n ≈ 1 (n = 4 in the van der Waals model), the value of the critical parameters $P_c \approx 10^5$ Pa, $T_c \approx 300$ K, as well as the frequency $\nu_{\text{bch}} \approx 10^{-2}$ Hz and the period $T_{\text{bch}} \approx 10^2$ s of a biochemical batch reaction. As a result, we obtain the following energy values per unit volume, which must be given to a system with biochemical reactions per unit time in order to achieve its stability limit:

(a) with the isothermal approach

$$P^* = (P - P_c)/T_{bch} \approx 10^{-2} W/m^3, \tag{8.35}$$

(b) with the isobaric approach

$$P^* = (\partial P/\partial T)_\rho \times (T - T_{bch})/T_{bch} \approx 4 \times 10^{-2} W/m^3. \tag{8.36}$$

Comparing the values of P^* from (8.34) with the values of P^* given in (8.35), (8.36), it should be concluded that sufficiently small EMF powers of the millimeter range can cause the transition of the system with biochemical reactions through the limit of stability and the formation of ordered coherent structures in it.

Criteria and principles for the formation of coherent structures in bioobjects under the action of EMF of the millimeter range. Let us formulate the criteria, the fulfillment of which is necessary for the formation of coherent structures in bioobjects under the influence of electromagnetic radiation with wavelengths corresponding to the range between the microwave and high-frequency ranges. The criteria have the form of the following inequalities:

(a) under isothermal conditions of the biochemical reaction in the presence of electromagnetic radiation of the millimeter range

$$\omega \varepsilon_0 \varepsilon'' E_0^2 / (P - P_c) \nu_{bch} > 1; \tag{8.37}$$

(b) under isobaric conditions of the biochemical reaction in the presence of electromagnetic radiation of the millimeter range

$$\omega \varepsilon_0 \varepsilon'' E_0^2 / (\partial P/\partial T)_\rho (T - T_c) \nu_{bch} > 1. \tag{8.38}$$

From the above formulas (4.37) and (4.38) it follows that the formation of coherent structures in systems with biochemical reactions under the influence of millimeter-wave electromagnetic radiation is facilitated by: (1) an increase in the frequency ω of electromagnetic oscillations; (2) an increase in the amplitude E_0 of the electrical component in the electromagnetic wave; (3) an increase in the imaginary part ε'' of the permittivity; (4) decrease in the frequency of periodic biochemical processes; (5) approximation of thermodynamic parameters (temperature, pressure) to their critical (bifurcation) values.

It should be noted that the presence of some invariants follows from the criteria (4.37) and (4.38), namely: $\omega\varepsilon''E_0^2 = const$. The presence of such a frequency-dependent invariant means that the same effect can be achieved at different frequencies, provided that the product of such characteristics of the electromagnetic field as the frequency ω and the square of the amplitude of the electric field strength $E_0{}^2$, by the characteristic of the biological object is imaginary. part of the permittivity ε'' will remain unchanged during the action.

The results obtained provide a basis for a deeper understanding of the physical and biophysical principles of the formation of coherent structures in systems with biochemical reactions under the action of millimeter-wave electromagnetic radiation. The primary act is the influence of EMR in the millimeter range on the Zakharyin-Ged zones, where biologically active points of the human body are located with a relatively lower electrical resistivity (relatively large electrical conductivity). This radiation is absorbed on the basis of certain molecular mechanisms, the detailed scheme of which should be based on a microscopic, that is, quantum mechanical, basis (in particular, during transitions between rotational sublevels of a certain vibrational level of the molecular spectra of biomolecules). Note that one of the interesting issues related to the problem of primary reception of millimeter-wave EMR in biological objects is the establishment of the frequency dependence of the electrodynamic characteristics $\varepsilon''(\omega)$ and $\tilde{\mu}''$ responsible for the absorption of EMR in biological tissues and fluids of the human body, as well as their extreme properties in the millimeter range of EMP.

As was mentioned above, the absorption of EMF leads to the emergence of a new type of ordering, which is characterized by a coherent nature of biochemical reactions (for example, reactions of formation and disappearance of mediator-receptor complexes). Coherence manifests itself in the presence of long-range and periodic in space connection between local values of the biochemical reaction order parameter, being its degree of completeness.

Spatial structures that arise in a system with biochemical reactions are characterized by anomalous susceptibility to external influences. As follows from the formulas of statistical non-equilibrium thermodynamics for kinetic coefficients, the electrical conductivity coefficient is determined by the spatio-temporal correlation function $<$ J (r, t) J (0,0) $>$ of electric charge flows (a so-called current correlator), which, in a reasonable approximation, can be represented through the product of the correlator function $<$ V (r,t) V (0,0) $>$ of velocities and the correlator function $< n_i$ (**r**, t) n_i (0,0) $>$ of concentrations:

$$\sigma \sim < J(r,t)J(0,0) > \sim < \upsilon(r,t)\upsilon(0,0) > \times < n_i(r,t)n_i(0,0) > . \tag{8.39}$$

It follows that the last factor on the right side of formula (8.39) is directly related to the susceptibility extremes: $\chi_i \sim < n_i(r,\ t)n_i(0,0) > .\chi_i \sim < n_i(r,\ t)n_i(0,0) > .$

8.3.4 Conclusions

This chapter discussed physical and mathematical ideas, as well as laws and principles that gives a fairly reliable scientific basis and makes it somewhat possible to try explaining the unusual human perception, which is currently considered "extrasensory". The consistent implementation of such a rather complex and risky task (what the author wrote about in the Abstract at the very beginning of this chapter) required the use of the following approaches:

1. The use of the "rule of three sigma", which is known in mathematical statistics (see Introduction, Sects. 8.3.1 and 8.3.2, as well as formula (8.7) from this chapter), explains the fundamental possibility of the existence of sufficiently rare random events obeying the normal law (Gaussian distribution). This mathematical distribution of random deviations from the mean value (mathematical expectation) is the most popular and widely used in statistical processing of various properties of human activity. For example, the Gaussian distribution is used in many European universities for quantitative assessment of students' knowledge based on the European Credit Transfer System (ECTS).
2. Section 8.2.2 decribes the results of experimental studies of the phenomenon of radiotaxis. Radiotaxis is the directed movement of particles, as well as cells, microorganisms, etc., under the influence of an external stimulus (external radiation). Using the method of photon-correlation spectroscopy allows to study the spectrum of scattered light and to measure the speed and direction of movement of the polymer particles of the suspension, which is caused by the action of radiation emanating from the "extrasensor's" palms. The results obtained in experiments on radiotaxis allows to conclude: the so-called "extrasensory" capabilities of a person are most likely explained by the action of ordinary physical fields of thermal radiation caused by the presence of temperature gradients.
3. The problem of the possibility of intuitive predictions on the basis of the so-called anticipatory solutions of the wave equations of Maxwell's electromagnetic theory is discussed. These solutions are usually not taken into account for the physical reasons that they contradict the principle of causality (see Sect. 8.2.3). As is known, the principle of causality requires that in the "cause-effect" sequence of events, the cause necessarily precedes the effect. This same problem is also discussed at the level of the basic equations of quantum mechanics in connection with the works of Paul Dirac and Richard Feynman, which predict the existence of the positron — the antiparticle of the electron. If the sign of time and simultaneously the sign of the charge of the electron are changed, then Dirac's equations [28, 29] remain unchanged, that is, invariant with respect to such transformations. This means that an electron moving forward in time has the same right to exist as a positron moving backward in time. Thus, the anticipatory solutions of Maxwell's wave equations are nothing more than an explanation for the movement of antimatter from infinitely distant points of the Universe to us.
4. The physical and biophysical mechanisms of the influence of weak external fields (in particular, electromagnetic fields of the millimeter range) on medical

and biological systems are studied (see Sect. 8.2.4). One of the main conclusions comes down to the following statement: transition of a system with biochemical reactions from a metastable to a stable state under weak external influences makes it possible to use an attractive hypothesis of the universality of various phase transitions and critical phenomena to study the influence of weak physical fields on processes in medical and biological systems that occur in a complex world outside and inside us.

5. Section 8.2.5 is devoted to the discussion of the problem of the relationship between true science and false science. The definition of true science is given as a sphere of human activity aimed at obtaining objective knowledge about nature and society, i.e., about the real world around and within us. Criteria and examples of true science are given. The main stages of knowledge of natural phenomena with the help of methods of true science, based on "Richard Feynman's Classification of the three stages of studying natural phenomena", which was proposed in [24, 25] and discussed in Chap. 2. This paragraph also defines false science, the main criteria of false science, the statements of a number of scientists (Niels Bohr, Petr Kapitsa, Mikhail Volkenstein) about the signs of false science, as well as three examples of false scientific theories and methods (Torsion method of treating diseases; The biological field; Long- term memory of water).
6. It is worth dwelling on the very interesting and, one might say, almost fantastic problem of quantum teleportation—the phenomenon of rapid transfer of quantum objects to macroscopic distances, which is discussed in Sect. 8.2.3. The great scientific importance of the problem of quantum deportation is emphasized by the awarding of the 2022 Nobel Prize in Physics to three physicists Alain Aspect (France), John Clauser (USA) and Anton Zeilinger (Austria). A theoretical physicist Michio Kaku believes that the practical implementation of quantum teleportation of real objects will take place in about 50 years.

After all, when we think a lot about incredible events, they become clear over time!

References

1. *Medical and Biological Physics*, Ed. A.V. Chalyi. Knyga Plus, Kyiv (2005)
2. *Proceedings of the International Symposium "Fundamental and Applied Aspects of Application of Microwave radiation in Medicine"*. Edited by Prof. S.P. Sitko, Kyiv (1989)
3. A.V. Chalyi, Non-equilibrium phase transitions in living system affected by law-intensive microwave radiation. Phys. Alive **1**(1), 81–92 (1993)
4. A.V. Chalyi, Fluctuation effects in stationary open systems. Phys. Alive **3**(1), 22–28 (1995)
5. W. Wien, MLA style: The Nobel Prize in Physics 1911. NobelPrize.org. Nobel Prize Outreach AB (2023). https://www.nobelprize.org/prizes/physics/1911/summary/
6. M. Planck, MLA style: The Nobel Prize in Physics 1918. NobelPrize.org. Nobel Prize Outreach AB (2023). https://www.nobelprize.org/prizes/physics/1918/summary/
7. Levchuk Yurij Mykolayovych / V.I. Nazarenko// Encyclopedia of Modern Ukraine / Eds.: I.M. Dzyuba, A.I. Zhukovskii, M.G. Zheleznyak et al.; Institute of Encyclopedic Research, National Academy of Sciences of Ukraine, Kyiv (2016)

8. Yu.M. Levchuk et al. Photon-correlation spectroscopy of albumen. Ukr. Biochem. J. **74**(5) (2002)
9. J. Mossbridge, P. Tressoldi and J. Utts, Predictive physiological anticipation preceding seemingly unpredictable stimuli: a meta-analysis. Front Psychol. **3**, 390 (2012)
10. S.L. Baumgart, M.S. Franklin, H.K. Jimbo et al., Prediction of truly random future events using analysis of prestimulus electroencephalographic data. AIP Conference Proceedings. AIP Publishing 1841(1) (2017)
11. J.A. Mossbridge, D. Radin, Precognition as a form of prospection: a review of the evidence. Psychol. Consciousness Theory Res. Practice **5**(1), 78–93 (2018)
12. I.O. Vakarchyk, *Quantum mechanics* (Ivan Franko Lviv National University, Lviv, 2012)
13. MLA style: Press release. NoblePrize.org. Nobel Prize Outreach AB 2024. https://www.nobelprize.org/prize/physics/2922/press-release/
14. Y.-N. Luo, H.-S. Zhong, M. Erhard, X.-L. Wang, L.-C. Peng, M. Krenn, X. Jiang, L. Li, N.-L. Liu, C.-Y. Lu, A. Zeilinger, J.-W. Pan, Quantum teleportation in high dimensions. Phys. Rev. Lett. (2019)
15. E. Schrödinger, *Abhandlungen zur Wellenmechanik*. Leipzig (1927)
16. E. Schrödinger, *Vier Vorlesungen uber Wellenmechanik*. Berlin (1928)
17. *Wolfgang Pauli—Biographical*. NobelPrize.org. Nobel Prize Outreach AB (2924). https://www.nobelprize.org/prizes/physics/1945/pauli/biographical/
18. W. Pauli, The connection between spin and statistics. Phys. Rev. **58**(8), 716–722 (1940)
19. M. Kaku. *Visions: How Science Will Revolutionize the 21st Century*. New York: Ancor (1998)
20. M. Kaku, *Physics of the future: how science will shape human destiny and our daily Livesby the year 2100* (Ancor, New York, 2012)
21. S.-I. Tomonago, J. Schwinger, R.P. Feynman, *NobelPrize.org. Nobel Prize Outreach AB* (2024)
22. L.B. Okun, The concept of mass. Phys. Today **42**, 31 (1989)
23. A.M. Gabovych, N. Gabovych, How to explain the non-zero mass of electromagnetic radiation consisting of zero-mass photons? Europ. J. Phys. **28**(4), 649 (2007)
24. R.P. Feynman, R.B. Leighton, M. Sands, *The Feynman Lectures on Physics*, vol. 1, Chapter 26, California Institute of Technology (2013)
25. A.V. Chalyi, Synergetic dialogue "physics-medicine": Hexagons in living and inanimate nature. J. Mol. Liquids **329**, 114249 (2021)
26. A.V. Chalyi, N.L. Grytsenko. Historical development of the concept of mass in physical theories of the XVII-XXI centuries. 1. Classical mechanics and the theory of relativity. Physics and astronomy at school, №3, 43–47 (2012)
27. A.V. Chalyi, N.L. Grytsenko, Historical development of the concept of mass in physical theories of the XVII-XXI centuries. 2. The law of mass conservation in the process of electron-positron annihilation and in the PET diagnostic method. Physics and astronomy at school, №6, 43–47 (2012)
28. P.A.M. Dirac, *The Principles of Quantum Mechanics* (Oxford University Press, Oxford, 1930)
29. P.A.M. Dirac, The quantum theory of the emission and absorption of radiation **114**(767), 247–65 (1927)
30. H. Haken, *Synergetics. An Introduction. Nonequilibrium Phase Transitions and Self-Organization in Physics, Chemistry and Biology*, Springer, Berlin, Heidelberg, New York (1978)
31. H. Haken, *Advanced Synergetics* (Springer-Verlag, Berlin, Heidelberg, New York, Tokyo, Instability Hierarchies of Self-Organizing Systems and Devices, 1983)
32. *"Mind from Matter" An Essay on Evolutionary Epistemology*, p. 167, (1986)
33. A.L. Mackay, *A Dictionary of Scientific Quotations*, p. 35 (1991)
34. A.V. Chalyi, N.V. Stuchinska, A.V. Melenevska, *Higher Mathematics for Students of Higher Medical and Pharmaceutical Educational Institutions*. Technique, Kyiv (2001)
35. A.V. Chalyi, L.M. Chernenko, Phase transitions in finite-size systems and synaptic transmission, in D. Beysens, N. Boccara, G. Forgacs (Eds.) *Dynamic Phenomena at Interfaces, Surfaces and Membranes* (New York, Nova Sci. Publishers, 1993)

36. A.V. Chalyi, A.N. Vasilev, E.V. Zaitseva, Synaptic transmission as a cooperative phenomenon in confined systems. Condensed Matter Phys. **20**(1), 13804 (2017)
37. L.D. Landau, E.M. Lifshits, *Electrodynamics of Continuous Media.* 2nd Edition, Volume 8 in Course of Theoretical Physics, Elsevier Ltd. (1984)

Index

A. V. Chalyi, *What is Medicine?*, https://doi.org/10.1007/978-3-031-64979-0